14795 BIFM

£26.95

Industrial Electronics

Industrial Electronics

■

Frank D. Petruzella

GLENCOE
McGraw-Hill

New York, New York   Columbus, Ohio   Mission Hills, California   Peoria, Illinois

Cover photograph: COMSTOCK Inc./Michael Stuckey. Safety section photographs: *Top left:* Charles Thatcher/Tony Stone Worldwide; *Top right:* Lou Jones/The Image Bank; *Lower right:* © Cindy Lewis.

ACKNOWLEDGMENTS

The *Basic Skills in Electricity and Electronics* series was conceived and developed through the talents and energies of many individuals and organizations.

The original, on-site classroom testing of the texts and manuals in this series was conducted at the Burr D. Coe Vocational Technical High School, East Brunswick, New Jersey; Chantilly Secondary School, Chantilly, Virginia; Nashoba Valley Technical High School, Westford, Massachusetts; Platt Regional Vocational Technical High School, Milford, Connecticut; and the Edgar Thomson, Irvin Works of the United States Steel Corporation, Dravosburg, Pennsylvania. Postpublication testing took place at the Alhambra High School, Phoenix, Arizona; St. Helena High School, St. Helena, California; and Addison Trail High School, Addison, Illinois.

Early in the publication life of this series, the appellation "Rainbow Books" was used. The name stuck and has become a point of identification ever since.

In the years since the publication of this series, extensive follow-up studies and research have been conducted. Thousands of instructors, students, school administrators, and industrial trainers have shared their experiences and suggestions with the authors and publishers. To each of these people we extend our thanks and appreciation.

Library of Congress Cataloging-in-Publication Data

Petruzella, Frank D.
    Industrial electronics / Frank D. Petruzella.
       p.   cm.
    Includes index.
    ISBN 0-02-801996-2
    1. Industrial electronics.   I. Title.
    TK7881.P48   1996
    621.3—dc20   95-36563
                                      CIP

Industrial Electronics

Send all inquiries to:
Glencoe/McGraw-Hill
936 Eastwind Drive
Westerville, Ohio 43081

ISBN 0-02-801996-2

Printed in the United States of America.

1 2 3 4 5 6 7 8 9   VH   01 00 99 98 97 96 95

Contents

Editor's Foreword

The McGraw-Hill *Basic Skills in Electricity and Electronics* series has been designed to provide entry-level competencies in a wide range of occupations in the electrical and electronic fields. The series consists of coordinated instructional materials designed especially for the career-oriented student. Each major subject area covered in the series is supported by a textbook, an activities manual, and an instructor's resource guide. All the materials focus on the theory, practices, applications, and experiences necessary for those preparing to enter technical careers.

There are two fundamental considerations in the preparation of materials for such a series: the needs of the learner and needs of the employer. The materials in this series meet these needs in an expert fashion. The authors and editors have drawn upon their broad teaching and technical experiences to accurately interpret and meet the needs of the student. The needs of business and industry have been identified through questionnaires, surveys, personal interviews, industry publications, government occupational trend reports, and field studies.

The processes used to produce and refine the series have been ongoing. Technological change is rapid, and the content has been revised to focus on current trends. Refinements in pedagogy have been defined and implemented based on classroom testing and feedback from students and instructors using the series. Every effort has been made to offer the best possible learning materials.

The widespread acceptance of the *Basic Skills in Electricity and Electronics* series and the positive responses from users confirm the basic soundness in content and design of these materials as well as their effectiveness as learning tools. Instructors will find the texts and manuals in each of the subject areas logically structured, well-paced, and developed around a framework of modern objectives. Students will find the materials to be readable, lucidly illustrated, and interesting. They will also find a generous amount of self-study and review materials to help them determine their own progress.

The publisher and editor welcome comments and suggestions from teachers and students using the materials in this series.

Charles A. Schuler
Project Editor

Basic Skills in Electricity and Electronics

Charles A. Schuler, Project Editor

BOOKS IN THIS SERIES

Introduction to Television Servicing by Wayne C. Brandenburg

Mathematics for Electronics by Harry Forster, Jr.

Electricity: Principles and Applications by Richard J. Fowler

Communication Electronics by Louis E. Frenzel, Jr.

Instruments and Measurements: A Text-Activities Manual by Charles M. Gilmore

Microprocessors: Principles and Applications by Charles M. Gilmore

Small Appliance Repair by Phyllis Palmore and Nevin E. André

Industrial Electronics by Frank D. Petruzella

Electronics: Principles and Applications by Charles A. Schuler

Digital Electronics by Roger L Tokheim

Preface

∎

This book has been written as a course of study that will introduce the reader to a broad range of industrial electronic circuits and equipment. The ever-increasing need to improve production output and reduce costs requires a technically literate workforce that can make use of modern manufacturing techniques. With this in mind, I have made every effort to present up-to-date material which reflects what industry is demanding of graduates and what instructors are teaching in their courses.

This textbook is intended for use in industrial training programs in technical schools and colleges and in industry. Although the reader needs to have a working knowledge of basic electrical theory and terminology before making use of this textbook, the mathematical requirements are minimal.

The text is comprehensive! It includes coverage of power distribution, power electronics, motor control, PLCs, and process control systems. With such a broad range of subjects, a conscientious effort has been made to establish cohesion so that the reader will see how all these subjects fit together. Component and system construction, operation, and installation are emphasized. Various practical applications are presented throughout the textbook as they relate to the modern industrial environment.

The textbook is logically structured, easy to read, and highly illustrated. It features a variety of worked examples and numerous self-tests that reinforce learning and build self-confidence. Performance objectives at the beginning of each chapter tell students exactly what is expected. Critical thinking questions at the end of each chapter build the independent problem-solving skills that are necessary for on-the-job troubleshooting.

Two companion pieces, the *Activities Manual* and the *Instructor's Resource Guide,* are available for use with this textbook.

The *Activities Manual* contains multiple-choice, true/false, completion, and matching-type questions for each chapter and serves as an excellent review of the material presented. In addition, the *Activities Manual* contains lab activities for each chapter. These practical assignments outline suggested experiments and testing applications. Special attention has been given to the use of readily available components for these experiments. The *Instructor's Resource Guide* contains answers to all the textbook review questions and to the *Activities Manual* objective questions and practical assignments.

Frank D. Petruzella

Safety

■

Electric and electronic circuits can be dangerous. Safe practices are necessary to prevent electrical shock, fires, explosions, mechanical damage, and injuries resulting from the improper use of tools.

Perhaps the greatest hazard is electrical shock. A current through the human body in excess of 10 milliamperes can paralyze the victim and make it impossible to let go of a "live" conductor or component. Ten milliamperes is a rather small amount of electrical flow: it is only *ten one-thousandths* of an ampere. An ordinary flashlight uses more than 100 times that amount of current!

Flashlight cells and batteries are safe to handle because the resistance of human skin is normally high enough to keep the current flow very small. For example, touching an ordinary 1.5-V cell produces a current flow in the microampere range (a microampere is one-millionth of an ampere). This much current is too small to be noticed.

High voltage, on the other hand, can force enough current through the skin to produce a shock. If the current approaches 100 milliamperes or more, the shock can be fatal. Thus, the danger of shock increases with voltage. Those who work with high voltage must be properly trained and equipped.

When human skin is moist or cut, its resistance to the flow of electricity can drop drastically. When this happens, even moderate voltages may cause a serious shock. Experienced technicians know this, and they also know that so-called low-voltage equipment may have a high-voltage section or two. In other words, they do not practice two methods of working with circuits, one for high voltage and one for low voltage. They follow safe procedures at all times. They do not assume protective devices are working. They do not assume a circuit is off even though the switch is in the OFF position. They know the switch could be defective.

As your knowledge and experience grow, you will learn many specific safe procedures for dealing with electricity and electronics. In the meantime:

1. Always follow procedures.
2. Use service manuals as often as possible. They often contain specific safety information.
3. Investigate before you act.
4. When in doubt, *do not act.* Ask your instructor or supervisor.

General Safety Rules for Electricity and Electronics

Safe practices will protect you and your fellow workers. Study the following rules. discuss them with others, and ask your instructor about any you do not understand.

1. Do not work when you are tired or taking medicine that makes you drowsy.
2. Do not work in poor light.
3. Do not work in damp areas or with wet shoes or clothing.
4. Use approved tools, equipment, and protective devices.
5. Avoid wearing rings, bracelets, and similar metal items when working around exposed electric circuits.
6. Never assume that a circuit is off. Double check it with an instrument that you are sure is operational.
7. Some situations require a "buddy system" to guarantee that power will not be turned on while a technician is still working on a circuit.
8. Never tamper with or try to override safety devices such as an interlock (a type of switch that automatically removes power when a door is opened or a panel removed).
9. Keep tools and test equipment clean and in good working condition. Replace insulated probes and leads at the first sign of deterioration.
10. Some devices, such as capacitors, can store a *lethal* charge. They may store this charge for long periods of time. You must be certain these devices are discharged before working around them.

11. Do not remove grounds, and do not use adaptors that defeat the equipment ground.

12. Use only an approved fire extinguisher for electrical and electronic equipment. Water can conduct electricity and may severely damage equipment. Carbon dioxide (CO_2) or halogenated-type extinguishers are usually preferred. Foam-type extinguishers may also be desired in some cases. Commercial fire extinguishers are rated for the type of fires for which they are effective. Use only those rated for the proper working conditions.

13. Follow directions when using solvents and other chemicals. They may be toxic, flammable, or may damage certain materials such as plastics.

14. A few materials used in electronic equipment are toxic. Examples include tantalum capacitors and beryllium oxide transistor cases. These devices should not be crushed or abraded, and you should wash your hands thoroughly after handling them. Other materials (such as heat shrink tubing) may produce irritating fumes if overheated.

15. Certain circuit components affect the safe performance of equipment and systems. Use only exact or approved replacement parts.

16. Use protective clothing and safety glasses when handling high-vacuum devices such as picture tubes and cathode-ray tubes.

17. Don't work on equipment before you know proper procedures and are aware of any potential safety hazards.

18. Many accidents have been caused by people rushing and cutting corners. Take the time required to protect yourself and others. Running, horseplay, and practical jokes are strictly forbidden in shops and laboratories.

Circuits and equipment must be treated with respect. Learn how they work and the proper way of working on them. Always practice safety; your health and life depend on it.

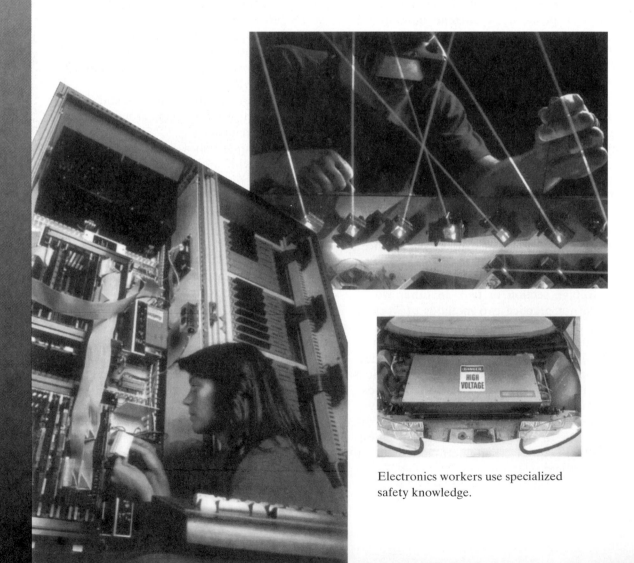

Electronics workers use specialized safety knowledge.

CHAPTER 1

Industrial Safety

■

CHAPTER OBJECTIVES

This chapter will help you to:

1. *Avoid* electrical hazards in the workplace.
2. *Identify* the electrical factors that determine the severity of an electric shock.
3. *List* the general safety precautions that must be observed when working with electrical equipment.
4. *Explain* the purpose and process involved in grounding an electrical system.
5. *Outline* the basic steps in a lockout procedure.
6. *Outline* the first-aid procedures for bleeding, burns, and electric shock.
7. *State* the purpose of the National Electrical Code.

Safety is the number one priority in any job. Every year, electrical accidents cause serious injury or death. Many of these casualties are young people just entering the workplace. They are involved in accidents that result from carelessness, from the pressures and distractions of a new job, or from a lack of understanding about electricity. This chapter is designed to develop an awareness of the dangers associated with electrical power and the potential dangers that can exist on the job or at a training facility.

■

1-1 ELECTRIC SHOCK

Often we think that serious electric shock can be caused only by high-voltage circuits. This is not so! More people are injured or killed by 120-V household voltage every year than in all other electrical-related accidents. If you walked away from your last electric shock, consider yourself lucky. Do not depend on luck. Work safely with electricity and live!

Electric shock occurs when a person's body becomes part of the electric circuit. The three electrical factors involved in an electric shock are resistance, voltage, and current.

▐▐▐➡ **YOU MAY RECALL** that electrical *resistance* (*R*) is the opposition to the flow of current in a circuit and is measured in ohms (Ω).

The lower the body resistance, the greater the potential electric shock hazard. Body resistance can be divided into external (skin resistance) and internal (body tissues and blood stream resistance). Dry skin is a good insulator; moisture lowers the resistance of skin, which explains why shock intensity is greater when the hands are wet. Internal resistance is *low* owing to the salt and moisture content of the blood. There is a wide degree of variation in body resistance. A shock that may be fatal to one person may cause only brief discomfort to another. Typical body resistance values are listed in Table 1-1. Body resistance can be measured with an *ohmmeter* (Fig. 1-1 on page 2).

Table 1-1	Skin Condition or Area and Its Resistance
Skin Condition or Area	Resistance Value
Dry skin	100,000 to 600,000 Ω
Wet skin	1000 Ω
Internal body—hand to foot	400 to 600 Ω
Ear to ear	about 100 Ω

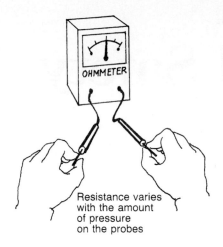

Resistance varies
with the amount
of pressure
on the probes

Fig. 1-1   Measuring body resistance.

▐▐▐▶ **YOU MAY RECALL**   that *voltage* (*V*) is the pressure that causes the flow of electric current in a circuit and is measured in units called volts (V).

The amount of voltage that is dangerous to life varies with each individual because of differences in body resistance and heart conditions. *Generally, any voltage above 30 V is considered dangerous.*

▐▐▐▶ **YOU MAY RECALL**   that *electric current* (*I*) is the rate of flow of electrons in a circuit and is measured in amperes (A).

The amount of current flowing through a person's body depends on the voltage and resistance. Body current can be calculated using the following formula:

$$\text{Current through body} = \frac{\text{Voltage applied to body}}{\text{Resistance of body}}$$

$$I \text{ (amperes)} = \frac{V \text{ (volts)}}{R \text{ (ohms)}}$$

or

$$I \text{ (milliamperes, mA)} = \frac{V \text{ (volts)}}{R \text{ (kilohms, kV)}}$$

$$1 \text{ A} = 1000 \text{ mA}$$

$$1 \text{ k}\Omega = 1000 \text{ }\Omega$$

The amount of current that passes through the body and the length of *time* of exposure are perhaps the two most reliable criteria of shock intensity. It doesn't take much current to cause a painful or even fatal shock. A current of 1 mA ($\frac{1}{1000}$ of an ampere) can be felt. A current of 10 mA will produce a shock of suffi-

cient intensity to prevent voluntary control of muscles, which explains why, in some cases, the victim of electric shock is unable to release his or her grip on the conductor while the current is flowing. A current of 100 mA passing through the body for one second or longer can be fatal. *Generally, any current flow above 0.005 A or 5 mA is considered dangerous.*

A flashlight cell can deliver more than enough current to kill a human being, yet it is safe to handle. This is because the resistance of human skin is high enough to limit greatly the flow of electric current. In lower voltage circuits, resistance restricts current flow to very low values. Therefore, there is little danger of an electric shock. Higher voltages, on the other hand, can force enough current though the skin to produce a shock. *The danger of harmful shock increases as the voltage increases.*

The *pathway* through the body is another factor influencing the effect of an electric shock. For example, a current from hand to foot, which passes through the heart and part of the central nervous system, is far more dangerous than a shock between two points on the same arm (Fig. 1-2).

Self-Test

Answer the following questions.

1. True or false? The higher the body resistance, the greater the potential electric shock hazard.
2. Moisture _____ the resistance.
3. True or false? Internal body resistance is very high because of bone structure.
4. Generally, any voltage above _____ V is considered dangerous.
5. Generally, any current above _____ mA is considered dangerous.

1-2 SAFETY IN THE WORKPLACE

Safety has become an increasingly important factor in the working environment. The electrical industry, in particular, regards safety to be unquestionably the top priority because of the hazardous nature of the business. A safe operation depends largely upon all plant personnel being informed and aware of potential hazards. *Obey all accident prevention signs* (Fig. 1-3)!

Many statistics show that 98 percent of all accidents are avoidable. With such a wide area

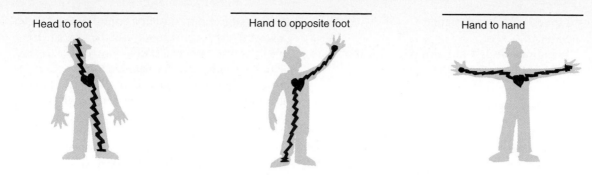

Fig. 1-2 **Typical electric current pathways that stop normal pumping of the heart.**

in which to make improvements, everyone can contribute something toward reducing accidents. Major causes are unsafe acts resulting from human error and material failure. It is the unsafe acts of individuals that cause 88 percent of all accidents. Material failure accounts for only 10 percent.

In addition to companies themselves, other groups promote worker safety. These include the United States government, insurance companies, and labor unions. In 1970, Congress created a regulatory agency known as the *Occupational Safety and Health Administration (OSHA)*. OSHA sets standards that regulate safety in plants. Its inspectors check on companies to make sure they are following these safety regulations. OSHA also inspects and approves safety products.

The following colors have been established by OSHA to designate certain cautions and dangers:

- *Red* is used to designate:
  Fire protection equipment and apparatus
  Portable containers of flammable liquids
  Emergency stop buttons and switches

- *Yellow* is used to designate:
  Caution and physical hazards
  Waste containers for explosive or combustible materials
  Caution against starting, using, or moving equipment under repair
  The starting point or power source of machinery

- *Orange* is used to designate:
  Dangerous parts of machines
  Safety starter buttons
  The exposed parts (edges) of pulleys, gears, rollers, cutting devices, and power jaws

- *Purple* is used to designate:
  Radiation hazards

Eye Protection
Must be Worn

Head Protection
Must be Worn

Hearing
Protection
Must be Worn

Hand Protection
Must be Worn

Breathing
Protection
Must be Worn

Foot Protection
Must be Worn

CAUTION!
Slippery Floor

CAUTION!
Fork Lift

DANGER!
Compressed Gases

No
Smoking

DANGER!
Flammable

DANGER!
Poison

Fire
Extinguisher

Eyewash

First Aid

Safety
Shower

Fig. 1-3 **Typical accident prevention signs.** *(Courtesy Safety Supply Canada)*

- *Green* is used to designate:
  Safety
  Locations of first-aid equipment (other than fire fighting equipment)

PERSONAL SAFETY ATTIRE

The clothing worn at work is important for personal safety. Appropriate attire should be worn for each particular job site and work activity (Fig. 1-4). The following points should be observed:

1. Hard hats, safety shoes, and goggles must be worn in areas where they are specified.
2. Safety earmuffs must be worn in areas that are noisy.
3. Clothing should fit snugly to avoid the danger of becoming entangled in moving machinery.
4. Metal jewelry should not be worn while working on energized circuits; gold and silver are excellent conductors of electricity.
5. Long hair should be confined or kept trimmed when working around machinery.

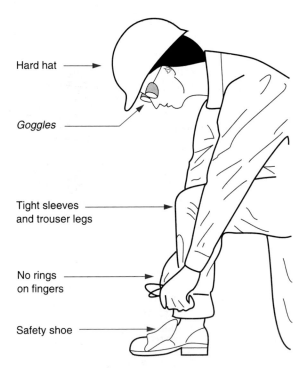

Hard hat

Goggles

Tight sleeves
and trouser legs

No rings
on fingers

Safety shoe

Fig. 1-4 Clothing and equipment used for personal safety. *(Courtesy Safety Supply Canada)*

1-3 GROUNDING

Electricity is the flow of electrons. The flow of electric current is something like the flow of water from the mountains to the ocean. Water always tries to find a way to the ocean. Electricity always tries to find a way to the ground. The route electricity takes is called its *path to ground.* If you are part of an electrical path to ground, electricity may pass through you. It could seriously burn or kill you. If you touch a live electrical wire while standing on the ground, or something that is in touch with the ground, you may become part of a path to ground.

Grounding refers to the deliberate connection of parts of a wiring installation to a common earth connection. In general, grounding guards against two hazards: *fire* and *shock*.

A fire hazard can occur when current leaks from a broken live wire or connection and reaches a point of zero voltage by some path other than the normal one. Such a path offers high resistance, so the current can generate enough heat to start a fire.

A shock hazard generally arises when there is little or no leaking current, but the potential for abnormal current flow exists. For example, if an exposed live wire touched the metal frame of an ungrounded piece of electrical equipment, the voltage of the live wire would charge the metal frame. If you then touched the charged metal frame, your body could provide a current path to ground, and you could suffer a serious shock.

Figure 1-5 illustrates grounding for protection. In order for this protection system to work, *both* the electric current-carrying conductor *system* and the circuit *hardware* (metal parts) must be grounded. In a properly grounded system, a direct short to ground fault produces a high current surge. This current blows a fuse or trips a circuit breaker to immediately open the circuit. Grounding has nothing to do with the operation of electrical equipment. Its sole purpose is the protection of life and property.

An ungrounded power tool can kill you! Always use properly grounded power tools. Use only those power tools with three-pronged plugs or double-insulated tools with two-pronged plugs (Fig. 1-6). Inspect cords and equipment often to make sure that ground pins are in safe condition.

The use of a grounding wire in a three-wire cord with a three-pronged plug cap and a grounding receptacle reduces the danger of shock, but it does not eliminate the danger completely. Sometimes a tool develops a ground fault that is not a solid or direct con-

IMPROPER GROUNDING IS AN INVITATION
TO DISASTER!

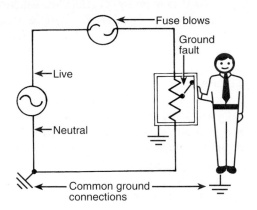

- Ground fault produces a
  short circuit which blows the fuse
- No shock is received from
  touching the metal frame

(a) Properly grounded circuit

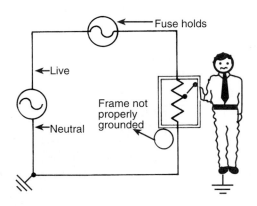

- Ground fault produces no
  abnormal current flow
- Fuse holds
- Shock is received by
  touching the metal frame and ground

(b) Improperly grounded circuit

Fig. 1-5 Grounding for protection.

nection between the live wire and the case. This can be caused by a partial breakdown of insulation or by moisture within the device. When this occurs, the ground-fault current may not be high enough to blow a 15-A fuse or trip a 15-A breaker. However, it may produce a high enough current to shock or electrocute anyone who comes into contact with the device. For example, Fig. 1-7 shows how electricity leaking out from frayed hot wires passes through the metal casing to the person holding the tool.

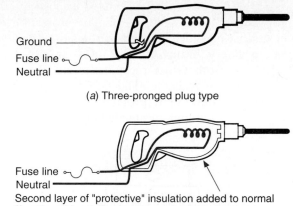

(a) Three-pronged plug type

Second layer of "protective" insulation added to normal "functional" insulation isolates motor and all current-carrying parts from contact with metal casing.

(b) Double-insulated two-pronged plug type

Fig. 1-6 Use properly grounded tools.

A *ground-fault circuit interrupter* (*GFCI*) is designed to minimize the probability of electrocution under the conditions just described. Under normal conditions the current in the hot wire and the neutral wire are absolutely identical. However, if wiring or a tool is defective and allows some current to leak to ground, a GFCI will sense the difference in the two wires. The GFCI is fast-acting; the unit will shut off the current or interrupt the circuit within $\frac{1}{40}$ second (s) after its sensor detects a leakage as small as 5 mA.

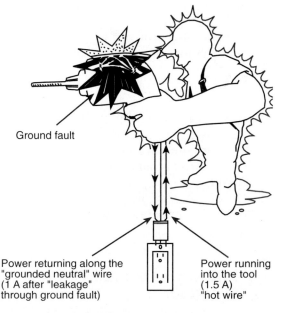

Ground fault

Power returning along the "grounded neutral" wire (1 A after "leakage" through ground fault)

Power running into the tool (1.5 A) "hot wire"

Fig. 1-7 Ground-fault current is not high enough to trip the breaker or blow the fuse.

A GFCI is *not* to be considered a substitute for grounding but as supplementary protection that senses leakage currents too small to operate ordinary branch circuit fuses or circuit breakers. Both GFCI receptacles and circuit breakers are available. The GFCI receptacle provides ground-fault protection to users of any electric equipment plugged into the receptacle (Fig. 1-8).

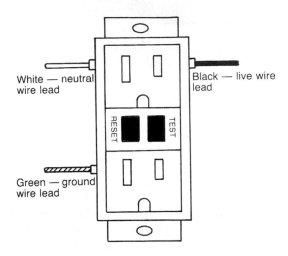

White — neutral wire lead

Black — live wire lead

RESET

TEST

Green — ground wire lead

Fig. 1-8 GFCI receptacle.

Self-Test

Answer the following questions.

6. The color _____ has been established by OSHA to designate dangerous parts of machines.
7. Safety earmuffs must be worn in _____ areas.
8. True or false? You can be suspended from your job for not following safe work practices.
9. Electricity always tries to find a return path to _____.
10. In general, grounding guards against two hazards: _____ and _____.
11. True or false? In a properly grounded system, a direct ground fault produces a short circuit.
12. Only _____-insulated power tools can be safely operated with two-pronged plugs.

1-4 LOCKOUT OF ELECTRICAL SOURCES

Electrical *lockout* and *tagout* (Fig. 1-9) refer to the process of padlocking the power source in the OFF position and indicating, on an appropriate card, the procedure that is taking place. This procedure is necessary so that no one will inadvertently turn the equipment to the ON position while it is being worked on. Lockout procedures involve a few basic, simple steps. Those steps may require five minutes of your time, but those five minutes are vital. Failure to lock-out properly could result in injury or even death.

"Lockout" means achieving a zero state of energy while equipment is being serviced. Proper lockout procedures are required for maintenance, repair, troubleshooting, adjusting, installation, or cleaning of electrical or mechanical equipment. Just pressing a stop button to shut down machinery won't provide you with security. Someone else working in the area can simply reset it. Even a separate automated control could be activated to override the manual controls.

It's essential for all interlocking or dependent systems to also be deactivated and de-energized. These could feed into the system being isolated, either mechanically or electrically. It's important to test the start button before resuming any work in order to verify that the energy source(s) have been isolated.

(Courtesy Safety Supply Canada.)

Fig. 1-9 Electrical lockout and tagout.

BASIC STEPS IN LOCKOUT PROCEDURE

Document all lockout procedures in a plant safety manual. This manual should be available to all employees and outside contractors working on the premises. Management should have policies and procedures for safe lockout and should also educate and train everyone involved in locking out electrical or mechanical equipment.

Identify the location of all switches, power sources, controls, interlocks, and other devices that need to be locked out in order to isolate the system. Review schematics of the system if they are available.

Stop all running equipment by using the controls at or near the machine.

Disconnect the switch. (Do not operate if the switch is still under load.) Stand clear of the box and face away while operating the switch with the left hand (if the switch is on the right side of the box).

Lock the disconnect switch in the OFF position. If the switch box is the breaker type, make sure the locking bar goes right through the switch itself and not just the box cover. Some switch boxes contain fuses, and these should be removed as part of the lockout process. If this is the case, use a fuse puller to remove them.

Use a tamper-proof lock with one key, which is kept by the individual who owns the lock. Combination locks, locks with master keys, and locks with duplicate keys are not recommended.

Tag the lock with the signature of the individual performing the repair and the date and time of the repair. There may be several locks and tags on the disconnect switch if more than one person is working on the machinery. The machine operator's (and/or the maintenance operator's) lock and tag will be present as well as the supervisor's.

Verify the isolation. Use a voltage test to determine that voltage is present at the line side of the switch or breaker. When all phases of outlet are dead with the line side live, you can verify the isolation.

Remove tags and locks when the work is completed. Each individual must remove his or her own lock and tag. If there is more than one lock present, the person in charge of the work is the last to remove his or her lock.

Before reconnecting the power, check that all guards are in place and that all tools, blocks, and braces used in the repair are removed. Make sure that all employees stand clear of the machinery.

Self-Test

Answer the following questions.

13. Electrical lockout refers to the process of _____ the power source in the OFF position.
14. True or false? Proper lockout procedures are required for cleaning mechanical equipment.
15. True or false? A disconnect switch should not be operated if the switch is still under load.
16. True or false? There should only be one lock and tag on the disconnect switch even if more than one person is working on the machinery.

1-5 GENERAL ELECTRICAL SAFETY PRECAUTIONS

With proper precautions, there is no reason for a technician to receive a shock. Receiving an electrical shock is a clear warning that proper safety measures have not been followed. To maintain a high level of electrical safety while you work, there are a number of precautions you should follow. Your individual job will have its own unique safety requirements. However, the following are given as essential basics.

- Never take a shock on purpose.
- Keep material or equipment at least 10 feet away from high-voltage overhead power lines.

- Do not close any switch unless you are familiar with the circuit that it controls and know the reason for its being open.
- When working on any circuit, take steps to ensure that the controlling switch is not operated in your absence. Switches should be padlocked open, and warning notices should be displayed.
- Avoid working on "live" circuits as much as possible.
- When installing new machinery, ensure that all metal frame work is efficiently and permanently grounded.
- Always treat circuits as "live" until you have proven them to be "dead." Presumption at this point can kill you. It is a good practice to take a meter reading before starting work on a dead circuit.
- Avoid touching any grounded objects while working on electrical equipment.
- *Remember that even with a 120-V control system, you may well have a higher voltage in the panel.* Always work so that you are clear of any of the higher voltages. (Even though you are testing a 120-V system, you are most certainly in close proximity to 240-V or 480-V power.)
- Don't reach into energized equipment while it is being operated. This is particularly important in high-voltage circuits.
- *Use good electrical practices even in temporary wiring for testing.* At times you may need to make alternate connections, but make them secure enough so that they are not in themselves an electrical hazard.
- When working on live equipment containing voltages over approximately 30 V, work with only one hand. Keeping one hand out of the way greatly reduces the possibility of passing a current through the chest.
- Discharge capacitors before handling them. Capacitors connected in live direct current (dc) circuits can store a lethal charge for a considerable time after the voltage to the circuits has been switched off. Discharging capacitors is done by shorting them. An insulated jumper probe with a built-in resistor can be used to safely discharge a capacitor (Fig. 1-10). The resistor limits the discharge flow of electrons to avoid a damaging current surge.

Complex industrial electrical and electronic equipment, as well as power distribution circuits, require on-going attention to safety. The following general safety precautions apply to meter measurements.

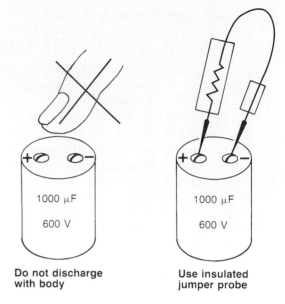

Do not discharge with body   **Use insulated jumper probe**

Fig. 1-10   Safely discharging a capacitor.

- *Never* use the ohmmeter on a live circuit.
- *Never* connect an ammeter in parallel with a voltage source.
- *Never* overload an ammeter or voltmeter by attempting to measure currents or voltages far in excess of the range-switch setting. If it is not an autoranging meter, always start on high settings so that you get a low-scale reading on your initial test; this will protect you and the meter in case the voltage (or amperage) is higher than anticipated.
- *Make certain* that any terminals you are measuring across are not accidentally shorted together, or to ground, by the test leads.
- *Check* meter test leads for frayed or broken insulation before working with them.
- *Avoid* touching the bare metal clips or tips of test probes.
- *Whenever possible,* remove voltage before connecting meter test leads into the circuit.
- *To lessen* the danger of accidental shock, disconnect meter test leads immediately after the test is completed.
- *On-line troubleshooting is recommended only for control voltages of 120 V or less.* This is never a technique for higher-voltage circuits, nor is it to be used on motors or any other high-current applications.
- Many meters have a HOLD button that captures a reading and displays it from memory after the probe has been removed from the circuit. Personal safety of the user is probably the most important reason for a "hold" function. A technician often is faced with awkward or cramped situations where he or

she needs to carefully watch the placement of the test probe while keeping hands clear of high voltages or moving machinery. In a high-energy environment, a probe slip can cause a short in the electrical circuit, resulting in serious injury to the user or damage to the meter, and may even cause the technician to accidentally fall into the work area.

A review of accidents involving electrical workers indicates that most deaths and injuries result from:

- Not maintaining safe limits of approach to live apparatus.
- Not taking adequate work protection or isolation.
- Not using safe work practices or following the safety rules.
- Using defective or poorly maintained tools and equipment.

Working with electricity should not cost an arm, a leg, . . . or your life! But sad to say for some it has.

Self-Test

Answer the following questions.

17. "Dead" circuits should be checked by _____ the voltage.
18. When installing new machinery, ensure that all metal frame work is _____.
19. True or false? When working on "live" equipment, probing with one hand reduces the possibility of passing a current through the chest.
20. Discharging capacitors is done by _____ them.
21. True or false? On-line troubleshooting is recommended for high-voltage circuits.
22. The _____ function of a meter should be used when taking measurements in awkward or cramped situations.
23. True or false? It is not possible to receive a shock from frayed or broken insulation on meter test leads.

1-6 FIRST AID

First aid is the immediate and temporary care given to the victim of an injury or illness. Its purpose is to preserve life, assist recovery, and prevent aggravation of the condition. A properly stocked first-aid kit should be readily available (Fig. 1-11 on page 10). If someone is hurt, *first* send for help immediately. A few basic first-aid procedures follow.

BLEEDING

To control bleeding, apply direct pressure on the wound using a clean pad or your hand. Raise the affected arm, leg, or head above heart level.

BURNS

For first-degree and minor second-degree burns, immerse the injured area in cold water or apply cold packs to relieve the pain. *Do not break blisters.* For second-degree burns with open blisters and all third-degree burns, no water or cold packs should be applied as this increases the likelihood of shock and infection. These serious burns should be treated with thick, clean bandages. No particles of charred clothing should be removed except by skilled medical practitioners. If the victim has suffered facial burns, he or she should be kept propped up and monitored for breathing difficulty. If only feet, legs, or arms are burned, they should be elevated above the level of the heart. For any serious burn, get medical help to the scene as soon as possible.

ELECTRIC SHOCK

To treat electric shock, turn the power off or use a dry board or stick to remove the electric contact from the victim. Do *not* touch the victim until he or she has been separated from the current. Begin first-aid procedures. Administer artificial respiration if the victim is not breathing. Keep the victim warm; position the victim so that the head is low and turned to one side to encourage flow of blood and to avoid an obstruction to breathing.

If breathing stops, you can help the victim by administering *artificial respiration*. The mouth-to-mouth method of artificial respiration is the most effective. In this method, you breathe air into the victim's lungs with your own mouth. The basic mouth-to-mouth method of artificial respiration is as follows (Fig. 1-12 on page 11):

1. Place the victim on his or her back immediately. *Turn the head and clear the throat* area of water, mucus, foreign objects, or food.

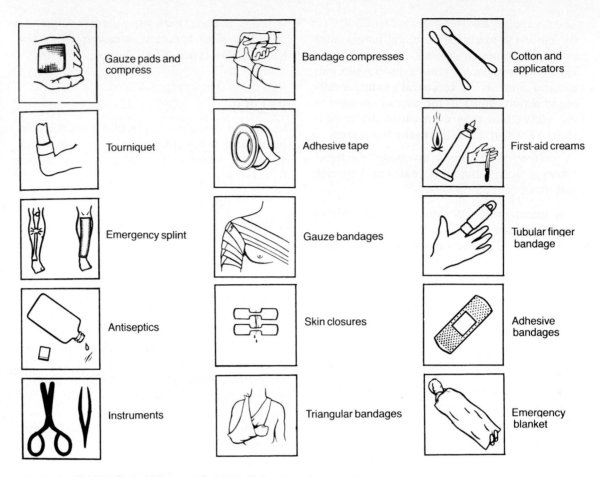

Fig. 1-11 Typical first-aid items. (*Courtesy Safety Supply Canada*)

2. *Tilt the victim's head back* to open the air passage.
3. *Lift the victim's jaw up* to keep the tongue out of the air passage.
4. *Pinch the victim's nostrils* closed to prevent air leakage when you blow.
5. *Seal your lips* around the victim's mouth.
6. *Blow into the victim's mouth* until you see the chest rise.
7. *Remove your mouth* to allow natural exhalation.
8. *Repeat* 12 to 18 times per minute, watching to see that the victim's chest rises and falls, until natural breathing starts.

1. Trigger the nearest fire alarm to alert all personnel in the workplace as well as the fire department.
2. If possible, disconnect the electric power source.
3. Use a carbon-dioxide or dry-powder fire extinguisher to put out the fire. Under no circumstances use water, as the stream of water may conduct electricity through your body and give you a severe shock.
4. Ensure that all persons leave the danger area in an orderly fashion.
5. Do not reenter the premises unless advised to do so by fire department personnel.

1-7 FIRE PREVENTION

Fire prevention is a very important part of any safety program. The chance of a fire occurring can be greatly reduced by *good housekeeping*. You should know where your fire extinguishers are located and how to use them.

In case of an electrical fire, the following procedures should be followed:

1-8 ELECTRICAL CODES AND STANDARDS

Two of the institutions responsible for safety are the National Fire Protection Association, which sponsored the National Electrical Code, and the National Board of Fire Underwriters, which created the Underwriters' Laboratories.

The *National Electrical Code (NEC)* is a set of guidelines describing procedures that minimize the hazards of electric shock, fires, and explosions caused by electrical installation. The text of the NEC is contained in chapters; each chapter is broken into individual articles. For example, the NEC gives tabulations of safe current-carrying capacities of various sizes and types of wire. This is a practical source of information as it includes the limitations of various types of insulations and the effect of various applications. The NEC is reprinted every three years. There are a number of changes each time a new edition is issued.

The NEC serves as a basis from which local governmental authorities write ordinances that deal with protecting the lives of people working with or using electricity or electrical devices. Local laws almost always refer to the NEC as the "standard minimum," sometimes making additions to it to meet local conditions. Local electrical inspectors and fire marshals enforce their own codes and can accept or refuse an installation according to local laws.

The NEC is *not* a textbook to be used as a basis of instruction. Rather, it is a *set of rules,* developed over many years, that has been found to provide safe and practical electrical installations. "Shall" and "should" are frequently used in the NEC. "Shall" pertains to the things that *must* be done to be accepted by the code; "should" are those things that are not required but should be done for minimum safety.

Throughout this text, reference will be made to the National Electrical Code. The electrical code pertinent to you is the one that governs electrical installations in your particular area.

Electric products are generally required to pass standardized tests for safe usage. One of the best-known testing organizations is the *Underwriters' Laboratories,* identified with the *UL* symbol shown in Fig. 1-13 on page 12. The various types of material used in electric wiring should be of a type listed by UL to ensure that an acceptable safety level is being maintained.

The purpose of the Underwriters' Laboratories is "to establish, maintain and operate laboratories for the investigation of materials, devices, products, equipment, construction, methods and systems with regard to hazards affecting life and property." Manufacturers

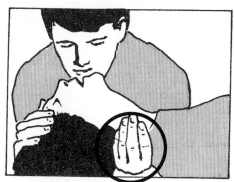

(*a*) Tilt head, clear throat, lift jaw

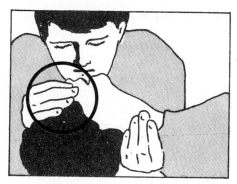

(*b*) Pinch nostrils

(*c*) Make a tight seal— blow into mouth

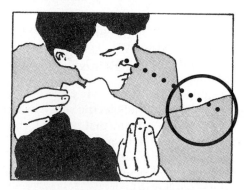

(*d*) Watch to see chest rise and fall— repeat 12 to 18 times per minute

Fig. 1-12   Artificial respiration, mouth-to-mouth method.

Fig. 1-13   The Underwriters' Laboratories label.

submit samples of their products to the UL labs. After vigorous testing to UL standards, the product, if it passes, is listed with UL and given a listing mark. The tests are only the *minimum* safety requirements; all products given a UL mark are not of equal quality. Listed merchandise may be recognized by the UL label, which is either attached to or stamped on the merchandise.

Standards established by the *National Electrical Manufacturers Association (NEMA)* assist users in proper selection of industrial control equipment. NEMA standards provide practical information concerning the rating, testing, performance, and manufacture of motor control devices such as enclosures, contactors, and starters.

Self-Test

Answer the following questions.

24. The purpose of first aid is to _____ life, _____ recovery, and prevent _____ of the injury.
25. True or false? When treating for shock, do not directly touch the victim unless he or she is separated from the power source.
26. If breathing stops, you can help the victim by knowing how to administer _____ .
27. True or false? Under no circumstances should water be used to put out an electrical fire.
28. What is the National Electrical Code?
29. What does the UL mark on a piece of electrical equipment mean?
30. List three motor control devices that are rated by NEMA.

SUMMARY

1. The three electrical factors involved in an electric shock are resistance, voltage, and current.
2. The lower the body resistance, the greater the potential electric shock hazard.
3. Body resistance may be divided into external (skin resistance) and internal (body tissues and blood stream resistance).
4. The higher the voltage, the greater the potential electric shock hazard.
5. Any voltage above 30 V is considered dangerous.
6. The higher the current, the greater the potential electric shock hazard.
7. Any current flow above 5 mA is considered dangerous.
8. The longer the time of exposure, the greater the potential electric shock hazard.
9. Electric shock current that passes through the heart is very dangerous.
10. A safe operation depends largely upon all plant personnel being informed and aware of potential hazards.
11. The Occupational Safety and Health Administration (OSHA) sets standards that regulate safety in plants.
12. Appropriate attire should be worn for each particular job site and work activity.
13. If you become part of an electrical path to ground, you could be seriously burned or even killed.
14. Grounding refers to the deliberate connection of parts of a wiring installation to a common earth connection.
15. Proper grounding guards against fire and shock hazards.
16. Electrical lockout is necessary so that someone will not inadvertently turn the equipment to the ON position while it is being worked on.
17. Receiving an electric shock is a clear warning that proper safety measures have not been followed.
18. Avoid working on "live" circuits and touching any grounded objects when probing live circuits.
19. Discharge capacitors before handling them.
20. Never overload meters by connecting them incorrectly or attempting to measure values far in excess of the range setting.

21. Check meter test leads for frayed or broken insulation before working with them.
22. If someone is hurt, send for help immediately.
23. To control bleeding, apply direct pressure on the wound.
24. For first-degree burns, immerse the injured area in cold water.
25. Administer artificial respiration if the victim is not breathing.
26. The chance of a fire occurring can be greatly reduced by good housekeeping.
27. Use a carbon-dioxide or dry-powder fire extinguisher to put out an electrical fire.
28. The National Electrical Code (NEC) is a set of rules that has been developed to provide safe and practical electrical installations.
29. The Underwriters' Laboratories (UL) do standardized tests of materials used in electric wiring to ensure that an acceptable safety level is being maintained.
30. Standards established by the National Electrical Manufacturers Association (NEMA) are used in the selection of motor control equipment.

CHAPTER REVIEW QUESTIONS

Answer the following questions.

1-1. Does the severity of an electric shock increase or decrease with each of the following changes?
   a. A decrease in the source voltage
   b. An increase in body current flow
   c. An increase in body resistance
   d. A decrease in the length of time of exposure

1-2. *a.* Calculate the theoretical body current flow (in amperes and milliamperes) of an electric shock victim who comes in contact with a 120-V energy source. Assume a total resistance of 1000 Ω (skin, body, and ground contacts).
   b. What effect would this amount of current have on the body?

1-3. List four items of personal safety attire often required on a job site.

1-4. State what each of the following abbreviations stands for:
   a. OSHA
   b. NEC
   c. UL
   d. NEMA

1-5. Explain how a properly grounded system provides protection against a direct ground fault.

1-6. A disconnect switch is to be pulled open as part of a lockout procedure. Explain the safe way to proceed.

1-7. Why is it essential for all interlocking or dependent systems to also be deactivated as part of the lockout procedure?

1-8. List five safety rules that apply to working on electrical equipment in general.

1-9. List five safety rules that apply specifically to meter connections.

1-10. List the four major causes of death and injury resulting from accidents involving electrical workers.

1-11. Outline the basic first-aid procedure for bleeding.

1-12. Outline the basic first-aid procedure for first-degree burns.

1-13. What important rescue procedure should be followed in case of an electrical accident involving a live electric circuit?

1-14. List the important steps to be followed when administering mouth-to-mouth artificial respiration.

1-15. How does the National Electrical Code compare to local laws dealing with electrical installations?

1-16. What is the purpose of Underwriters' Laboratories?

CRITICAL THINKING QUESTIONS

1-1. Technician A comes in contact with a live wire and receives a fatal shock. Technician B comes in contact with the *same* live wire and receives only a mild shock. What are some reasons why this might occur?

1-2. Every year, electrical accidents cause serious injury or death. Many of these casualties are young people just entering the workplace. To what do you think most of these accidents can be attributed?

1-3. You have been assigned the task of explaining the company lockout procedure to new employees. Outline what you would consider to be the most effective way of doing this.

1-4. Obtain a copy of the National Electrical Code (NEC). What are the different types of information it contains?

Answers to Self-Tests

1. false
2. lowers
3. false
4. 30
5. 5
6. orange
7. noisy
8. true
9. ground
10. fire; shock
11. true
12. double
13. padlocking
14. true
15. true
16. false
17. metering
18. grounded
19. true
20. shorting
21. false
22. hold
23. false
24. preserve; assist; aggravation
25. true
26. artificial respiration
27. true
28. A set of rules that has been found to provide safe and practical electrical installations.
29. The equipment has passed minimum safety requirements as determined by the Underwriters' Laboratories.
30. enclosures, contactors, and starters

CHAPTER 2

Understanding Industrial Electrical Diagrams

∎

CHAPTER OBJECTIVES

This chapter will help you to:

1. *Recognize* symbols frequently used on industrial electrical diagrams.
2. *Read* and *construct* ladder diagrams.
3. *Interpret* information found on wiring diagrams.
4. *Become familiar with* the terminology used in motor control circuits.

This chapter introduces the different types of electrical diagrams that are necessary for working with control circuits. In order to facilitate making and reading electrical drawings, certain standard symbols are used. An electrical circuit is drawn as either a ladder diagram or a wire diagram. This chapter will help you to understand the relationship between electrical symbols and the ladder diagram. Terms and practical applications are also presented.

∎

2-1 ELECTRICAL SYMBOLS

The symbols used to represent industrial electrical and electronic components can be considered a form of technical shorthand. The use of these symbols tends to make circuit diagrams less complicated and easier to read and understand.

In industrial control systems, symbols and related lines show how the parts of a circuit are connected to one another. Unfortunately, not all electrical and electronic symbols are standardized. You will find slightly different symbols used by different manufacturers. Also, symbols sometimes look nothing like the real thing, so we have to learn what the symbols mean. Table 2-1 on page 16 lists abbreviations commonly used in industry today.

2-2 LADDER DIAGRAMS

A *ladder* (or line or elementary) *diagram* is a *schematic* representation of an electrical circuit. In this type of schematic, the two power lines connect to the power source, and the various circuits connect across them like rungs in a ladder (Fig. 2-1 on page 16). Notice that the ladder diagram is a schematic representation of the circuit; it is not a physical representation. The electrical components and conductors are arranged according to their electrical function in the circuit—that is, schematically. Simplicity is the purpose of the schematic layout of the ladder diagram. Diagram complexity is greatly reduced by indicating each circuit as a single vertical line.

On some diagrams you will see both heavy and light conductor lines (Fig. 2-2 on page 17). The *heavy* lines are used for high-current-carrying conductors such as the main lines and motor leads. The *lighter* lines are used for control circuits such as the switches, timers, and relays. Conductors may cross each other but make no electrical contact; this is represented by intersecting lines with no dot. Conductors that may make contact are represented by a heavy dot at the junction.

From page 15:

Ladder diagram

On this page:

Power circuit

Control circuit

Off-the-shelf position

De-energized position

Table 2-1 Commonly Used Abbreviations for Electrical Terms and Devices

ALM	alarm	MTR	motor
AC	alternating current	M	motor starter
ARM	armature	NEG	negative
AUTO	automatic	NEUT	neutral
BAT	battery	NC	normally closed
BKR	breaker	NO	normally open
CAP or C	capacitor	OHM	ohmmeter
CKT	circuit	OL	overload relay
CONT	control	PH	phase
CR	control relay	PL	pilot light
CEMF	counter-electromotive force	PLS	plugging switch
CT	current transformer	POS	positive
DIO or D	diode	PWR	power
DC	direct current	PS	pressure switch
DIR	direction	PRI	primary
DISC	disconnect switch	PB	pushbutton
D	down	REC	rectifier
DWG	drawing	RES or R	resistor
DS	drum switch	REV	reverse
DB	dynamic braking	RH	rheostat
EMF	electromotive force	SSW	safety switch
FLD	field	SEC	secondary
FS	float switch	SS	selector
FLS	flow switch	SCR	semiconductor-
FTS	foot switch		controlled
FWD	forward		rectifier
FREQ	frequency	1PH	single-phase
FU	fuse	SOL	solenoid
GEN	generator	SW	switch
GRD	ground	TEMP	temperature
HP	horsepower	THS	thermostat switch
IC	integrated circuit	3PH	three-phase
INTLK	interlock	TD	time delay
LT	lamp	TR	time-delay relay
LS	limit switch	T, TRANS	transformer
MAN	manual	UV	under voltage
MEM	memory	U	up

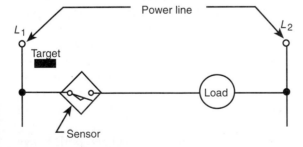

Fig. 2-1 Basic ladder diagram.

Most schematics can be broken down into two basic parts: the *power circuit* and the *control circuit*. The purpose of having a power and control circuit is to provide control of the machine without using devices (except contactors and wiring) that must carry many amperes. By using devices such as contactors, we can control a motor or other loads that draw a large amount of amperage with a control system that could use a lower voltage and much less amperage. The power circuit provides the main power and the power to the motors, whereas the control circuit provides the control.

All switches and relay contacts can be classified as normally open or normally closed. The positions drawn on diagrams are the electrical characteristics of each device as it would be found when purchased and not connected in any circuit. This is sometimes referred to as the *off-the-shelf position*. It is important to understand this because it may also represent the de-energized position in a circuit. The *de-energized position* refers to the component position when the circuit is de-energized, or no power is present on the circuit. This point of reference is often used as the starting point in the analysis of the operation of the circuit.

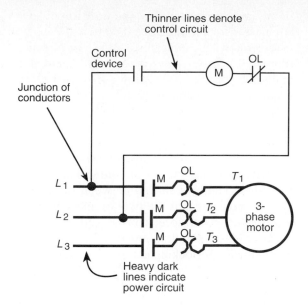

Fig. 2-2 **Conductor lines and connections.**

In order to identify the relay coil and its several contacts, we put a letter or letters in the circle that represents the coil (Fig. 2-3). Each contact that is operated by this coil will have the coil letter or letters written next to the symbol for the contact. Sometimes, when there are several contacts operated by one coil, a number is added to the letter to indicate the contact number. Although there are standard meanings of these letters, most diagrams provide a key list to show what the letters mean; generally they are taken from the name of the device.

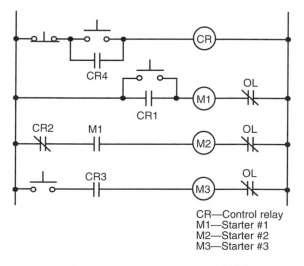

CR—Control relay
M1—Starter #1
M2—Starter #2
M3—Starter #3

Fig. 2-3 **Identification of contacts and coils.**

The *load* is the electrical device in the line or ladder diagram that uses the electrical

power from L_1 to L_2. Control relays, solenoids, and pilot lights are examples of loads. At least one load device must be included in each *rung* (individual circuit) of the diagram. Without a load device, the control devices would be switching an open circuit to a short circuit between L_1 and L_2.

All loads have one side connected to L_2 (Fig. 2-4). Generally, no more than one load is placed in any one circuit line between L_1 and L_2. When more than one load must be connected in the line diagram, the loads must be connected in parallel. This will ensure that the full line voltage from L_1 and L_2 will appear across each load. If the loads are connected in series, neither will receive the entire line voltage necessary for proper operation.

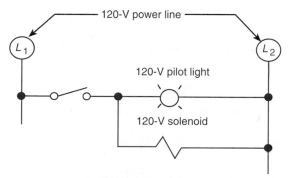

Fig. 2-4 **Connection of additional loads.**

▐▶ **YOU MAY RECALL** that in a series circuit the applied voltage is divided among each of the loads. In a parallel circuit the voltage across each branch is the same and is equal in value to the applied voltage.

Loads are operated by control devices such as switches, pushbuttons, limit switches, and pressure switches. Control devices are connected between L_1 and the load. All additional STOP or OFF control devices must be wired in *series*. All additional START or ON control devices must be wired in parallel (Fig. 2-5 on page 18).

The ladder diagram uses numbers to help you locate electrical devices, numbered wire locations, and schematic locations. Each *line* or *rung* is marked (lines 1, 2, 3, etc.) starting with the top line and reading down. A line can be defined as a complete path from L_1 to L_2 that contains a load. Figure 2-6 on page 18 illustrates the marking of each line in a line diagram with three separate lines. The circled numbers identify the lines on the physical drawing; you will not find these numbers anywhere in the

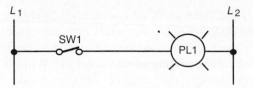

(a) Control device properly connected between L_1 and the load

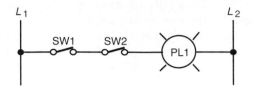

(b) Additional OFF control device wired in series

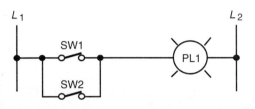

(c) Additional ON control device wired in parallel

Fig. 2-5   Connection of additional control devices.

electrical panel. They are used only as an aid in reading the diagram. However, you will find a number on the end of each wire throughout the electrical panel. These numbers correspond to the wire identification numbers on the ladder diagram. Thus, on line 3 (Fig. 2-6) you can see three wire numbers: 1, 6, and 2.

Wires common to each other are usually designated by a single number on the diagram. For example, wire 2 (lines 1 and 3 shown in Fig. 2-6) has a single number designation. Yet you know that all the wires common to this numbered wire are also number 2 wires. On the right side of the ladder diagram next to the common control wire is a series of descrip-

tions. The descriptions tell the function of that circuit by the output device it controls.

A broken line indicates a mechanical function. *It is not an electrical conductor.* Do not make the mistake of reading a broken line as a part of the electrical circuit. In Fig. 2-7 the vertical broken lines on the forward and reverse pushbuttons indicate that their normally closed and normally open contacts are mechanically connected. Thus, pressing the button will open the one set of contacts and close the other. The broken line between the F and R coils indicates that the two are mechanically interlocked. Therefore, coils F and R cannot close contacts simultaneously because of the mechanical interlocking action of the device.

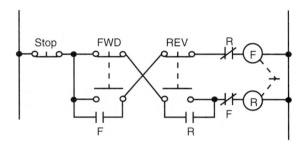

Fig. 2-7   Mechanical functions.

If a control circuit is supplied from a grounded circuit, it must be connected so that an accidental ground in the control circuit will not start the motor or make the stop button or control inoperative. Figure 2-8(a) shows the control transformer properly grounded. When the circuit is operational, the entire circuit to the left of coil M is the *ungrounded* circuit (it is the "hot" leg). Therefore, *any short to ground in the ungrounded circuit will blow the control transformer fuse.*

What would happen if the circuit was grounded at L_1 [Fig. 2-8(b)]? A short to ground any place in the control circuit would

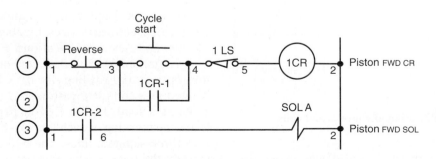

Fig. 2-6   Line and wire numbers.

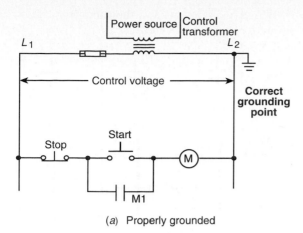

(a) Properly grounded

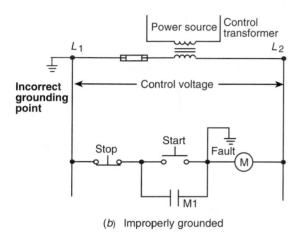

(b) Improperly grounded

Fig. 2-8   Grounding for a control circuit supplied from a grounded circuit.

activate the output device. Can you imagine the danger that would be involved? A short circuit to ground would not blow the fuse; rather, it would activate the circuit, starting the motor unexpectedly. Pressing the stop button would not de-energize the M coil. Equipment damage and personnel injuries would be very likely. Clearly, output devices must be directly connected to the grounded side of the circuit.

▐▐▐➤ **YOU MAY RECALL**   that there is only one fuse on the circuit side of a control transformer. There is a fuse on the "hot" side but *not* on the grounded neutral side. There is an important safety factor in this design. If a fuse was placed on the grounded side and it opened—and the fuse on the hot side remained conductive—the circuit would be inoperative but the conductors would remain live. Although the circuit could not function, it presents a hazard to anyone working on it. *In practice, a fuse is never placed on the grounded side of the circuit.*

Looking at the whole circuit in a ladder diagram schematic at once can be very confusing. Try breaking the schematic down into smaller pieces and applying some basic rules:

- Read the schematic from left to right.
- Read the schematic from top to bottom.
- Notice that on the schematic diagram all devices are shown in their de-energized state.
- When the coil that controls a set of contacts is energized, change the contacts on your schematic to their energized state.
- Frequently, relay contacts on the line you are reading are controlled by other relays, switches, and so on, on other lines.

Self-Test

Answer the following questions.

1. True or false? Symbols always look like the real thing.
2. List the commonly used abbreviation for each of the following electrical terms:
   a. alternating current
   b. control relay
   c. ground
   d. manual
   e. positive
   f. integrated circuit
   g. secondary
   h. three-phase
3. A ladder diagram is a(n) _____ representation of an electrical circuit.
4. In a ladder diagram, the electrical components are arranged according to their electrical _____ in the circuit.
5. True or false? The heavier lines on electrical diagrams usually indicate high-current-carrying conductors.
6. On electrical diagrams, conductors that cross each other and make contact are represented by a heavy _____ at the junction.
7. The circuits on most motor control schematics can be broken down into being part of the _____ circuit or the _____ circuit.
8. True or false? All electrical components are normally drawn in their off-the-shelf position.
9. True or false? A pressure switch would be classified as a load device.
10. In a(n) _____ circuit, the voltage across each branch is the same.
11. All additional START or ON control devices must be wired in _____.

12. On a ladder diagram, a line can be defined as a complete _____ from L_1 to L_2 that contains a load.
13. True or false? Wires common to each other are usually designated by a single number on the diagram.
14. True or false? Broken lines on a diagram are usually used to indicate control circuits.
15. True or false? If a control circuit is supplied from a grounded circuit, it makes no difference what side of the line is grounded.
16. True or false? A fuse is never placed on the grounded side of the circuit.

2-3 WIRING DIAGRAMS

A *wiring diagram* is intended to show the actual *connection* and *physical location* of all component parts in a circuit. Coils, contacts, motors, and the like are shown in the actual position that would be found on an installation. Since wiring connections and terminal markings are shown, this type of diagram is helpful when wiring the device, or tracing wires when troubleshooting.

Frequently, there will be a wiring diagram inside a magnetic motor starter cover. Figure 2-9 illustrates a typical wiring diagram for a motor starter. The diagram shows, as closely as possible, the actual location of all of the component parts of the device. The open terminals (marked by an open circle) and arrows represent connections made by the user. Note that bold lines denote the power circuit, and thin lines are used to show the control circuit. Conventionally, in ac magnetic equipment, black wires are used in power circuits, and red wires are used in control circuits.

For smaller circuits it is advantageous to use physical rather than schematic diagrams. It is simpler to identify terminals and wire locations from a physical diagram. Occasionally, for equipment with simple circuits, you may find that the entire unit is represented with only a wiring diagram. Figure 2-10 shows a typical wiring diagram for an electric chain hoist.

A *conduit layout diagram* indicates the start and the finish of the electrical conduits and shows the approximate path taken by any conduit in progressing from one point to another. Integrated with a drawing of this nature is the conduit and cable schedule, which tabulates each conduit as to number, size, function, service, and also includes the number and size of wires to be run in the conduit. Such a drawing is essential for the initial installation of any

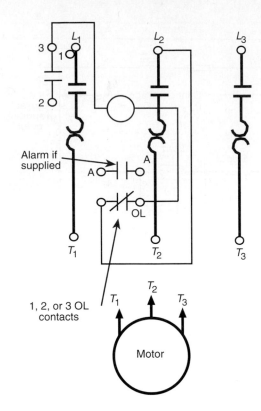

Fig. 2-9   Typical wiring diagram for a motor starter.

electrically operated equipment. A typical conduit layout is shown in Fig. 2-11.

A *diagram of connections* attempts to show the physical connections and wiring involved in the construction of any piece of electrical equipment. Insofar as is practicable, these diagrams are laid out in such a way that upon looking at the equipment and referring to the drawing, if one point on the equipment is known the other points can be readily determined by inspection. A typical diagram of connections is shown in Fig. 2-12 on page 22. This is a diagram of connections of a typical combination magnetic line starter complete with control transformer. From this sketch it can be seen that the diagram approximates quite closely the actual appearance, especially with regard to terminals for outgoing connections.

To assist in installing and tracing wires in complex installations, a *plan of circuits diagram* is frequently employed. This drawing rarely attempts to show complete details of panel board or equipment wiring and is chiefly used to indicate wiring from terminal boards, panel boards, or other equipment to ultimate devices. The example previously used, showing a line starter motor and pushbutton station, reduced to the form used on a plan of circuits, is shown in Fig. 2-13 on page 22. The auxiliary

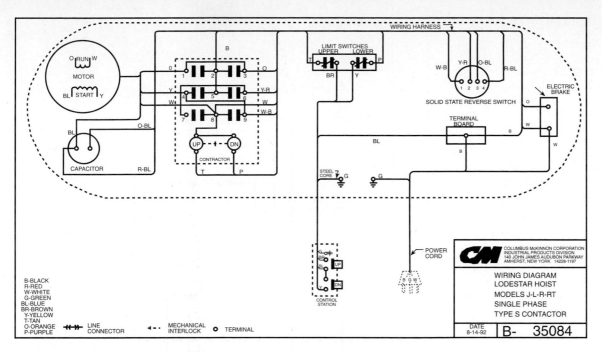

Fig. 2-10 Typical wiring diagram for an electric chain hoist. (*Source: CM Hoist Div., Columbus McKinnon Corporation*)

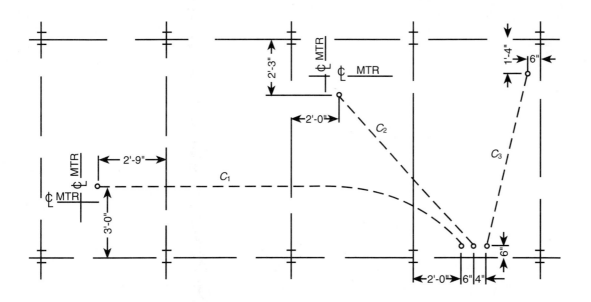

Conduit No.	Conduit Size	Conductor		From	To	Service	Remarks
		Size	Type				
C_1	1"	3-#6	TW	Panel #1	Pump #1	Power	
C_2	$1\frac{1}{4}$"	3-#4	TW	Panel #1	Pump #2	Power	
C_3	1"	4-#12	TW	Panel #1	Pushbutton Station	Control	

Fig. 2-11 Conduit layout diagram.

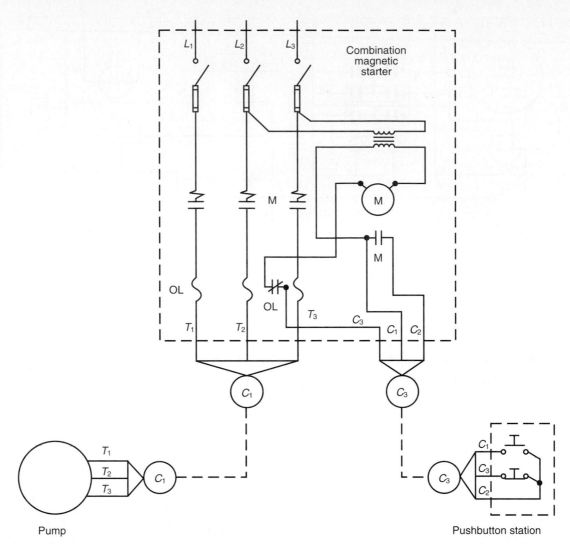

Fig. 2-12   Diagram of connections.

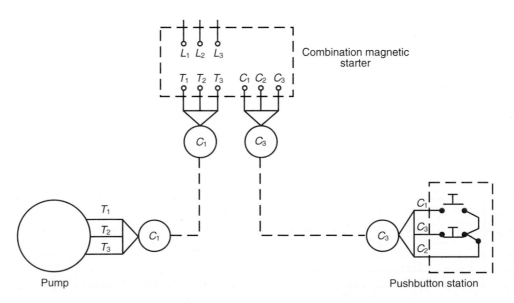

Fig. 2-13   Plan of circuits diagram.

equipment is still shown in detail, but the line starter is now reduced to show only incoming and outgoing connections. The internal connections of the line starter are omitted.

You read a wiring diagram in much the same way as you read a ladder diagram; that is, the circuit functions are represented by their appropriate symbols. However, you need to follow conductor lines rather than a single horizontal circuit line. This adds to the potential confusion when reading a wiring diagram. The wiring diagram is a useful circuit representation. Its use, however, is generally limited to small, specific circuit functions.

2-4 SINGLE-LINE AND BLOCK DIAGRAMS

A *single-line diagram* permits greater simplification of a circuit by omitting auxiliary functions. The single-line drawing, like the one shown in Fig. 2-14, is used by many motor control equipment manufacturers as a road map in the study of motor control installations. The installation is reduced to the simplest possible form, yet it still shows the essential requirements and equipment in the circuit.

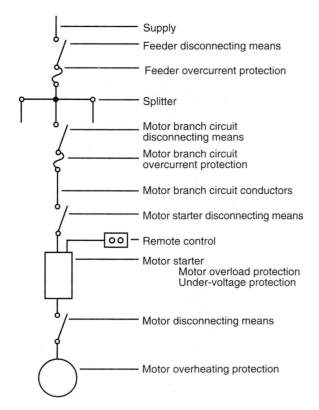

Fig. 2-14 Single-line diagram of a motor installation.

Supply
Feeder disconnecting means
Feeder overcurrent protection
Splitter
Motor branch circuit disconnecting means
Motor branch circuit overcurrent protection
Motor branch circuit conductors
Motor starter disconnecting means
Remote control
Motor starter
    Motor overload protection
    Under-voltage protection
Motor disconnecting means
Motor overheating protection

Single-line diagrams are often found showing major switching and switchgear design, where a considerable amount of information needs to be shown and yet the main object is to make the layout as simple as possible. Figure 2-15 on page 24 shows the single-line diagram of a small power distribution system.

A *block diagram* represents the major functional parts of complex electronic and electrical systems by blocks. Individual components and wires are *not* shown. Instead, each block represents electrical circuits that perform specific functions in the system. The functions the circuits perform are written in each block. Arrows connecting the blocks indicate the general direction of current paths. Figure 2-16 on page 24 shows the block diagram of a solid-state crane control system.

Self-Test

Answer the following questions.

17. A wiring diagram is intended to show the actual _____ of all components.
18. What other information is often integrated with a conduit layout diagram?
19. True or false? A plan of circuits diagram normally attempts to show the complete details of internal equipment wiring.
20. True or false? A single-line diagram reduces a circuit to its simplest form.
21. True or false? No wires or components are normally shown on block diagrams.

2-5 MOTOR CONNECTIONS AND TERMINOLOGY

Most industrial machinery is driven by electric motors. Industries would cease to function without properly designed, installed, and maintained motor control systems.

In general, motors are classified according to the type of power used (ac or dc) and the motor's principle of operation. Figure 2-17 on page 25 shows the symbols used to identify the basic parts of a direct current (dc) motor. The rotating part of the motor is referred to as the *armature;* the stationary part of the motor is referred to as the *stator,* which contains the *series field winding* and the *shunt field winding.* In dc machines A_1 and A_2 always indicate the armature leads, S_1 and S_2 indicate the series field leads, and F_1 and F_2 indicate the shunt field leads.

Single-line diagram

Block diagram

Armature

Stator

Series field winding

Shunt field winding

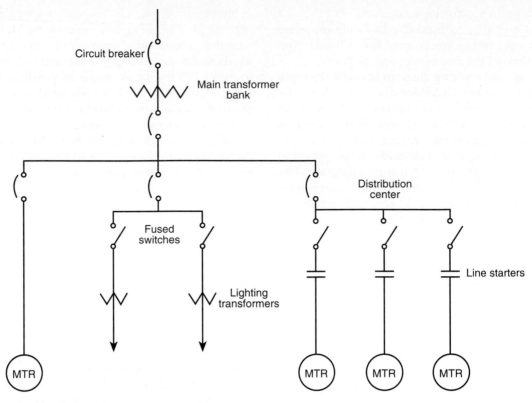

Fig. 2-15   Single-line diagram of a power distribution system.

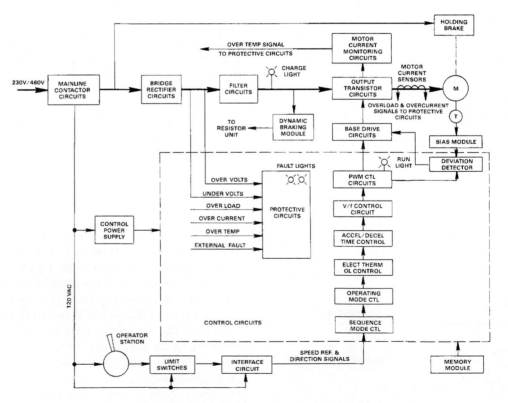

Fig. 2-16   Block diagram of a solid-state crane control system. (*Courtesy Harnischfeger Corporation*)

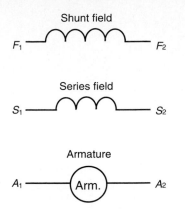

Shunt field

F_1 ⸺◠◠◠⸺ F_2

Series field

S_1 ⸺◠◠⸺ S_2

Armature

A_1 ⸺(Arm.)⸺ A_2

Fig. 2-17   Parts of a dc motor.

It is the kind of excitation provided by the field and nothing more that makes the difference between one type of dc motor and another; the construction of the armature has nothing to do with the type of motor.

There are three general types of dc motors classified according to the method of field excitation.

- A *shunt motor* (Fig. 2-18) uses a comparatively high resistance field winding of many turns of fine wire, usually connected in parallel (in shunt with the armature).
- A *series motor* (Fig. 2-19 on page 26) uses an extremely low resistance field winding of very few turns of heavy wire connected in series with the armature.
- A *compound motor* (Fig. 2-20 on page 26) uses a combination of shunt field (many turns of fine wire) in parallel with the armature, and series field (few turns of heavy wire) in series with the armature.

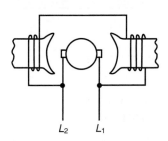

Counterclockwise		Clockwise	
Line 1 F_1–A_1	Line 2 F_2–A_2	Line 1 F_1–A_1	Line 2 F_2–A_1
F_1 ◠◠◠ F_2		F_1 ◠◠◠ F_2	
L_1 A_1 (A) A_2 L_2		L_1 A_2 (A) A_1 L_2	

Fig. 2-18   Standard dc shunt motor connections for counterclockwise and clockwise rotation.

The standard direction of shaft rotation for motors is counterclockwise facing the end opposite the drive end. The purpose of applying markings to the terminals of motors according to a standard is to aid in making connections to other parts of the electric power system and to avoid improper connections, which may result in unsatisfactory operation or damage. Terminal markings are normally used to tag only those terminals to which connections must be made from outside circuits.

The use of alternating current (ac) motors is more widespread than dc motors. The rotating part of an ac motor is referred to as the *rotor;* the stationary part of the motor is referred to as the stator. A squirrel-cage rotor consists of a laminated steel core with conductors cast into the rotor slots. All ac motors are classified according to the type of ac power used: either *single-phase* or *three-phase* (polyphase).

The majority of single-phase ac motors are constructed in fractional horsepower sizes for 120- to 240-V, 60-hertz (Hz) power sources. Types of single-phase motors include the split-phase, capacitor, shaded-pole (Fig. 2-21 on page 27).

Some characteristics of a split-phase motor are:

- It is the simplest and most common single-phase motor.
- The starting winding produces a phase difference to start the motor and is switched out by a centrifugal switch as running speed is approached.
- Its design is low in cost.
- Starting torque is limited to applications such as furnace fans and some power tools.
- Typical sizes range up to about ½ horsepower.
- The motor can be reversed by reversing the leads to the starting winding or main winding, but not to both.

A typical dual-voltage split-phase motor, which would be either 120 or 240 V, has two run windings instead of one. The two run windings are connected in series to the supply lines for high voltage, and the two windings are connected in parallel for low voltage. The start winding is connected across the supply lines for low voltage and at one line to the midpoint of the run windings for high voltage. To reverse a dual voltage split-phase motor, interchange the two start winding leads.

Advantages to using a permanent capacitor motor are listed on the next page.

Shunt motor

Series motor

Compound motor

Rotor

Single-phase motor

Three-phase motor

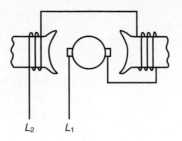

Counterclockwise			Clockwise		
Line 1 A_1	Tie A_2–S_1	Line 2 S_2	Line 1 A_2	Tie A_1–S_1	Line 2 S_2
L_1 A_1 A A_2 S_1 S_2 L_2			L_1 A_1 A A_2 S_1 S_2 L_2		

Fig. 2-19 Standard dc series motor connections for counterclockwise and clockwise rotation.

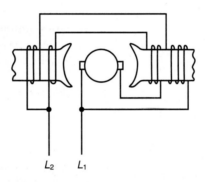

Counterclockwise			Clockwise		
Line 1 F_1–A_1	Tie A_2–S_1	Line 2 F_2–S_2	Line 1 F_1–A_2	Tie A_1–S_1	Line 2 S_2–F_2
F_1 F_2 L_1 A_1 A A_2 S_1 S_2 L_2			F_1 F_2 L_1 A_2 A A_1 S_1 S_2 L_2		

Fig. 2-20 Standard dc compound (cumulative) motor connections for counterclockwise and clockwise rotation. For differential compound connection, reverse S_1 and S_2.

- It uses a capacitor permanently connected in series with one of the stator windings to achieve a compromise between good starting torque and good running characteristics.
- This design is lower in cost than other capacitor motors that incorporate capacitor switching systems.
- It achieves better starting torque and running characteristics than a split-phase motor.

- Its smoother, more efficient running compares more favorably with three-phase operation.

   Now let's look at some of the characteristics of a shaded-pole motor.

- The shaded portion of the pole is isolated from the rest of the pole by a copper conductor that forms a single turn around it.

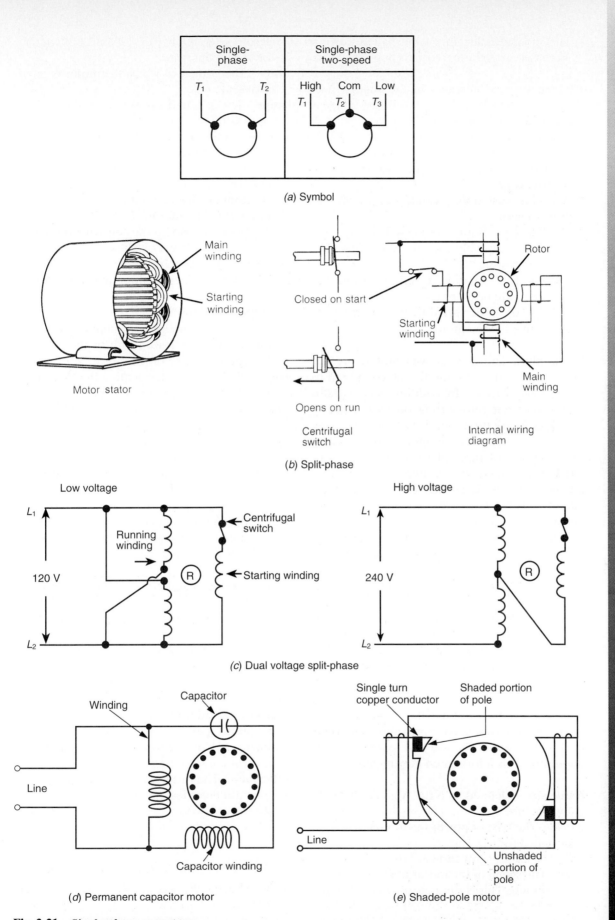

Single-phase	Single-phase two-speed
T_1 T_2	High Com Low T_1 T_2 T_3

(a) Symbol

Main winding

Starting winding

Motor stator

Closed on start

Opens on run

Centrifugal switch

Rotor

Starting winding

Main winding

Internal wiring diagram

(b) Split-phase

Low voltage

L_1

Running winding

Centrifugal switch

120 V

Ⓡ

Starting winding

L_2

High voltage

L_1

240 V

Ⓡ

L_2

(c) Dual voltage split-phase

Winding

Capacitor

Line

Capacitor winding

(d) Permanent capacitor motor

Single turn copper conductor

Shaded portion of pole

Line

Unshaded portion of pole

(e) Shaded-pole motor

Fig. 2-21   Single-phase ac motors.

- The magnetic flux in the unshaded portion increases with the current through its winding.
- Magnetic flux increases in the shaded portion; however, it is delayed by current induced in the copper shield.
- The magnetic field sweeps across the pole face from the unshaded portion to the shaded portion, developing a torque in the squirrel cage.
- It is a low-cost motor, with low efficiency and power factor.
- Shaded-pole motors are used where low torque is acceptable and are usually less than $\frac{1}{10}$ horsepower.
- Direction of rotation is always from the unshaded area toward the shaded area.
- The speed can be varied by varying the voltage applied to the motor.

Single-phase large horsepower motors are not used because they are inefficient, compared to three-phase motors. In addition, single-phase motors are not self-starting on their running windings, as are three-phase motors.

Large horsepower ac motors are usually three-phase. All three-phase motors are constructed internally with a number of individually wound coils. Regardless of how many individual coils there are, the individual coils will always be wired together (series or parallel) to produce three distinct windings, which are referred to as phase A, phase B, and phase C (Fig. 2-22).

The *motor nameplate* contains important information about the connection and use of the motor. When a motor is selected for an installation, the horsepower, speed, and heat rating must be known. When it is installed, the voltage and full-load current of dc motors must be known. For ac motors the voltage, full-load current, frequency, phases, and locked rotor current must be known.

The motor nameplate provides important information about the use and replacement of the motor. This information may include:

frame size   Defined by NEMA and indicated by a number.

phase   The type of power the motor was designed for.

hp   The horsepower rating of the motor.

rpm   The speed of the motor in revolutions per minute. This is running, not synchronous, speed.

cycles   The frequency of the power at which the motor operates.

volts   The voltage at which the motor is rated to operate.

amps   The full-load current at rated load, rated voltage, and rated frequency.

degree Centigrade/rise   The permissible temperature increase (in degrees Centigrade) above ambient temperature of a motor operating at rated load and speed.

serial number   The manufacturer's code number for the motor.

type   Code describing the construction of the motor and the power it runs on.

housing   Motor enclosure.

service factor   An indication that when the motor is operated at rated voltage and frequency, it may be safe to overload the motor to the horsepower obtained by multipying the service factor by the horsepower rating of the motor.

code   NEMA code designating the locked-rotor kilovolt ampere (kVA) per horsepower; for example, M allows from 10 to 11.2 kVA per horsepower.

hours   The period of time when the motor can operate without overheating. Small motors are generally rated "continuously."

Terminology is of the utmost importance when attempting to understand the electrical motor control. Common motor control terms include the following:

across-the-line   A method of motor starting. Connects the motor directly to the supply line on starting or running. (Also called *full voltage*.)

automatic starter   A self-acting starter. Completely controlled by the master or pilot switches or some other sensing device.

auxiliary contact   Contact of a switching device in addition to the main circuit contacts. Operated by the main contacts.

contactor   A device to repeatedly establish or interrupt an electric power circuit.

efficiency   The ratio of mechanical output power to electrical input power:

$$\frac{\text{Output}}{\text{Input}}$$

Note: The list of common motor terms continues on page 30.

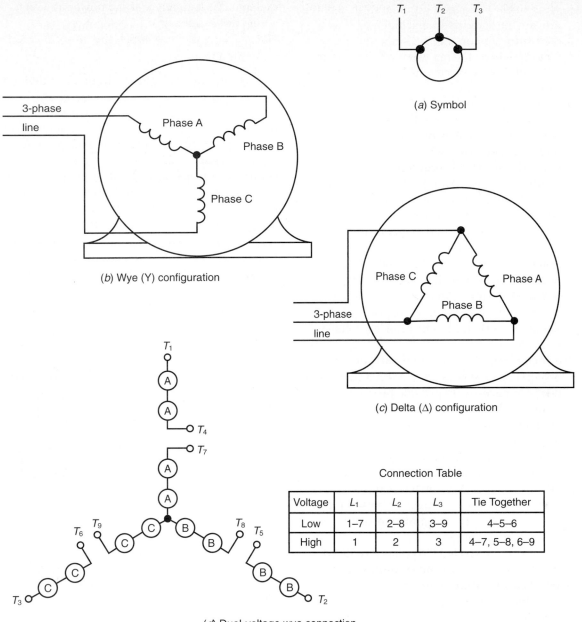

(a) Symbol

(b) Wye (Y) configuration

(c) Delta (Δ) configuration

Connection Table

Voltage	L_1	L_2	L_3	Tie Together
Low	1–7	2–8	3–9	4–5–6
High	1	2	3	4–7, 5–8, 6–9

(d) Dual-voltage wye connection

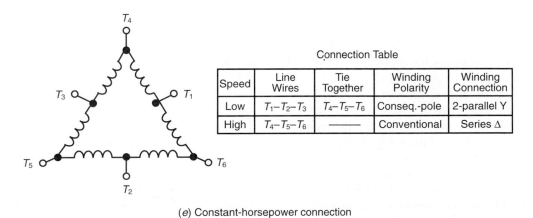

Connection Table

Speed	Line Wires	Tie Together	Winding Polarity	Winding Connection
Low	T_1–T_2–T_3	T_4–T_5–T_6	Conseq.-pole	2-parallel Y
High	T_4–T_5–T_6	———	Conventional	Series Δ

(e) Constant-horsepower connection

Fig. 2-22　Three-phase ac motors.

jogging (inch) Momentary operation. Small movement of a driven machine.

locked-rotor amperes The amperage draw of a motor when starting where the magnetic flux and the rotor are at the same position or locked.

low voltage protection (LVP) Magnetic control only; nonautomatic restarting. A three-wire control. A power failure disconnects service; when power is restored, manual restarting is required.

low voltage release (LVR) Magnetic control only; automatic restarting. A two-wire control. A power failure disconnects service; when power is restored, the controller automatically restarts.

magnetic contactor A contactor that is operated electromechanically.

multispeed starter An electric controller with two or more speeds (reversing or nonreversing) and full or reduced voltage starting.

overload relay Running overcurrent protection. Operates on excessive current. It does not necessarily provide protection against a short circuit. It causes and maintains interruption of the device from a power supply.

plugging Braking by reverse rotation. The motor develops retarding force.

pushbutton A master switch that is a manually operable plunger or button for actuating a device, assembled into pushbutton stations.

reduced voltage starter Applies a reduced supply voltage to the motor during starting.

relay Used in control circuits and operated by a change in one electrical circuit to control a device in the same circuit or another circuit. Ampere rated.

remote control Controls the function initiation or change of electrical device from some remote point.

selector switch A manually operated master switch for rotating the motion for an actuating device. Assembled into pushbutton master stations.

slip The difference in speed between the rotation of the magnetic flux in the stator and the rotation of the rotor shaft.

starter An electric controller. A start-, stop-, and protect-connected motor.

timer A pilot device, also considered a timing relay, that provides an adjustable time period to perform its function. It is motor driven, solenoid actuated, and electronic.

torque The turning twisting force. There are two types of torque that are considered when looking at motors: starting torque and running torque.

Magnetic motor control circuits are divided into two basic types: the two-wire control circuit, and the three-wire control circuit. *Two-wire control* uses a maintained-contact type of pilot device to provide *low voltage release (LVR)*. Figure 2-23 shows a typical two-wire control circuit. Notice that as long as the two-wire control device is closed, power can be supplied to the coil of the controller. If the motor is stopped by a power interruption, the two-wire control device will not open. Since the control device does not open, the motor will restart when power is restored to the system.

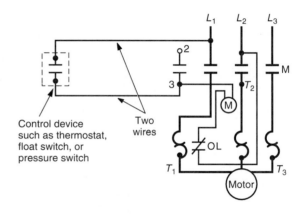

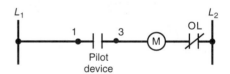

Fig. 2-23 Two-wire control.

Two-wire control circuits are used in applications, such as blower fans and pumps, where the automatic restarting characteristic is desirable and there is no danger of a person's being injured if the equipment should suddenly restart after a power failure.

Three-wire control uses a momentary-contact pilot device and a holding circuit contact to provide *low voltage protection (LVP)*. This means that the starter will drop out when there is a voltage failure, but it will *not* pick up automatically when voltage returns. The most common example of a three-wire control circuit is probably the start-stop, pushbutton control shown in Fig. 2-24.

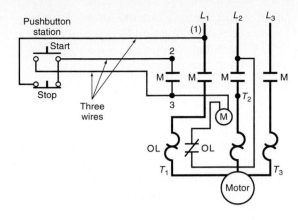

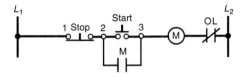

Fig. 2-24 Three-wire control.

The control circuit is completed through the stop button and also through a holding contact (2–3) on the starter. When the starter drops out, the holding contact opens and breaks the control circuit until the start button is pressed to restart the motor. In the event of a power failure, these circuits are designed to protect against automatic restarting when the power returns. This type of protection should be used where accidents or damage might result from unexpected starts.

Self-Test

Answer the following questions.

22. Motors are classified according to the type of power used as being _____ or _____.

23. What part of the motor is indicated by each of the following lead designations?
    a. A_1 and A_2
    b. S_1 and S_2
    c. F_1 and F_2
24. List the three general types of dc motors.
25. True or false? The standard direction of shaft rotation for motors is clockwise facing the drive end.
26. Terminal markings are usually used to tag only those terminals to which connections must be made from _____ circuits.
27. True or false? The use of dc motors is more widespread than the use of ac motors.
28. The rotating part of an ac motor is referred to as the _____; the stationary part is called the _____.
29. True or false? Single-phase ac motors are constructed in large horsepower sizes.
30. True or false? Three-phase motors are not self-starting on their running windings.
31. The three distinct windings of a three-phase motor are referred to as: _____, _____, and _____ .
32. Information required to identify a motor is contained on the motor _____.
33. State the correct motor terminology for each of the following:
    a. Ratio of output to input
    b. The turning twisting force
    c. The amperage draw of a motor when starting
    d. Applies reduced voltage to a motor during starting
    e. Provides an adjustable time period to perform a function
34. Two-wire control circuits use a _____-contact type of pilot device.
35. True or false? Three-wire control circuits are designed to protect against automatic restarting when power returns after a power failure.

SUMMARY

1. The use of symbols tends to make diagrams less complicated.
2. Before an electrical diagram can be examined, it is necessary to understand what each symbol and abbreviation represents.
3. A ladder diagram is a schematic representation of a circuit with components arranged according to their function.
4. On diagrams, heavy lines are used for high-power main and motor circuits, whereas lighter lines are used for low-power control circuits.
5. Most schematics can be broken down into two basic parts: the power circuit and the control circuit.

6. Active components are drawn in their de-energized or deactivated position on diagrams.
7. Relay contacts are identified by the same letter that represents the coil.
8. At least one load must be included in each rung of a ladder diagram.
9. When more than one load is connected in a rung of a ladder diagram, the loads are connected in parallel.
10. Additional STOP or OFF control devices are wired in series; additional START or ON control devices are wired in parallel.
11. Ladder diagrams use numbers to help you locate devices and wire locations.
12. Wires common to each other are usually designated by a single number on the diagram.
13. A broken line indicates a mechanical function.
14. Grounded control circuits must be wired so that any short to ground in the entire circuit will blow the control transformer fuse.
15. Schematic ladder diagrams are read from top to bottom, with each rung being read from left to right.
16. Wiring diagrams show the actual connection and physical location of all component parts.
17. Conduit layout diagrams, diagram of connections, and plan of circuits diagrams are all associated with wiring diagrams.
18. A single-line diagram shows the circuit in its simplest form.
19. Block diagrams are used to represent the major functional parts of complex circuits.
20. Motors are classified according to the type of power used and their principle of operation.
21. The three general types of dc motors are the shunt, series, and compound motor.
22. The rotating part of an ac motor is referred to as the rotor; the stationary part is called the stator.
23. All ac motors are classified as either single-phase or three-phase.
24. Single-phase motors are limited to small horsepower applications.
25. The windings on a three-phase motor are called phase A, phase B, and phase C.
26. A motor nameplate contains information about the connection and the use of the motor.
27. Two-wire control uses a maintained-contact type of pilot device to provide low voltage release.
28. Three-wire control uses a momentary-contact pilot device and a holding circuit contact to provide low voltage protection.

CHAPTER REVIEW QUESTIONS

Answer the following questions.

2-1. What difficulty is encountered when standard symbols are not used in electrical diagrams?

2-2. Identify the following abbreviations:

a. PH   e. MTR
b. SOL   f. TD
c. PB   g. BAT
d. RES   h. PWR

2-3. Compare the way a circuit is represented in a ladder diagram and a wiring diagram.

2-4. How is the power circuit distinguished from the control circuit on a schematic diagram?

2-5. All make-and-break devices (switches, contact points, pushbuttons, etc.) are represented on ladder diagrams in a given operating position. What is that position?

2-6. What do the letter designations "NO" and "NC" stand for? Explain their meanings.

2-7. Explain how a relay coil with several contacts is identified on a ladder diagram.

2-8. *a.* At least one load device must be included between L_1 and L_2 in each rung of a ladder diagram. Why?

b. If more than one load must be connected in the line diagram, how are the loads connected?

2-9. How is each line in a line diagram marked to distinguish that line from all other lines?

2-10. What does a broken line on a ladder diagram indicate?

2-11. Why is the output device always the last component in the circuit, with one terminal connected to the grounded (common) control wire?

2-12. In what order is a ladder diagram read?

2-13. What information is contained on the conduit and cable schedule?

2-14. What does a plan of circuits diagram show?

2-15. How does a ladder diagram differ from a single-line diagram?

2-16. What is the major function of the block diagram?

2-17. Name the internal parts of a dc motor.

2-18. How are ac motors classified according to the type of ac power used?

2-19. Compare the operation of a two-wire and a three-wire control.

2-20. Name the control element commonly used to step down a high-voltage power source (e.g., 480 Vac) to a lower-control voltage source (e.g., 120 Vac)?

CRITICAL THINKING QUESTIONS

2-1. What would happen if two identical loads were connected in series with a 120-Vac power source?

2-2. Pilot devices do not consume power. Why?

Answers to Self-Tests

1. false
2. a. AC
   b. CR
   c. GRD
   d. MAN
   e. POS
   f. IC
   g. SEC
   h. 3PH
3. schematic
4. function
5. true
6. dot
7. power
   control
8. true
9. false
10. parallel

11. parallel
12. path
13. true
14. false
15. false
16. true
17. location
18. a conduit and cable schedule
19. false
20. true
21. true
22. ac or dc
23. a. armature
    b. series field
    c. shunt field
24. shunt, series, and compound
25. true

26. outside
27. false
28. rotor
    stator
29. false
30. false
31. phase A, phase B, and phase C
32. nameplate
33. a. efficiency
    b. torque
    c. locked rotor amperes
    d. reduced voltage starter
    e. timer
34. maintained
35. true

CHAPTER 3

Transformers and Power Distribution Systems

∎

CHAPTER OBJECTIVES

This chapter will help you to:

1. *Outline* the method used in transmitting power over long distances.
2. *Explain* the methods and purpose of protecting electric distribution systems.
3. *Discuss* the causes and effects of power line disturbances.
4. *Explain* the theory of operation of various types of transformers.
5. *Describe* the operating characteristics of single-phase and three-phase transformer systems.
6. *Identify* the function and parts of a unit substation.

The electrical power that is created in a generating station must go through many stages of distribution before it can be used by electrical loads. In this chapter we will study how the transformer enables us to transmit electric power over great distances and to distribute it safely in factories.

∎

3-1 POWER DISTRIBUTION SYSTEMS

The *central-station system* of power generation and distribution enables power to be produced at one location for immediate use at another location many miles away. Transmitting large amounts of electric energy over fairly long distances is accomplished most efficiently by using *high voltages*. High voltages are used in transmission lines to *reduce* the required amount of *current flow*. Example 3-1 illustrates how this is accomplished. The power transmitted in a system is proportional to the voltage multiplied by the current. If the voltage is raised, the current can be reduced to a small value while still transmitting the same amount of power. Reducing the necessary current flow at high voltage means the size and cost of conductor wires required to handle the entire output of a generating plant may be no larger than the lower voltage leads serving industrial customers. Reducing the current also minimizes the amount of power lost in the lines.

There are certain limitations to the use of high-voltage transmission systems. The higher the transmitting voltage, the more difficult and expensive it becomes to safely insulate between line wires, as well as from line wires to ground. For this reason the voltages in a typical high-voltage grid system are reduced in stages as they approach the area of final use.

EXAMPLE 3-1

If 10,000 W are to be transmitted, a current of 100 A would be required if the voltage used were only 100 V.

$$100 \text{ V} \times 100 \text{ A} = 10,000 \text{ W}$$

On the other hand, if the transmission voltage is stepped up to 10,000 V, a current flow of only 1 A would be needed to transmit the same amount of power.

$$10{,}000 \text{ V} \times 1 \text{ A} = 10{,}000 \text{ W}$$

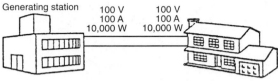

(a) Transmission at 100 V

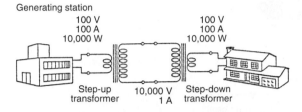

(b) Transmission at 10,000 V

Much control, protection, transformation, and regulation take place in a power distribution system. Figure 3-1 on page 36 illustrates the typical stages through which the distribution system must go in delivering the power required to industrial users.

Three-phase power is usually supplied to commercial, agricultural, and industrial customers. As load requirements increase, the use of single-phase power is no longer practical or economical. In terms of the conductors necessary to delivery a given amount of power, three-phase systems can deliver 1.73 times more power for the same amount of wire than a single-phase system. In addition, three-phase systems permit the design of more efficient electric equipment. Three-phase motors, for example, are more efficient than comparable single-phase motors. For that reason, larger size ac motors are available *only* in three phase. Three-phase primary circuits, however, may have single-phase branches tapped from them (Fig. 3-2 on page 37).

In a typical power distribution system, the flow of current is not constant. Loads vary with the hour of the day and the day of the week. Because the voltage drop in parts of the system is determined by the amount of current flow, a continuous change in the voltage measured at the customer's service could occur. Automatic voltage regulators are used to try to maintain constant voltage levels. Electric power companies generally allow variations of ±10 percent from a nominally stated voltage.

A significant reduction in power line voltage for extended periods is commonly referred to as a *brownout*. Intentional brownouts are sometimes used by the power utility to force a reduction in power consumption. For example, with a 10 percent reduction in voltage, the power demand will automatically be lowered by 19 percent.

PROTECTION

Grounding is one of the most important aspects of an electrical distribution system.

⮕ **YOU MAY RECALL** that the purpose of grounding is to protect life from the danger of electric shock and property from damage.

In order for grounding to be effective, both earth grounding and equipment grounding are required. *Earth grounding* establishes the zero voltage reference for an electrical distribution system. It also protects the electrical system and equipment from superimposed voltages from lightning and contact with higher voltage systems. Moreover, the earth ground prevents the buildup of potentially dangerous static charge in a building. *Equipment grounding* effectively interconnects all non-current-carrying conductive surfaces, such as equipment enclosures, raceways, and conduits to earth ground.

Lightning arresters (also referred to as surge arresters) are used where there is a danger of lightning strikes or voltage surges. The lightning arrester works on the spark gap principle, like the spark plug in an automobile. One side of the arrester is connected to *ground;* the other side is connected to the wire that is to be protected. Under normal circuit voltage conditions, these two points are insulated by the air gap between them. When lightning strikes the line, the resulting high voltage ionizes the air and produces a low impedance discharge path to ground. Specially designed arresters are available for use on overhead power lines as well as on signal circuits such as telephone circuits (Fig. 3-3 on page 38).

An *overcurrent* is either an overload current or a short-circuit current. The overload current is an excessive current relative to normal operating current, but one that is confined to the normal conductive paths provided by the conductors as well as other components and loads

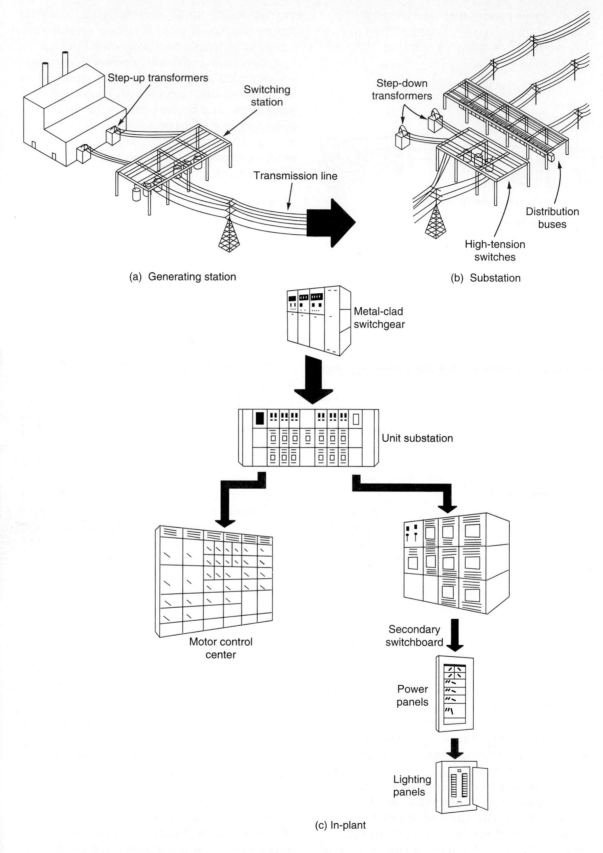

(a) Generating station

(b) Substation

(c) In-plant

Fig. 3-1 Stages in the delivery of power to an industrial user.

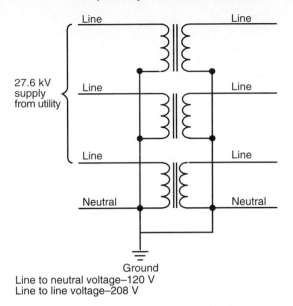

Three-phase wye-connected transformer

27.6 kV supply from utility

Line to neutral voltage–120 V
Line to line voltage–208 V

Fig. 3-2   Typical 208-V three-phase wye-connected service.

of the distribution system. As the name implies, a short-circuit current is one that flows outside the normal conducting paths.

Overloads are most often between one and six times the normal current level. Usually, they are caused by harmless temporary surge currents that occur when motors are started up or transformers are energized. Such overload currents, or transients, are normal occurrences. Because they are of brief duration, any temperature rise is trivial and has no harmful effect on the circuit components. (It is important that proactive devices do not react to overloads.) Continuous overloads can result from defective motors, overloaded equipment, or too many loads on one circuit. Such sustained overloads are destructive and must be cut off by protective devices before they damage the distribution system or system loads. Because they are of relatively low magnitude compared to short-circuit currents, however, removal of the overload current within a few seconds will generally prevent equipment damage. A sustained overload current results in overheating of conductors and other components and will cause deterioration of insulation, which may eventually result in severe damage and short circuits if not interrupted.

Whereas overload currents occur at rather modest levels, the *short circuit*, or fault current, can be many hundreds of times larger than the normal operating current. A high-level fault may be 50,000 A (or larger). If not cut off within a matter of a few thousandths of a second, damage and destruction can become rampant. There can be severe insulation damage, melting of conductors, vaporization of metal, ionization of gases, arcing, and fires. Simultaneously high-level short-circuit currents can develop huge magnetic-field stresses. The magnetic forces between bus bars and other conductors can be many hundreds of pounds per linear foot; even heavy bracing may not be adequate to keep them from being warped or distorted beyond repair.

Overcurrent protection is fundamental to the safe operation of all medium- and high-voltage distribution systems that are used in industrial plants. The *fuse* is a reliable overcurrent protective device. A fusible link or links encapsulated in a tube and connected to contact terminals comprise the fundamental elements of the basic fuse. Electrical resistance of the link is so low that it simply acts as a conductor. When destructive currents occur, however, the link very quickly melts and opens the circuit to protect conductors and other circuit components and loads. Figure 3-4(b) on page 39 shows a typical high-voltage fuse used in a power distribution system. The fuse is hinged to allow for easy removal. This feature, however, does not allow the fuse to be used as a disconnecting means for the circuit.

The fuse selection for a particular installation must meet predetermined frequency, voltage, and current requirements. Fuses are available for both 25- and 60-Hz frequency systems. The *voltage rating* of a fuse is the *highest* voltage at which it is designed to safely interrupt the current. (It may be used at any voltage at or below the rated voltage without affecting its operating characteristics.)

The *continuous-current rating,* also known as *ampere rating,* of a fuse represents the maximum amount of current the fuse will carry without blowing. (When subjected to a current above its ampere rating, the fuse will open the circuit after a predetermined period of time.)

The *interrupting-current rating,* also known as *short-circuit rating,* of a fuse is the maximum current it can safely interrupt at rated voltage (no explosion or body rupture).

Circuit breakers (Fig. 3-5 on page 40) are *switches* that automatically open electrical circuits when an overload condition occurs. As with other equipment, circuit breakers are divided into those rated for 1000 V and less (low-voltage) and those rated for more than

Overloads

Short circuit

Fuse

Voltage rating

Continuous-current rating (ampere rating)

Interrupting-current rating (short-circuit rating)

Circuit breakers

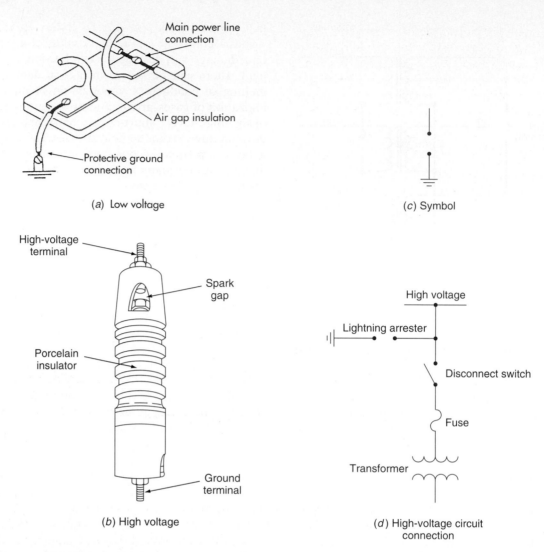

(a) Low voltage

(c) Symbol

(b) High voltage

(d) High-voltage circuit connection

Fig. 3-3 Lightning arresters.

1000 V (medium- and high-voltage). Low-voltage breakers are universally operated in air, so it is not necessary to designate them as air circuit breakers. Medium- and high-voltage breakers, on the other hand, use mediums other than air in which to open the circuit; therefore, they must be designated as air, oil, gas, or vacuum.

In the oil circuit breaker, the contacts are immersed in oil contained in a metal tank. Instead of being quenched in oil, the arc of the air circuit breaker breaks in a blast of air. The breaker may be opened or closed either by a hand-operated lever or automatically. A number of tripping methods are used for overcurrent protection.

The reliable and safe distribution of electric power is dependent upon the use of protective devices for making and breaking circuits, detecting the presence of faulted currents, and

for isolating a faulty circuit with the least disturbance and damage to the system and human life. *Coordination* is the process of selecting protective devices so that there is a minimum of power interruption in case of a fault or overload. For a particular situation, a value of high-voltage fuse should be selected which ensures that other protective devices between the fuse and the loads can react to a given condition in less time. Coordination studies require that the time-current characteristic charts of different protective devices be compared and that the selection of the proper devices be made accordingly.

To obtain selective coordination, the correct ratio must be maintained between the ampere rating of a main fuse and that of the feeder fuse, and between the feeder fuse and the branch circuit fuse. If the ratios are correct and a fault occurs, only the fuse nearest that fault will open and isolate the faulted circuit. Other branches of the distribution system will be unaffected; the fault will not cause more than one fuse to open. Examples of ratios of fuse ampere ratings that provide selective coordination are shown in Fig. 3-6 on page 40.

POWER QUALITY

The voltage produced by the utility electricity generators has a sinusoidal waveform with a frequency of 60 cycles per second, or 60 Hz. Any variation to the voltage waveform, in magnitude or in frequency, can cause problems with the operation of electrical equipment. Under these circumstances, the variation is called a *power line disturbance* (Fig. 3-7 on page 41). Note in Fig. 3-7(*c*) that if a load that is adversely affected by power variation is not present, no quality problem is experienced.

An electrical disturbance source can be coupled with other circuitry by various means.

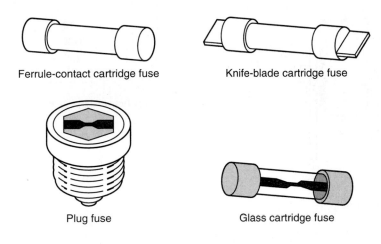

Ferrule-contact cartridge fuse

Knife-blade cartridge fuse

Plug fuse

Glass cartridge fuse

(*a*) Low-voltage fuses

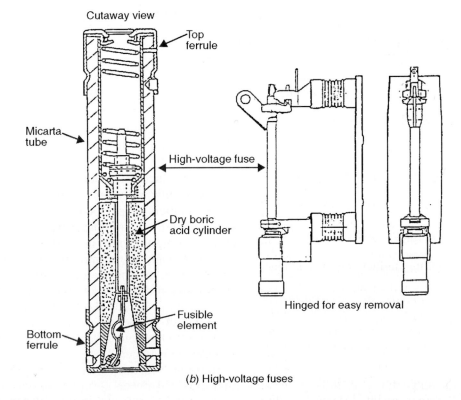

Cutaway view

Top ferrule

Micarta tube

High-voltage fuse

Dry boric acid cylinder

Fusible element

Bottom ferrule

Hinged for easy removal

(*b*) High-voltage fuses

Fig. 3-4 Fuses.

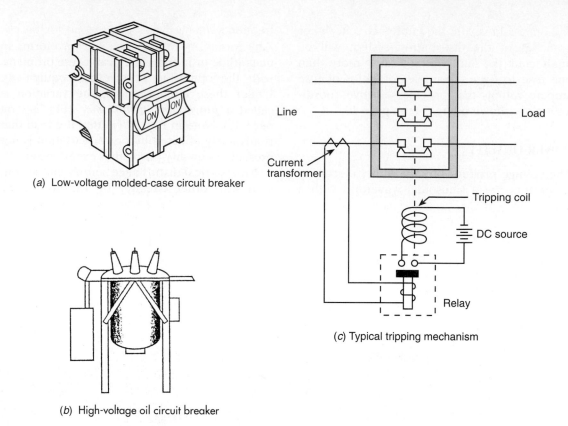

(a) Low-voltage molded-case circuit breaker

(b) High-voltage oil circuit breaker

(c) Typical tripping mechanism

Fig. 3-5   Circuit breakers.

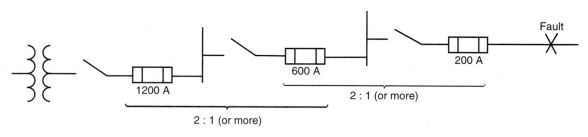

Fig. 3-6   Selective coordination of ratios of fuse ampere ratings.

These include:

- *Conductive coupling,* directly by means of wires.
- *Capacitive coupling,* capacitively via an electric field.
- *Inductive coupling,* inductively via a magnetic field.
- *Radiation,* by means of electromagnetic waves. *Electromagnetic interference* (*EMI*) and *radio frequency interference* (*RFI*) are terms used to describe this type of disturbance.

Of increasing importance is the effects of *harmonics* on power quality. Harmonics are the by-products of modern electronics. They are especially prevalent wherever there are large numbers of personal computers, adjustable speed drives, and other types of *nonlinear* loads that draw current in short pulses.

Stated simply, harmonics are electrical energy at a frequency other than 60 Hz—specifically, multiples of 60 Hz. Thus, the second harmonic is 120 Hz, the third is 180 Hz, the seventh is 420 Hz, and so on (Fig. 3-8). Utilities do *not* generate harmonics; user loads generate harmonics.

Harmonics are created by nonlinear loads that draw current in abrupt pulses rather than in a smooth sinusoidal manner. These pulses

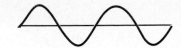

(a) Pure sinusoidal ac voltage waveform

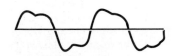

(b) Distorted ac voltage waveform

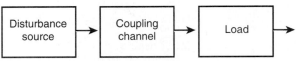

(c) Three elements needed to produce a problematic power line disturbance; source, coupling channel, and receptor

Fig. 3-7 Power line disturbance.

cause distorted current wave shapes which, in turn, cause harmonic currents to flow back into other parts of the power system. The diode/capacitor power supply found in computers is a typical example of a single-phase nonlinear load. The incoming ac voltage is diode rectified and is then used to charge a large filter capacitor, which draws a pulse of current only during the peak of the wave. During the rest of the wave, when the voltage is below the capacitor residual, the capacitor draws no current. This process is repeated over and over.

Harmonics, unlike transients, represent a steady-state problem because they are present as long as the harmonic-generating equipment is in operation. In many cases, harmonics will not have detrimental effects on equipment

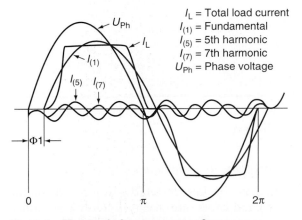

I_L = Total load current
$I_{(1)}$ = Fundamental
$I_{(5)}$ = 5th harmonic
$I_{(7)}$ = 7th harmonic
U_{Ph} = Phase voltage

Fig. 3-8 Harmonic frequency waveform.

operation. If the harmonics are very severe, however, or if loads are highly sensitive, a number of problems may arise. Table 3-1 on page 42 lists some problems caused by harmonics.

To prevent harmonic currents from causing damage or power outages, it is important to survey the electrical power distribution to determine whether harmonic currents are present. Once the problem has been identified, several alternative corrective actions are available. Special filters can be installed to screen out harmonics. Depending on the source and nature of the harmonic current, this solution may correct the problem. Other corrective actions include balancing the load in each phase of a three-phase, 4-wire system; adding extra neutral wires to each phase; and de-rating or replacing transformers.

EMERGENCY POWER SUPPLY SYSTEMS

When power in an electrical system is disrupted for an extended period, life-threatening conditions may occur. Continuous lighting and power are essential in places of public assembly, theaters, hotels, sports arenas, health-care facilities, and the like. Power interruption in an industry can result in loss of critical data and damage to process control systems. Engine-driven alternators are commonly used to provide emergency power in the event of a power failure. Prime movers for these on-site alternators may be powered by gasoline, diesel, or gaseous fuel engines. Figure 3-9 on page 43 shows the block diagram of a typical *standby power supply (SPS)* system with a transfer arrangement used to automatically energize the emergency system upon failure of the normal current supply. Standby power supply systems are off-line or standby. *Uninterruptible power supply (UPS)* systems, however, are on-line or continuous.

On-line systems are typically more reliable and effective. Both systems may be either rotary or static. Rotary systems employ rotating machines; static systems use solid-state components. A properly selected UPS/SPS system is the only product, other than a generating unit, that can protect critical loads against power interruptions exceeding 0.5 s.

Figure 3-10 on page 43 shows the block diagram of a static, uninterruptible power supply. During normal operation, the incoming voltage is rectified, supplied to a battery, fed to an inverter, and output as ac power. If commercial

Table 3-1   Problems Caused by Harmonic Effects

Equipment	Harmonic Effects	Results
Transformers	• Voltage harmonics cause higher transformer voltage and insulation stress; normally not a significant problem.	• Transformer heating • Reduced life • Increased copper and iron losses • Insulation stress • Noise
Capacitors	• Capacitor impedance decreases with increasing frequency, so capacitors act as sinks where harmonics converge. • Supply system inductance can resonate with capacitors at some harmonic frequencies, causing large currents and voltages to develop.	• Heating of capacitors owing to increased dielectric losses • Short circuits • Fuse failure • Capacitor explosion
Motors	• Increased losses may occur.	• Motor heating • Reduced efficiency • Mechanical vibration and noise • Reduced life • Insulation stress
Circuit breakers	• May not operate properly in the presence of harmonic currents.	• A peak-sensing electronic trip circuit breaker responds to the peak of current waveform. As a result, the circuit breaker won't always respond properly to harmonic currents. Because the peak of the harmonic current is usually higher than normal, this type of circuit breaker may trip prematurely at a low current. If the peak is lower than normal, the breaker may fail to trip when it should.
Neutral conductors	• In a three-phase, 4-wire system, odd multiples of third harmonics, known as *triplens,* add together in the neutral conductor. • In systems with many single-phase, nonlinear loads, the neutral current can actually exceed the phase current.	• Overheating of the neutral conductor • Higher than normal voltage between the neutral conductor and ground • Overheating of neutral bus bars and connecting lugs
Electronic and computer-controlled equipment	• Electronic controls are often dependent on the zero crossing, or on the voltage peak, for proper control; however, harmonics can significantly alter these parameters, thus adversely affecting operation.	• Malfunction of control and protection equipment • Premature equipment failure • Erratic operation of static drives and robots

power is interrupted, the battery provides power to drive the inverter: therefore, the UPS system is always the final source of power for the load. The critical computer load is connected without interruption to a bypass source through a fast-acting static transfer switch in the event of UPS failure or overload.

Self-Test

Answer the following questions.

1. True or false? Relatively large diameter conductors are used in electrical utility transmission systems.

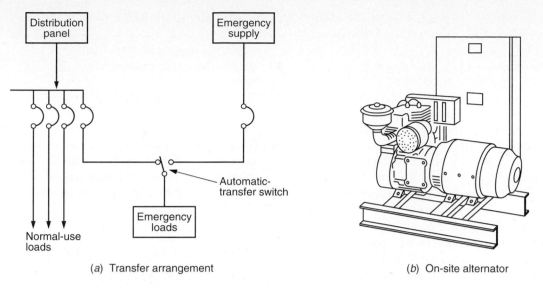

(a) Transfer arrangement

(b) On-site alternator

Fig. 3-9   Typical standby power supply system.

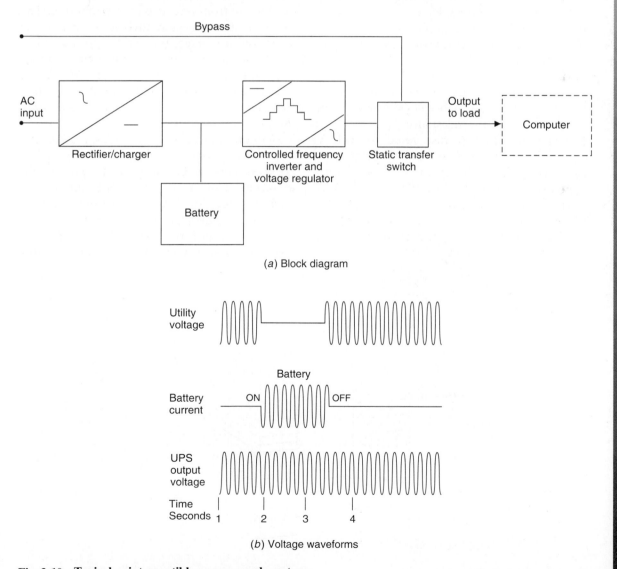

(a) Block diagram

(b) Voltage waveforms

Fig. 3-10   Typical uninterruptible power supply system.

2. High voltages are used in transmission lines to reduce the required amount of _____.
3. Substations contain step-down _____ that reduce the voltage.
4. The frequency of the power supplied by the utility is normally _____.
5. True or false? Single-phase power is usually supplied to an industrial customer.
6. True or false? A nominally stated utility voltage of 240 V could drop to 220 V and still be within an acceptable variation.
7. True or false? Brownouts are sometimes originated by power companies in order to reduce the load on their systems.
8. In order for grounding of an electrical distribution system to be effective, both _____ grounding and _____ grounding is required.
9. True or false? Lightning arresters work on the spark gap principle.
10. True or false? It is important that protective devices react to overloads of brief duration.
11. True or false? Short-circuit currents can be more damaging than overload currents.
12. True or false? A high-voltage fuse element normally has low resistance.
13. True or false? A fuse rated for 600 V should not be used on a 240-V circuit.
14. True or false? Normally, the continuous-current rating of a fuse is the same as its interrupting-current rating.
15. With proper _____, only the protective device nearest the fault will open should a fault occur.
16. Power line disturbance refers to any variation in _____ or _____ to the generated sinusoidal waveform.
17. Harmonics are created by _____ loads.
18. True or false? One of the harmful effects of harmonics is lowering of the system power factor.
19. True or false? Harmonic currents can be reduced by balancing the load in each phase of a three-phase 4-wire system.
20. True or false? Standby power supply systems are on-line or continuous.

3-2 TRANSFORMER OPERATION AND APPLICATIONS

TRANSFORMER PRINCIPLE

A transformer is a static device used to transfer energy from one ac circuit to another. This transfer of energy may involve an increase or decrease in voltage, but the frequency will be the same in both circuits. When the transformation takes place with an increase in voltage, it is called a *step-up transformer*. When voltage is decreased, it is called a *step-down transformer*.

Without transformers the widespread distribution of electrical power would be impractical. Transformers makes it possible to generate power at a convenient voltage, step up to a very high voltage for long-distance transmission, and then step down for practical distribution.

A basic transformer (Fig. 3-11) consists of two windings, or coils, wound around an iron

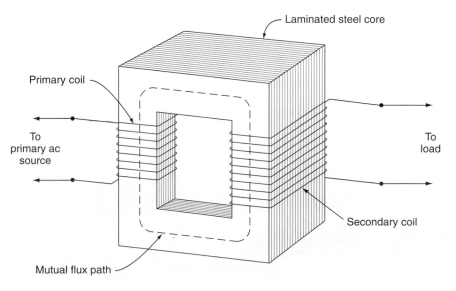

Fig. 3-11 Basic transformer.

core. The varying current of ac power is required to produce a varying magnetic flux in the iron core so that electrical energy from one winding is transferred to the other. The winding that receives the power from the supply is called the *primary winding,* whereas the winding that delivers power to the load is called the *secondary winding.* The ac frequency of the primary induces the same frequency in the secondary. In addition to changing voltage, transformers may be used to isolate circuits, regulate voltage or current, and for metering and protective circuits.

The principle of operation of a transformer is based on *mutual induction.* Mutual induction occurs when the magnetic field surrounding one conductor cuts across another conductor, inducing a voltage in it. This effect may be increased by forming the conductors into coils and winding them on a common magnetic core.

When the primary coil of a transformer is connected to an alternating voltage, there will be a current in the primary coil called the *exciting current.* This exciting current sets up an alternating flux that links the turns and induces a voltage in both windings. The voltage of self-induction induced in the primary is a counter voltage, opposite in polarity and almost equal in magnitude to the applied voltage. This high counter voltage limits the exciting current to a very low value. The voltage induced in the secondary winding is a result of mutual inductance. Because a typical power transformer has a flux linkage of almost 100 percent, the same voltage will be induced in each turn of each coil. The total induced voltage, therefore, will be directly proportional to the turns on the coil (Fig. 3-12).

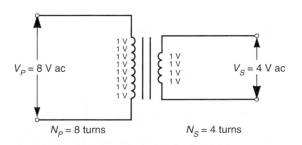

Fig. 3-12 Transformer-induced voltages.

EXAMPLE 3-2

A transformer is wound with eight turns on the primary coil and four turns on the secondary coil (Fig. 3-12). If 8 V ac is applied to the primary, the self-induced voltage in each primary turn would be equal to 8 ÷ 8, or 1 V. Because each turn of the secondary has the same voltage induced, the secondary voltage would be equal to 4×1, or 4 V.

Transformers are wound on cores made from stacks, or *laminations,* of sheet steel. The core ensures a good magnetic linkage between the primary and secondary windings. *Eddy currents* are caused by the alternating current that induces a voltage in the core of the transformer itself. Because the iron core is a conductor, it produces a current by the induced voltage. By laminating the core, the paths of the eddy currents are reduced considerably, as seen in Fig. 3-13, thereby reducing heat and power loss. Eddy currents are prevented from flowing from lamination to lamination by a thin coating of insulating material on the flat surfaces of the lamination. The eddy currents that do exist are very small and represent wasted power dissipated as heat in the core.

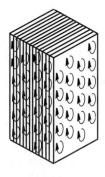

(*a*) Eddy currents in a solid piece of metal

(*b*) Laminations reduce the path of circulating eddy currents

Fig. 3-13 Eddy currents in steel core.

POWER, VOLTAGE, AND CURRENT RELATIONSHIPS

Transformers increase the voltage, decrease the voltage, or allow the voltage to remain the same between the primary and secondary windings without significant power loss. Transformer power output equals transformer power input minus the internal losses and is the product of voltage and current.

There is no gain or loss in energy in an *ideal* transformer. Energy is transferred from the primary circuit to the secondary circuit. This

Primary winding

Secondary winding

Mutual inductance

Exciting current

Laminations

Eddy currents

means that the volts multiplied by the amperes of the primary circuit equal the volts multiplied by the amperes of the secondary circuit. In other words, in the ideal transformer, the power output must equal the power input. There is some power loss in a practical transformer, but the average efficiency of a transformer is over 90 percent. *Efficiency losses* result from the ohmic resistance (copper losses) of the windings; *core losses*, which are caused by the induction of eddy currents in the core material; and *hysteresis*, or molecular friction, that is caused by changes in the polarity of the applied current.

The maximum power rating of a transformer is often included on the transformer's nameplate. Transformer power is rated in volt-amperes (VA) and *not* in watts. This is because transformer power is *apparent power*. It does not convert the power to heat; it merely transfers the power from a source to a load.

▶ **YOU MAY RECALL** that apparent power (VA) is the total power that an ac circuit is apparently using. It is the total power supplied to the circuit from the source, including *real* (watts) and *reactive* (VAR) power.

The ratio of the number of turns in the primary to the number in the secondary is the *turns ratio* of a transformer:

$$\text{Turns ratio} = \frac{N_p}{N_s}$$

where N_p = number of turns in primary
N_s = number of turns in secondary

In the ideal transformer, the voltage induced in each turn of the secondary is the same as the self-induced voltage of each turn in the primary. The voltage that is self-induced in each turn of the primary equals the voltage applied to the primary divided by the number of turns in the primary. Thus, the voltage ratio of a transformer is equal to its turns ratio. This can be written as:

$$\text{Turns ratio} = \text{voltage ratio}$$
$$\frac{N_p}{N_s} = \frac{V_p}{V_s}$$

where N_p = number of turns in primary
N_s = number of turns in secondary
V_p = primary voltage
V_s = secondary voltage

The voltage steps up or down across the transformer in proportion to the turns ratio. For

example, if the number of secondary turns is twice the number of primary turns, the secondary voltage will be twice the primary voltage. If the number of primary turns is twice the number of secondary turns, however, the secondary voltage will be half the primary voltage.

Transformers are classified as step-up or step-down in relation to their effect on voltage. A step-up transformer is one in which the secondary-coil output is greater than the primary-coil input voltage. This type of transformer has more turns in the secondary coil than in the primary coil. The ratio of the primary to secondary turns determines the input-to-output voltage ratio of the transformer.

EXAMPLE 3-3

If a step-up transformer has 50 turns for the primary winding and 100 turns for the secondary winding (see the diagram below), the turns ratio would be:

$$\text{Turns ratio} = \frac{N_p}{N_s} = \frac{50}{100} = \frac{1}{2} \text{ or } 1{:}2$$

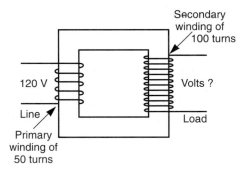

If V_p is stepped up by a factor of 2, the secondary voltage V_s will be equal to 2×120, or 240 V.

A *step-down transformer* is one in which the secondary-coil output voltage is less than the primary-coil input voltage. This type of transformer has *less* turns in the secondary coil than in the primary coil. Again, the ratio of the primary to secondary determines the input-to-output voltage ratio of the transformer.

EXAMPLE 3-4

If a step-down transformer has 100 turns for the primary winding and 5 turns for the secondary winding (see the diagram below), the turns ratio would be:

$$\text{Turns ratio} = \frac{N_p}{N_s} = \frac{100}{5} = \frac{20}{1} \text{ or } 20{:}1$$

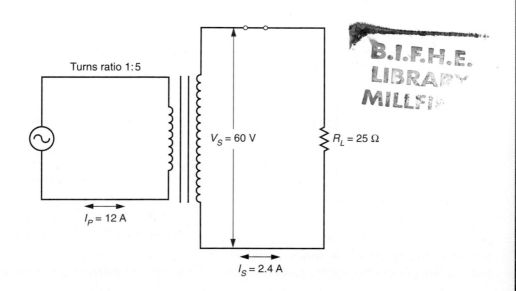

$V_P = 240\,V$ $V_S = ?$
$N_S = 5$
$N_P = 100$

If the primary voltage is 240 V and V_p is stepped down by a factor of 20, the secondary voltage V_s is equal to 240 ÷ 20, or 12 V.

A transformer automatically adjusts its input current to meet the requirements of its output or load current. Thus, when no current is being used from the secondary winding, no current flows in the primary except excitation current. Excitation current is the very small amount of current that is necessary to maintain the magnetic circuit. For self-regulation, a transformer depends on *counter-electromotive force (cemf)* generated in its primary winding by its own magnetism and an opposing magnetism produced by the current drawn by the load on the secondary winding.

If the secondary circuit of the transformer becomes shorted, the high current that results generates a great opposition to the primary winding flux. As a result, the cemf of the primary is made to drop very low; primary current increases dramatically also. It is for this reason that a fuse placed in series with the primary winding protects both the primary and secondary circuits from excessive current.

By Ohm's law, the amount of secondary current equals the secondary voltage divided by the resistance in the secondary circuit (a negligible coil resistance assumed).

When a transformer steps up the voltage, the secondary-coil current is correspondingly less than the primary-coil current, so the power (voltage multiplied by current) is the same in both windings. The ratio of primary to secondary current is *inversely proportional to the voltage or turns ratio.*

$$V_s \times I_s = V_p \times I_p$$

or

$$\frac{I_s}{I_p} = \frac{V_p}{V_s}$$

EXAMPLE 3-5

If a step-up transformer with a 1:5 turns ratio has a secondary voltage of 60 V across a 25-Ω resistive load (see the diagram below), I_s would be:

$$I_s = \frac{V_s}{R_L} = \frac{60\,V}{25\,\Omega} = 2.4\,A$$

With a turns ratio of 1:5, the current ratio is 5:1. Therefore, I_p would be:

$$I_p = 5 \times I_s = 5 \times 2.4$$
$$A = 12$$

Turns ratio 1:5

$V_S = 60\,V$ $R_L = 25\,\Omega$

$I_P = 12\,A$

$I_S = 2.4\,A$

EXAMPLE 3-6

If a step-down soldering gun transformer has a turns ratio of 200:1 and a secondary heating current of 400 A (see below), the primary current would be:

$$I_p = \frac{I_s}{200} = \frac{400\ A}{200} = 2\ A$$

If the primary is operated from a 120-V source, the secondary voltage would be:

$$V_s = \frac{V_p}{200} = \frac{120\ V}{200} = 0.6\ V$$

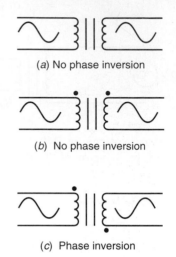

Isolation transformers are used to eliminate direct electric connections between the primary and secondary circuits without changing the voltage and current ratings. Figure 3-14 illustrates this application. The transformer's turns ratio is 1:1. Using the isolation transformer, the load is isolated from the voltage source, so there is no chance that the chassis might accidentally become hot owing to improper plug placement.

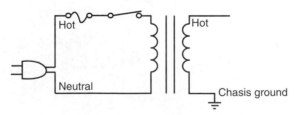

Fig. 3-14   Isolation transformer.

TRANSFORMER POLARITY

Transformers can reverse the *phase* of the input voltage by winding the secondary coil in the opposite direction of the primary coil. Figure 3-15 illustrates the method used to identify the relative phase relationship of transformer inputs and outputs on schematic diagrams. Notice that when the schematic does not contain "dots," no phase inversion takes place.

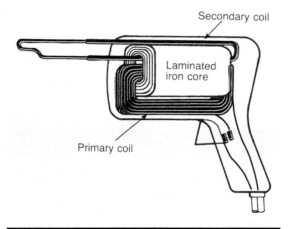

(a) No phase inversion

(b) No phase inversion

(c) Phase inversion

Fig. 3-15   Transformer phase inversion.

The *polarity* of a transformer can be shown by means of dots on the primary and secondary terminals. This type of marking is used on instrument transformers. On power transformers, however, the high-voltage winding leads are marked H_1 and H_2 and the low-voltage winding leads are marked X_1 and X_2 (Fig. 3-16). By convention, H_1 and X_1 have the same polarity, which means that when H_1 is instantaneously positive, X_1 also is instantaneously positive. These markings are used in establishing the proper terminal connections when connecting single-phase transformers in parallel, series, or three-phase configurations.

In practice, the four terminals on a transformer are mounted in a standard way so that the transformer has either *additive* or *subtractive* polarity. A transformer is said to have additive polarity when terminal H_1 is diagonally opposite terminal X_1. Similarly, a trans-

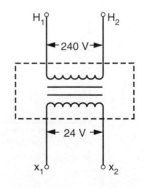

Fig. 3-16   Transformer terminal markings.

former has subtractive polarity when terminal H_1 is adjacent to terminal X_1 (Fig. 3-17).

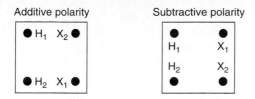

Fig. 3-17   Additive and subtractive polarity depends upon the location of the H and X terminals.

Subtractive polarity is standard for all single-phase transformers above 200 kVA, provided the high-voltage winding is rated above 8660 V. All other transformers have additive polarity.

EXAMPLE 3-7

It is sometimes necessary when transformer leads are unmarked to make a polarity test to identify and mark the leads. One method is shown in the diagram below. First, the H_1 and H_2 leads should be marked. Next, a jumper is connected between the H_1 lead and the low-voltage lead adjacent to it, and a voltmeter is connected between H_2 and the other low-voltage lead. A low voltage is then applied to the H_1 and H_2 leads, and the voltmeter reading is recorded. If the voltmeter reading is greater than the applied voltage, the transformer is additive and X_1 will be the lead on the right. If the voltmeter reading is less than the applied voltage, the transformer is subtractive and X_1 is on the left.

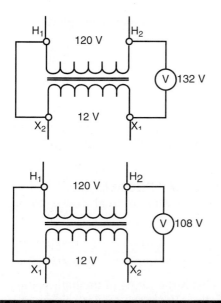

TRANSFORMER TYPES AND RATINGS

Many different types of transformers are used in industrial circuits. In most cases, these transformers are classified according to their applications. Unfortunately standards vary slightly, and some of the following terms may be interchangeable.

Power transformer. This generally refers to larger transformers that are used to change voltage and current levels to meet circuit power requirements. In most cases power transformers are designed for operation at ac line frequencies of 50-60 Hz. Two types of transformers are shown in Figs. 3-18 and 3-19. The *power supply transformer* in Fig. 3-18 is used to change the ac line voltage into another more suitable voltage. The *oil-cooled distribution transformer* in Fig. 3-19 is used to step-down utility system voltages to standard industrial voltages.

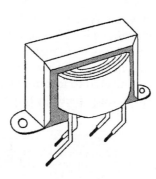

Fig. 3-18   A power supply transformer is used to change the ac line voltage into another more suitable voltage.

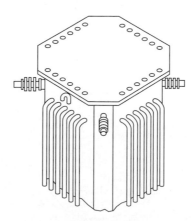

Fig. 3-19   An oil-cooled distribution transformer is used to step down utility system voltages to standard industrial voltages.

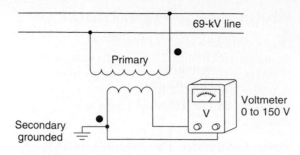

Fig. 3-20 A potential transformer is used to step down the system voltage to the voltage rating of the instrument.

Instrument transformers. These are small transformers used in conjunction with test and measurement instruments. A *potential transformer* (Fig. 3-20) is used to step-down the system voltage to the voltage rating of the instrument. A *current transformer* (Fig. 3-21) supplies the instrument with a small current that is proportionate to the main current. Current transformers are also used in conjunction with large overcurrent and overload devices. A very high voltage, capable of producing a fatal shock, can build up in the secondary winding when it is open. For this reason, the secondary leads should always be connected to an ammeter or kept short-circuited if the meter is removed.

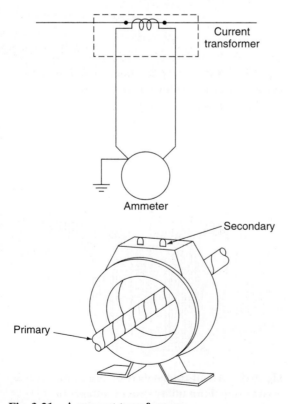

Fig. 3-21 A current transformer.

Autotransformers. This transformer has a single tapped winding common to both primary and secondary circuits. It is primarily used where a small increase or decrease in voltage is required. Autotransformers are

- Used on long distribution lines to boost voltage and compensate for line drop.
- Adjustable for variable voltage.
- Used on large three-phase motor starters to reduce current and torque during the motor starting period.

An autotransformer, like the one shown in Fig. 3-22, is used where several voltages are required. When it is not necessary to have electrical isolation between the primary and secondary circuits, and when direct connection between them is permissible, autotransformers offer the benefits of smaller size, lower weight, and lower cost.

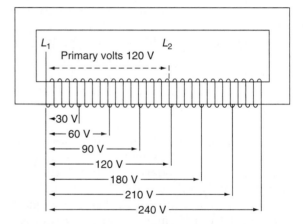

Fig. 3-22 An autotransformer.

Typical transformer nameplate ratings include the following:

kVA. Transformers are rated in kilovolt-ampere or volt-ampere rather than kilowatts, because the power factor of the load determines the transformer output in kilowatts. Because load power factor varies, it is not practical to rate transformers in kilowatts.

▆▶ **YOU MAY RECALL** that *power factor* (*PF*) is the ratio of real power to apparent power:

$$PF = \frac{\text{watts}}{\text{volt-ampere}} = \frac{\text{real power}}{\text{apparent power}}$$

H.V. This is the rating of the high-voltage winding. If it is the primary circuit, it is maximum voltage applied. If it is the secondary, it is the output voltage at full load.

L.V. This is the rating of the low-voltage winding. If it is the primary circuit, it is the maximum applied voltage. If it is the secondary, it is the output voltage at full load.

Cycles. This is the frequency rating in cycles per second, or Hz.

Polarity. This will tell you whether the transformer is additive or subtractive.

%Z. This is the transformer impedance expressed as a percentage of the total circuit impedance. It is important when transformers are connected in parallel that percent impedance be the same for proper sharing of the load. Percent impedance is also used to calculate short-circuit currents for circuit breaker and fuse applications.

▶ **YOU MAY RECALL** that *impedance* (*Z*) is the total opposition to current flow in an ac circuit and may include resistance *(R),* inductive reactance *(X_L),* and capacitive reactance *(X_C).*

°C rise. This is the temperature rise above ambient with the transformer fully loaded. If the temperature rises more than this, it indicates an overload or a faulty cooling system. Many transformers are provided with recording thermometers to keep a check on the system.

▶ **YOU MAY RECALL** that *efficiency* of mechanical or electrical equipment is usually stated as the ratio of power output to power input.

$$\frac{\text{Output}}{\text{Input}} \times 100 = \% \text{ efficiency}$$

It may also be stated as the input equals the output plus the losses.

Power input = power output + power losses

Since the losses in power transformers are very low, the efficiency of most transformers is approximately 97 percent. The two main losses affecting the efficiency of a transformer are copper and core losses. These losses appear as *heat.* The wire of which the windings are made must be large enough to carry the intended current, and its insulation must be able to withstand the intended voltage.

A transformer's power rating includes its ability to dissipate the heat caused by the losses. Transformers are broadly classified as *wet* or *dry.* In wet types, oil or some other liquid serves as a heat-transfer medium and insulation. The windings are immersed in oil, which acts as an insulator as well as a cooling agent. Vertical pipes, around the outside of a steel housing and enclosing the transformer, connect to the upper and lower parts of the housing and permit the oil to be cooled by natural convection and gravity circulation of the oil. Very large oil-cooled transformers may have forced circulation, using a pump to circulate the oil. Dry types use air or inert gas in place of liquid. Sometimes fans are used for forced-air cooling on transformers in order to provide additional power for load peaks and to lower oil and copper temperatures where the surrounding air is hot.

Self-Test

Answer the following questions.

21. True or false? Movement of the coils within a transformer produces the transformation of voltage.

22. True or false? Without transformers the widespread distribution of electrical power would be impractical.

23. The transformer winding that delivers power to the load is known as the _____ winding.

24. The principle of operation of a transformer is based on _____ induction.

25. The _____ current sets up an alternating flux in the core of a transformer.

26. Eddy currents are caused by the alternating current's inducing a voltage in the _____ of the transformer.

27. Transformer power output equals transformer power input minus the internal _____.

28. True or false? Transformer power is rated in volt-amperes instead of watts.

29. True or false? The voltage ratio of a transformer is equal to its turns ratio.

30. True or false? A step-up transformer is one in which the secondary coil current is greater than the primary coil current.

31. True or false? If the secondary circuit of a transformer becomes overloaded, the primary circuit will not be affected.

32. The turns ratio of an isolation transformer is always _____.

33. The low-voltage winding leads on a single-phase power transformer are marked _____ and _____.

34. True or false? A transformer is said to have additive polarity when terminal H_2 is diagonally opposite terminal X_2.
35. True or false? A distribution transformer would be classified as a current transformer.
36. True or false? Potential transformers are always step-down transformers.
37. True or false? An autotransformer does not provide electrical isolation between the primary and secondary circuits.
38. Impedance is the total opposition to current flow in an ac circuit and may include _____, _____, and _____.
39. True or false? An efficiency rate of 80 percent is average for most transformers.
40. True or false? In wet-type transformers, oil serves as a heat-transfer medium and insulation.

3-3 TRANSFORMER CONNECTIONS AND SYSTEMS

SINGLE-PHASE TRANSFORMER SYSTEMS

Transformers, like batteries and other power sources, may be connected into series or parallel arrangements. When transformers are grouped together in order to operate and share load, it is necessary to comply with the following requirements:

- Their voltage ratings must be equal.
- Their impedance ratios (percent impedance) must be equal.
- Their polarities must be determined and connections made accordingly.

Transformers are seldom connected in series. When they are, however, their current ratings must be large enough to carry the maximum current of the load. For the most efficient operation, their current ratings must be equal. Figure 3-23 shows typical examples of two single-phase transformers connected in series.

Parallel operation of transformers results in an increased kilovolt-ampere rating. When additional loads are to be added to a circuit, it is sometimes more practical to connect another transformer in parallel rather than to change the existing one. Two transformers of equal ratings connected in parallel will carry twice the kilovolt-ampere rating of either one. Figure 3-24 shows a typical example of two single-phase transformers connected in parallel. (*Note:* Particular attention must be paid to the polarity of each transformer so that terminals

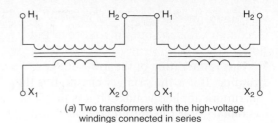

(a) Two transformers with the high-voltage windings connected in series

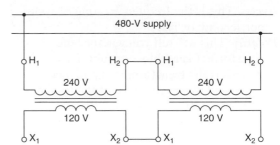

(b) Two transformers with series connections on both primary and secondary windings

Fig. 3-23   Series connection of transformers.

having the same polarity are connected together. An error in polarity produces a short circuit as soon as the transformers are excited.)

There are two types of single-phase distribution: *single-phase 2-wire distribution* and *single-phase 3-wire distribution*. Single-phase 2-wire consists of a hot wire and a neutral; single-phase 3-wire consists of two hot wires and a neutral. (*Note:* Although a single-phase 3-wire has two hot wires, it is still called single-phase *not two-phase*). A 240/120 V single-phase 3-wire is the type of distribution used to feed our houses. Loads can be connected between hot wires or from hot to neutral. The *neutral wire* is the conductor that is tapped off the center of the secondary of the transformer. The two hot wires are 180 electrical degrees apart; thus, the neutral only carries the *difference* in amperage between the two hot wires.

Figure 3-25 shows a dual-voltage distribution transformer connected for single-phase 3-wire and single-phase 2-wire. As already mentioned, the 3-wire system is used in residential wiring. It provides 120 V for lighting and small appliances and 240 V for heavy loads such as electric heating, ranges, and dryers [see Fig. 3-25(a)]. For safety reasons, the middle conductor, called neutral, is grounded; this allows for a maximum potential from line to ground of 120 V. If a ground occurs in either line 1 or line 2, the overcurrent device will open the circuit. If a ground occurs in line 1 or line 2 in an ungrounded system, 240 V would be present from the other line to ground. An ungrounded

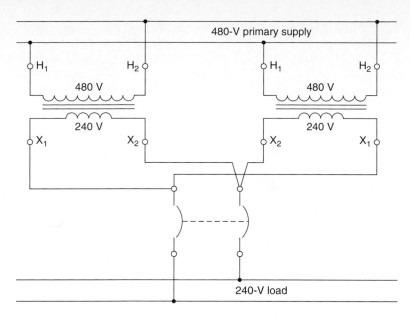

Fig. 3-24 Parallel connection of transformers.

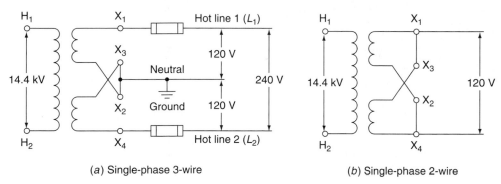

(a) Single-phase 3-wire

(b) Single-phase 2-wire

Fig. 3-25 Single-phase distribution.

system would require two grounds to open the circuit. Figure 3-25(b) shows the distribution transformer connected to give only 120 V. The resulting output voltage is only 120 V, but the power rating of the transformer remains unchanged.

EXAMPLE 3-8

The 50-kVA transformer shown below has a balanced resistive load. There is a 50-A load from both L_1 and L_2 to the neutral. There is also a

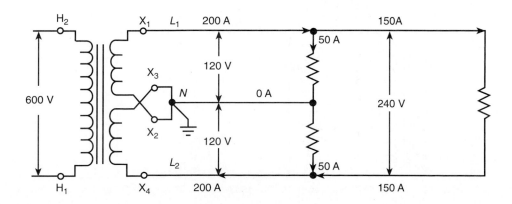

150-A load across L_1 and L_2. Current values are shown by the arrows.

a. What is the total current in the three lines? L_1 and L_2 both carry the 240-V load plus one of the 120-V loads, so the total in L_1 and L_2 is 50 A + 150 A or 200 A. The neutral conductor carries the difference between L_1 and L_2.

$$200 \text{ A} - 200 \text{ A} = 0 \text{ A}$$

b. What is the primary current (neglect losses)?

$$V_p I_p = V_s I_s$$

$$I_p = \frac{V_s I_s}{V_p}$$

$$= \frac{240 \times 200}{600}$$

$$= 80 \text{ A}$$

c. What is the load on the transformer in kilovolt-amperes?

$$kVA = \frac{VI}{1000}$$

$$= \frac{240 \text{ V} \times 200 \text{ A}}{1000}$$

$$= 48 \text{ kVA}$$

d. What is the percent load on the transformer?

$$\% \text{ load} = \frac{kVA \text{ load}}{kVA \text{ rating}} \times 100$$

$$= \frac{48 \text{ kVA}}{50 \text{ kVA}} \times 100$$

$$= 96\%$$

EXAMPLE 3-9

This diagram shows the same 50-kVA transformer with an unbalanced resistive load. The load from L_1 to neutral is 100 A, from L_2 to neutral is 50 A, and from L_1 to L_2 is 100 A. Current values are shown by the arrows.

a. What is the total current in the three lines? L_1 carries the 240-V load plus the 100-A, 120-V load.

$$100 \text{ A} + 100 \text{ A} = 200 \text{ A in } L_1$$

L_2 carries the 240-V load plus the 50-A, 120-V load.

$$100 \text{ A} + 50 \text{ A} = 150 \text{ A in } L_2$$

The neutral carries the unbalanced portion of the load.

$$200 \text{ A} - 150 \text{ A} = 50 \text{ A in the neutral}$$

b. What is the total load in kilovolt-amperes? To find the total load, calculate the individual loads in kilovolt-amperes and add them together.

$$kVA = kVA_1 + kVA_2 + kVA_3 + \text{etc.}$$

$$kVA = \frac{VI_1}{1000} + \frac{VI_2}{1000} + \frac{VI_3}{1000}$$

$$= \frac{(120 \text{ V} \times 100 \text{ A})}{1000} + \frac{(120 \text{ V} \times 50 \text{ A})}{1000}$$

$$+ \frac{(240 \text{ V} \times 100 \text{ A})}{1000}$$

$$= 12 \text{ kVA} + 6 \text{ kVA} + 24 \text{ kVA}$$

$$= 42 \text{ kVA}$$

c. What is the primary current?

$$I_p = \frac{kVA \times 1000}{V}$$

$$= \frac{42 \times 1000}{600 \text{ V}}$$

$$= 70 \text{ A}$$

d. What is the percent load on the transformer?

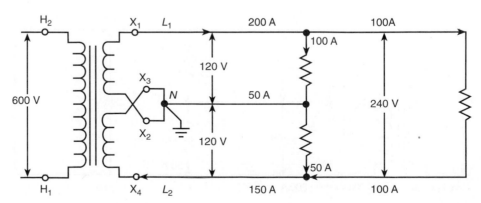

$$\% \text{ load} = \frac{\text{kVA load}}{\text{kVA rating}} \times 100$$

$$= \frac{42 \text{ kVA}}{50 \text{ kVA}} \times 100$$

$$= 84\%$$

Note: The total output rating of the 50-kVA transformer is 208 A $\left(\dfrac{50 \text{ kVA}}{240 \text{ V}} \right)$. Although the total transformer is only 84 percent loaded, coil $X_1 - X_2$ is almost 100 percent loaded. This shows the necessity of balancing single-phase 120-V loads as much as possible in order to protect the transformer windings and use the full capacity of the transformer.

THREE-PHASE TRANSFORMER SYSTEMS

Large amounts of power are generated using a *three-phase system.* The generated voltage will be stepped up and down many times before it reaches the loads in your home or plant. This transformation can be accomplished by using *wye-* or *delta-connection* transformers or a combination of the two along with differing voltage ratio transformers.

Three-phase transformers may be inherently three-phase, having three primary windings and three secondary windings mounted on a three-legged core. The same result, however, can be achieved by using three single-phase transformers connected together to form a so-called three-phase transformer bank. It is important to balance the load on a transformer bank so that any one transformer will not be overloaded.

The two types of three-phase distribution commonly used are three-phase *3-wire* and three-phase *4-wire.* The *three-phase 3-wire delta system* (Fig. 3-26) is used for *balanced loads* and consists of three transformer windings connected end to end. These three windings are connected in delta, and the phases are 120 electrical degrees apart. The phase-to-phase voltage is the same as the coil voltage. Two voltages most commonly used in industry are 208-V and 600-V three-phase 3-wire. The current output of the three-phase 3-wire delta transformer is the coil current multiplied by 1.73, which gives you line current. The number 1.73 is the square root of 3 and is used because the windings are 120 electrical degrees apart.

The other commonly used three-phase distribution is *three-phase 4-wire.* There are two types of three-phase 4-wire distribution: the *wye* or *star three-phase 4-wire,* and the *three-phase delta 4-wire.* The wye *three-phase 4-wire system* (Fig. 3-27 on page 56) has three phases connected at a common point, which is called the neutral. Because of this, none of the windings is affected by the other windings. Therefore, the wye three-phase 4-wire system is used for *unbalanced loads.* The phases are 120 electrical degrees apart; however, they have a common point, so the phase-to-phase voltage is equal to the phase-to-neutral voltage multiplied by 1.73. The line current is equal to the phase current. Two of the most commonly used voltages are 120 V/208 V three-phase 4-wire and 347 V/600 V three-phase 4-wire. The neutral of the wye three-phase 4-wire connection usually is bonded to ground to prevent phase-to-phase voltage at the neutral.

The *three-phase delta 4-wire transformer* is used to supply a small single-phase lighting load and three-phase power load at the same time. They come in sizes up to 150 kVA

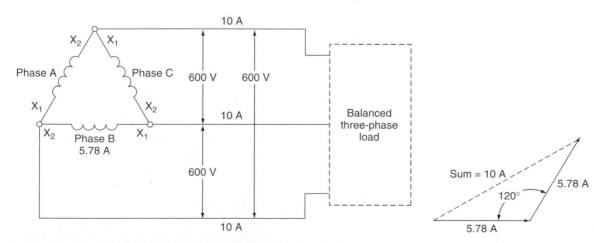

Fig. 3-26 Three-phase delta 3-wire connection.

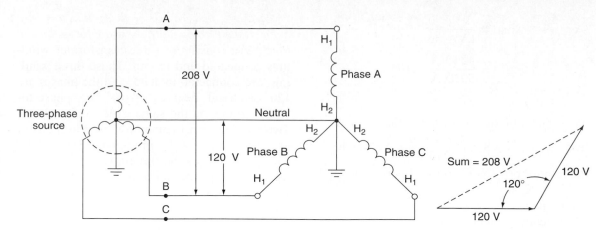

Fig. 3-27   Wye three-phase 4-wire connection.

and are built in such a way that 5 percent of the rated kilovolt-amperes of the transformer is taken from the center tap on the single-phase winding. Thus, the three-phase capacity is reduced by 25 percent. There are three voltages produced by this connection. The transformer connection shown in Fig. 3-28 is

for a three-phase delta 4-wire 120/208/240 V connection. Three-phase 240 V is obtained from the three phases, 208-V single-phase is obtained from phase A to the neutral tap, and 120-V single-phase is obtained from phase B or C to the neutral tap. To get phase A to neutral voltage, multiply the phase-to-phase voltage by 0.87 (cosine of 30): $240 \times 0.87 = 208$ V.

Figure 3-29 shows a *3-wire delta-to-delta three-phase step-down transformer bank*. The three single-phase transformers have both primary and secondary windings connected in delta. If the primary has 2400 V and the transformer ratio is 10:1, the secondary voltage in relation to the primary voltage is determined solely by the ratio, since delta line and phase voltages are the same.

Figure 3-30 on page 58 shows the diagram of the *4-wire wye-wye three-phase connection* that is commonly used for supplying three-phase and single-phase service. Single-phase transformers with 2400-V primaries, when connected wye in a three-phase bank, require 4160 line volts to give 2400 phase volts.

$$V_{\text{ph}} = \frac{V_{\text{line}}}{1.73}$$

$$= \frac{4160}{1.73}$$

$$= 2400 \text{ V}$$

If the primary has 2400 V and the transformer ratio is 20:1, the secondary phase voltage will be 120 V and the line voltage will be 208 V (120 V × 1.73). Three-phase loads are supplied at 208 V. The voltage for single-phase loads is 208 V or 120 V. Three single-phase transformers of the same power ratings are used in most wye-to-wye systems. The capacity

Voltages available:
3Φ—240 V from lines A, B, and C
1Φ—240 V from any two of the lines A, B, and C
1Φ—120 V from either B to *N* or C to *N*
1Φ—208 V from line A to *N*

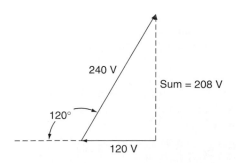

Fig. 3-28   Three-phase delta 4-wire connection.

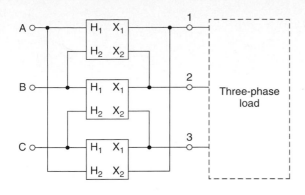

(a) Block diagram. The incoming lines (source) are A, B, C and the outgoing lines (load) are 1, 2, 3

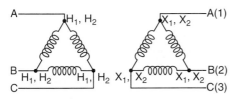

(b) Schematic symbol

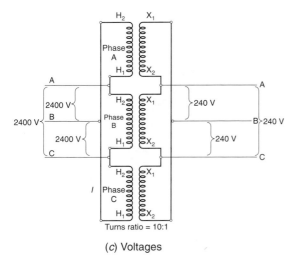

Turns ratio = 10:1

(c) Voltages

Fig. 3-29 Three-wire delta-to-delta three-phase transformer bank.

Self-Test

Answer the following questions.

41. True or false? Transformers of different voltage ratings may be grouped together.
42. Parallel operation of transformers results in an increased _____ rating.
43. _____-V single-phase _____-wire is the type of distribution system that is used to feed homes.
44. Three-phase transformers are normally connected in _____ or _____.
45. In a delta-connected transformer, the phase and line _____ are equal.
46. In a wye-connected transformer, the phase and line _____ are equal.
47. The common connection in a wye-connected transformer is called the _____ point.
48. True or false? A wye-connected three-phase 4-wire transformer has a phase-to-phase voltage of 600 V. The phase to neutral would be 300 V.
49. True or false? The three phases of a three-phase system are 120 electrical degrees apart.

3-4 SUBSTATIONS

Substations serve to change the line voltage by means of transformers and to *regulate* it. Figure 3-31 on page 59 shows the block diagram of an electronically operated tap switching transformer used to regulate voltage. The tap switches regulate output voltage by changing the ratio of primary windings to secondary windings in response to fluctuations in input voltage or load. The circuit uses a comparator in conjunction with solid-state switches (SCRs or triacs) to automatically select the appropriate taps to compensate for fluctuations. Voltage is regulated not continuously but in steps. Switching occurs when line voltage passes through zero, so transients are not created.

A *network* consists of substations, transmission lines, and generating stations that are connected together. A network allows excess power from one region to be fed to another region in response to demand. During the periods when the demand for electrical power drops, power stations shut down some generators. At times of peak demand, auxiliary equipment is set in operation. Networking results in a decrease in the electricity-generating capacity needed by any one area. This produces lower costs for all users.

of transformers is rated in kilovolt-amperes. The total kilovolt-ampere capacity of a transformer bank is found by adding the individual kilovolt-ampere ratings of each transformer in the bank.

To help maintain a balanced transformer bank, always try to connect loads so as to divide them evenly among the three transformers. This will naturally be done when connecting a three-phase motor, since a three-phase motor will draw the same amount of current from each line. Care should then be taken to balance the single-phase loads.

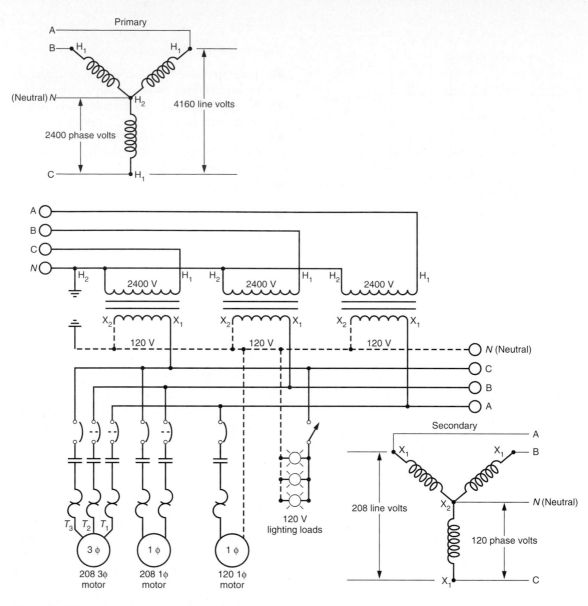

Fig. 3-30   Four-wire wye-to-wye three-phase transformer bank.

Substations also provide a safe point in the distribution system for disconnecting the power in the event of trouble, as well as a convenient place to take measurements and check the operation of the distribution system.

Interconnecting substations serve to tie together different power systems. This enables power exchanges between substations to increase, thus stabilizing the overall network. Substations provide a switching point where different connections may be made between various transmission lines. Figure 3-32 shows two substations interconnected to a network.

Secondary unit substations form the heart of an industrial plant's or commercial building's

electrical distribution. They receive the electric power from the electric utility and step it down to the utilization voltage level of 600 V nominal or less for distribution throughout the building. Unit substations offer a coordinated and integrated switch-gear and transformer package, factory assembled and tested, which requires a minimum amount of labor for installation at the site.

The diagram for a typical unit substation is shown in Fig. 3-33 on page 60. The unit substation is completely enclosed on all sides with sheet metal (except for the required ventilating openings and viewing windows) so that no live parts are exposed. Access within the

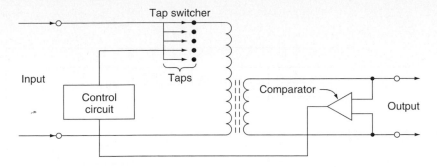

Fig. 3-31 Tap switching transformer.

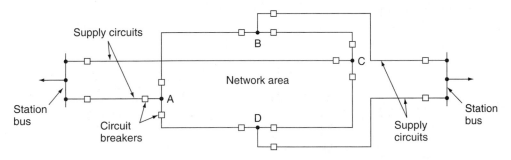

Fig. 3-32 Interconnecting substations.

enclosure is provided only through *interlocked doors* or bolted-on removable panels. The basic unit substation consists of the three following compartments:

1. High-voltage primary switch-gear section
    - This section incorporates the terminations for the primary feeder cables and primary switch-gear, all housed in one metal-clad enclosure.
    - A load-interrupter switch and fuse assembly are widely used for primary switch-gear.
    - The load-interrupter switch can interrupt currents only up to its continuous current rating, normally either 600 A or 1200 A.
    - Lightning protection should be provided if the unit substation is supplied from an overhead distribution system.
2. Transformer section
    - This section houses the transformer for stepping down the primary voltage to the low-voltage utilization level.
    - Dry-type, air-cooled transformers are universally used because they do not require any special fireproof vault construction.
3. Low-voltage distribution section
    - This switchboard section provides the protection and control for the low-voltage feeder circuits.

- It may consist of fusible switches, molded-case circuit breakers, draw-out-type power circuit breakers, or any combination of these devices.
- It may contain metering for the measurement of voltage, current, power, power factor, or energy.
- The secondary switch gear is intended to be tripped out in the event of overload or faults in the secondary circuit fed from the transformer; the primary gear should trip if a short circuit or ground fault occurs in the transformer itself.

Before attempting to do any work on a unit substation, the *load* should be disconnected from the transformer and locked open. Then the transformer primary should be disconnected, locked out, and grounded temporarily if over 600 V.

3-5 IN-PLANT DISTRIBUTION

Within the plant, electrical feeder systems are required to safely deliver energy to the location of each piece of equipment without any component overheating or unacceptable voltage drops (Fig. 3-34 on page 61). This connecting wiring

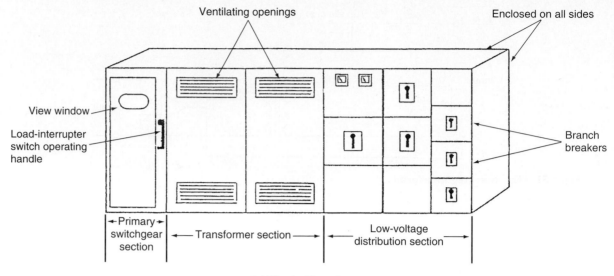

Ventilating openings

Enclosed on all sides

View window

Load-interrupter switch operating handle

Branch breakers

←Primary→ switchgear section

←— Transformer section —→

Low-voltage distribution section

(a) Physical layout

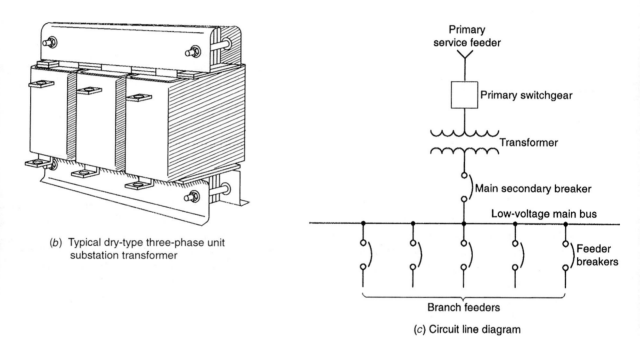

(b) Typical dry-type three-phase unit substation transformer

Primary service feeder

Primary switchgear

Transformer

Main secondary breaker

Low-voltage main bus

Feeder breakers

Branch feeders

(c) Circuit line diagram

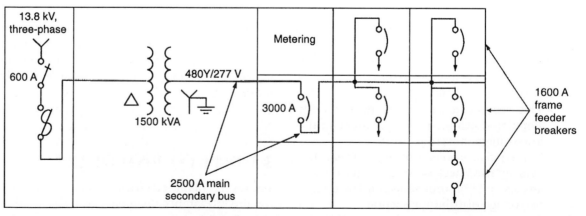

13.8 kV, three-phase

600 A

Metering

480Y/277 V

1500 kVA

3000 A

1600 A frame feeder breakers

2500 A main secondary bus

(d) Unit substation line diagram

Fig. 3-33   Secondary unit substation.

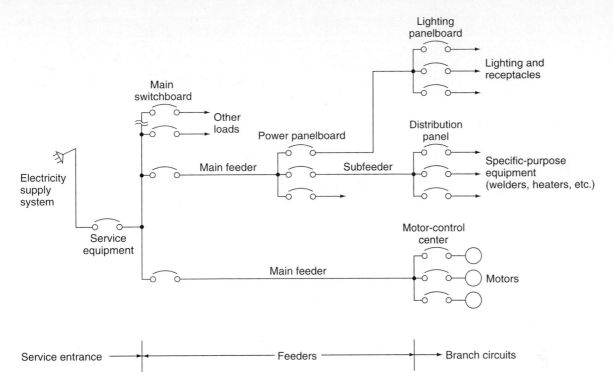

Fig. 3-34 In-plant distribution.

within a building can be divided into the following sections:

1. Service entrance
   - This section includes conductors for delivering energy from the electricity supply system to the premises being served.
   - The conductors are terminated near their point of entrance into the building in the service equipment.
   - In the case of a plant or large premises, the electrical power is usually supplied by the electric utility at a medium-voltage level; this requires a unit substation with a transformer to step down the voltage to the utilization level.
2. Feeders
   - This section includes conductors for delivering the energy from the service equipment location to the final branch-circuit overcurrent device; this protects each piece of utilization equipment.
   - Main feeders originate at the service equipment location, and subfeeders originate at panelboards or distribution centers at locations other than the service equipment location.
3. Branch circuits
   - This section includes conductors for delivering the energy from the point of

the final overcurrent device to the utilization equipment.
   - Each feeder, subfeeder, and branch circuit conductor in turns needs its own properly coordinated overcurrent protection in the form of a circuit breaker or fused switch.

Correct selection of conductors for feeders and branch circuits must take into account ampacity, short circuit, and voltage drop requirements. The ampacity rating of conductors in a raceway depends on the conductor material, gauge size, and temperature rating; the number of current-carrying conductors in the raceway; and the ambient temperature. The *ampacity* of a conductor refers to the maximum amount of current it can safely carry without becoming overheated. Manufacturers usually list the correct wire and fuse sizes in the installation instructions.

The National Electrical Code contains tables that list the ampacity for approved types of conductor size, insulation, and operating conditions. These tables are a practical source of information for specific circuit installations. Table 3-2 on page 62 is a typical listing of the ampacity of copper wires.

Short-circuit conditions can impose tremendous stresses on an electrical system. In the case of feeders, the resulting high short-circuit currents can cause the conductor temperature

Table 3-2 Ampacity of Copper Wires

Wire size	Not More Than Three Conductors in Conduit, Cable, or Buried Directly in the Earth		Single Conductors in Free Air		
	Type TW	Types RH, RHW, THW	Type TW	Types RH, RHW, THW	Weather-proof
	A	B	C	D	E
14	15	15	15	15	30
12	20	20	20	20	40
10	30	30	30	30	55
8	40	50	60	70	103
6	55	65	80	95	130
4	70	85	105	125	163
2	95	115	140	170	219
1/0	125	150	195	230	297
2/0	145	175	225	265	344
3/0	165	200	260	310	401

to rise very rapidly. The *short-circuit current rating* of a conductor depends on the conductor material, the temperature limitations of the insulation, and the duration of the short circuit.

It is very important to have the correct voltage at the outlet that serves a piece of utilization equipment. Most equipment is voltage sensitive. The National Electrical Code recommends a maximum voltage drop of 3 percent for any one branch circuit or feeder, with a maximum voltage drop from the service entrance to the utilization outlet of 5 percent. The *line voltage drop* can be easily measured when the equipment is operating by reading the voltage at the final overcurrent device and subtracting it from that of the voltage read at the equipment. If we read 240 V at the supply and 230 V at the equipment when it is operating, then there is a voltage drop of 10 V in the circuit (Fig. 3-35). The actual voltage drop in a length of wire is equal to the product of the current and resistance of the wire:

$$V_{drop} = I \times R_{wire}$$

where V_{drop} = voltage drop in volts (V)
I = current in amperes (A)
R_{wire} = wire resistance in ohms (V)

Line voltage drops are kept low by keeping the resistance of the line wires low. The resistance of a length of wire decreases as its diameter increases and increases as its length increases. For convenience, wire sizes are usually referred to by an equivalent gauge number rather than by the actual diameter. The *American Wire Gauge (AWG) table* is the standard used and consists of 40 wire sizes ranging from AWG 40 (the smallest) up to AWG 0000 (Table 3-3). Note that the *larger* the gauge number, the *smaller* the diameter of the conductor.

Excessive line voltage drop can be caused by using wires that are too small in diameter for the circuit's current requirements and length. When sizing circuit conductors for extremely long circuits, the expected voltage drops are estimated before the installation. If necessary, wire diameter sizes are increased above the rated current capacity that is required for the circuit to keep the line voltage loss within acceptable limits.

Current flow through a wire also causes a *power loss* that results from the conductor's resistance. Power loss in a wire is equal to the square of the current multiplied by the resistance of the wire:

$$P = I^2 \times R_{wire}$$

where P = power in watts (W)
I = current in amperes (A)
R_{wire} = wire resistance in ohms (Ω)

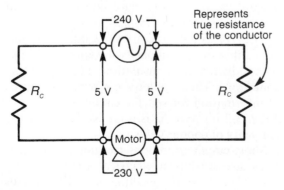

Fig. 3-35 Measuring line voltage drop.

Table 3-3 AWG Wire Sizes

AWG (B and S) Gauge	Standard Metric Size (mm)	Diameter in mils	Cross-sectional Area		Ohms per 1000 ft at 20° C (68° F)	Lbs per 1000 ft	Ft per lb
			Circular mils	Square in.			
0000	11.8	460.0	211,600	0.1662	0.04901	640.5	1.561
000	10.0	409.6	167,800	0.1318	0.06180	507.9	1.968
00	9.0	364.8	133,100	0.1045	0.00793	402.8	2.482
0	8.0	324.9	105,500	0.08289	0.09827	319.5	3.130
1	7.1	289.3	83,690	0.06573	0.1239	253.3	3.947
2	6.3	257.6	66,370	0.05213	0.1563	200.9	4.977
3	5.6	229.4	52,640	0.04134	0.1970	159.3	6.276
4	5.0	204.3	41,740	0.03278	0.2485	126.4	7.914
5	4.5	181.9	33,100	0.02600	0.3133	100.2	9.980
6	4.0	162.0	26,250	0.02062	0.3951	79.46	12.58
7	3.55	144.3	20,820	0.01635	0.4982	63.02	15.87
8	3.15	128.5	16,510	0.01297	0.6282	49.98	20.01
9	2.80	114.4	13,090	0.01028	0.7921	39.63	25.23
10	2.50	101.9	10,380	0.008155	0.9989	31.43	31.82
11	2.24	90.74	8324	0.006467	1.260	24.92	40.12
12	2.00	80.81	6530	0.005129	1.588	19.77	50.59
13	1.80	71.96	5178	0.004067	2.003	15.68	63.80
14	1.00	64.08	4107	0.003225	2.525	12.43	80.44
15	1.40	57.07	3257	0.002558	3.184	9.858	101.4
16	1.25	50.82	2583	0.002028	4.016	7.818	127.9
17	1.12	45.26	2048	0.001609	5.064	6.200	161.3
18	1.00	40.30	1624	0.001276	6.385	4.917	203.4
19	0.90	35.89	1288	0.001012	8.051	3.899	256.5
20	0.80	31.96	1022	0.0008023	10.15	3.092	323.4
21	0.71	28.46	810.1	0.0006363	12.80	2.452	407.8
22	0.63	25.35	642.4	0.0005046	16.14	1.945	514.2
23	0.56	22.57	509.5	0.0004002	20.36	1.542	648.4
24	0.50	20.10	404.0	0.0003173	25.67	1.223	817.7
25	0.45	17.90	320.4	0.0002517	32.37	0.9699	1,031.0
26	0.40	15.94	254.1	0.0001996	40.81	0.7692	1,300
27	0.355	14.20	201.5	0.0001583	51.47	0.6100	1639
28	0.315	12.64	159.8	0.0001255	64.90	0.4837	2067
29	0.280	11.26	126.7	0.00009953	81.83	0.3836	2607
30	0.250	10.03	100.5	0.00007894	103.2	0.3042	3287
31	0.224	8.928	79.70	0.00006260	130.1	0.2413	4145
32	0.200	7.950	63.21	0.00004964	164.1	0.1913	5227
33	0.180	7.080	50.13	0.00003937	206.9	0.1517	6591
34	0.160	6.305	39.75	0.00003122	260.9	0.1203	8310
35	0.140	5.615	31.52	0.00002476	329.0	0.09542	10,480
36	0.125	5.000	25.00	0.00001964	414.8	0.07568	13,210
37	0.112	4.453	19.83	0.00001557	523.1	0.06001	16,660
38	0.100	3.965	15.72	0.00001235	659.6	0.04759	21,010
39	0.090	3.531	12.47	0.000009793	831.8	0.03774	26,500
40	0.080	3.145	9.888	0.000007766	1049.0	0.02993	33,410

Naturally, it is desirable to hold line power loss to a minimum. The larger the conductors, the smaller the power of I^2R losses will be.

A compromise is generally reached in which the voltage drop will be held within acceptable limits and in which there is an economic balance between the cost of the conductors and the cost of the power lost.

All conductors installed in a building must be properly protected, usually by installing them in raceways. Raceways provide space, support, and mechanical protection for conductors, and they minimize the hazards from electric shocks and electrical fires. Commonly used types of raceways (Fig. 3-36) are described on the next page.

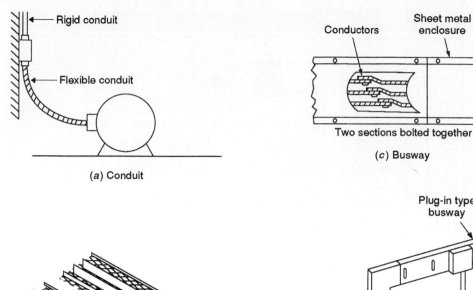

Fig. 3-36   Common types of raceways.

1. Conduits
   - Conduits available in rigid and flexible, metallic and nonmetallic types.
   - Conduit runs must be properly supported and have sufficient access points to facilitate the installation of the conductors.
   - Conduits must be large enough to accommodate the number of conductors based, generally, on a 40 percent fill ratio.
2. Cable trays
   - Cable trays are used to support feeder cables where a number of them are to be run from the same location.
   - They consist of heavy feeder conductors run in troughs or trays.
3. Low-impedance busways (bus duct)
   - The busways are used in buildings for high-current feeders.
   - They consist of heavy bus bars enclosed in ventilated ducts.
4. Plug-in busways
   - These busways are used for overhead distribution systems.
   - They provide convenient power tap-offs to the utilization equipment.

A *switchboard* is a piece of electrical equipment to which a large block of electric power is delivered from the substation and broken down into smaller blocks for distribution throughout the building. In the case of a unit substation, it is the low-voltage distribution section. Figure 3-37 illustrates a combination service entrance switchboard. The switchboard has the space and mounting provisions required by the local power company for metering their equipment and incoming power. In addition, other meters such as ammeters and voltmeters may be built into the meter compartment. The switchboard also controls the power and protects the distribution system through the use of switches, fuses, circuit breakers, and protective relays. Switchboards that have more than six switches or circuit breakers must include a *main switch* to protect or disconnect all circuits.

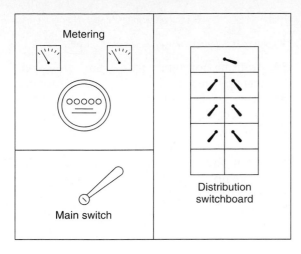

Fig. 3-37 Combination service entrance switchboard.

A *panelboard* is a wall-mounted distribution cabinet containing a group of circuit breaker or fuse protective devices for lighting, convenience receptacles, or power distribution branch circuits (Fig. 3-38). The panelboard is usually supplied from the switchboard and further divides the power distribution system into smaller parts. Panelboards make up the part of the distribution system that provides the last centrally located protection for the final power run to the load and its control circuitry.

Often, an industrial installation will require that many motors be controlled from a central location. For example, the incoming power, control circuitry, required overload and overcurrent protection, and any transformation of power are combined into one convenient cen-

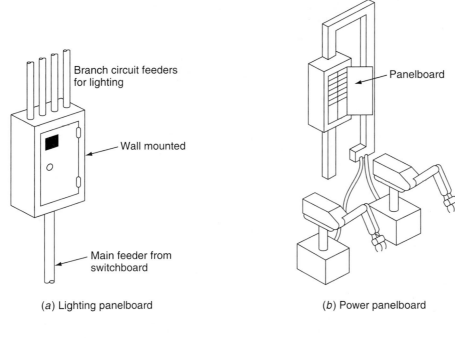

(a) Lighting panelboard

(b) Power panelboard

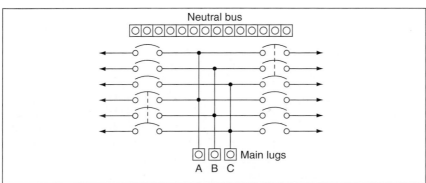

(c) Internal wiring for a three-phase 4-wire panelboard with circuit breakers

Fig. 3-38 Panelboard.

ter. This center is called the *motor control center* (Fig. 3-39). It is a compact, floor-mounted assembly, composed principally of combination starters. It consists of one or more vertical sections, with each section having a number of spaces for motor starters. The number of spaces is determined by the horsepower (hp) ratings of the individual starters. Thus, a starter that will control a 10-hp motor will take up less room than a starter that will control a 100-hp motor. The motor control center is a modular structure designed specifically for plug-in-type control units and motor control.

Self-Test

Answer the following questions:

50. Transmission line voltages are often regulated by means of _____ transformers.
51. True or false? Substations are never interconnected.
52. True or false? A 600-V power line would be classified as low voltage.
53. What are the three compartments that make up a unit substation?
54. True or false? Overload currents cause the conductor temperature to rise very rapidly.

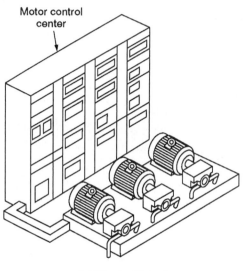

(a) Physical layout

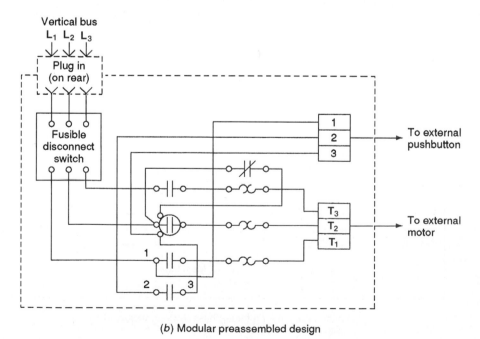

(b) Modular preassembled design

Fig. 3-39 Motor control center.

55. True or false? The ampacity of a wire may be reduced when it is installed in a raceway.

56. True or false? The ampacity of a conductor is the same as its current-carrying capacity.

57. According to the National Electrical Code, the voltage drop for any one branch circuit should not exceed _____ percent.

58. True or false? Resistance increases as the length of the conductor increases.

59. True or false? If the load current remains constant, the effect of changing the feeder size from No. 8 to No. 6 AWG is a decrease in line drop.

60. True or false? Wire diameters in standard wire tables include insulation thickness.

61. Bus duct is used in building for high-_____ feeders.

62. True or false? Switchboards are part of the plant distribution system that take a large block of electrical power and break it down into smaller blocks.

63. True or false? A switchboard is usually supplied with power from a panelboard.

64. True or false? The branch circuit is that portion of the distribution system between the final overcurrent device and the outlet or load.

65. A motor control center is composed principally of motor _____.

3-6 ELECTRIC POWER AND ENERGY

Electric energy is the energy carried by moving electrons. Whenever current exists in a circuit, there is a conversion of electric energy into other forms of energy. For example, current flow through a lamp filament converts electric energy into light and heat energy. *Electric power* can be defined as the *rate* at which electric energy is converted. For example, a 150-W light bulb converts energy at twice the rate of a 75-W bulb (Fig. 3-40).

▌▶ **YOU MAY RECALL** that the *watt* (W) is the basic unit of electric power. A summary of formulas for power measurement in dc and ac circuits follows:

DC Circuit

$P = V \times I$

$p = V^2/R$

$P = I^2R$

Single-Phase AC Circuits

$P = V \times I \times PF$

Three-Phase AC Circuits

$P = V_{\text{line}} \times I_{\text{line}} \times 1.73 \times PF$

where P = power in watts (W)
V = voltage in volts (V)
I = current in amperes (A)
R = resistance in ohms (Ω)
PF = power factor

For small industrial applications, the watt is too small a unit, so the *kilowatt* (*kW*), which is 1000 W, is used as the standard unit.

For large industrial applications, the unit kilowatt is too small for measurement, so the *megawatt* (*MW*), which is 1000 kW or 1,000,000 W, is used as the measurement unit.

Wattmeters take into account all factors—current, voltage, and power factor—at the same time and indicate the true power being consumed. Wattmeters are available in both analog and digital configurations. As shown in

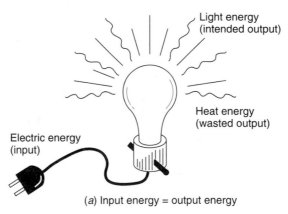

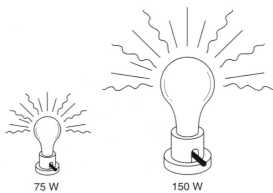

(a) Input energy = output energy

(b) The 150-W light converts energy at twice the rate of the 75-W light

Fig. 3-40   Electric power and energy.

Fig. 3-41, dc and single-phase power is the simplest to measure. The wattmeter has four terminals: two for voltage and two for current. The voltage terminals are hooked across the line, and the current leads are hooked in series with the load.

Whereas many of the loads throughout a plant are single-phase, their operating power comes from one phase of a polyphase distribution system. Normally a polyphase wattmeter is used to measure power in a three-phase circuit.

Power measurements in ac circuits present difficult measurement problems, but in polyphase circuits they are even more challenging. Because the flow of current in an ac circuit constantly changes direction because of the alternating polarity of the voltage, the power in an ac circuit is not as easy to measure as in a dc circuit.

Real power (watts), sometimes called *true power* or *average power,* in an ac circuit is the electric power that is actually converted into heat or work. In the dc circuit, we learned that power in watts is equal to the voltage multiplied by the current. The same is true of the *resistive ac circuit.* In Fig. 3-42, however, we see that *V* and *I* are always varying in an ac circuit. The power in such a circuit is the average of all the instantaneous values of voltage multiplied by the corresponding instantaneous values of current. The *true power,* or heating effect, is the average of these areas and can be shown to equal the product of the voltage *(V)* multiplied by current *(I).* The unit used to measure real power is the watt; real power is measured with a wattmeter.

Reactive power is the power that a capacitor or inductor *seems* to be using. In a sense, however, capacitors and inductors *do not use power;* they store energy. Inductive reactance takes power from the source to build its magnetic field, but then when the magnetic field is collapsing, it returns the power to the source. The effect is that no net power from the source is consumed. Similarly, capacitive reactance takes power from the source to charge its plates. During discharge the power is returned to the source. Again, the effect is that no net power is consumed from the source. Resis-

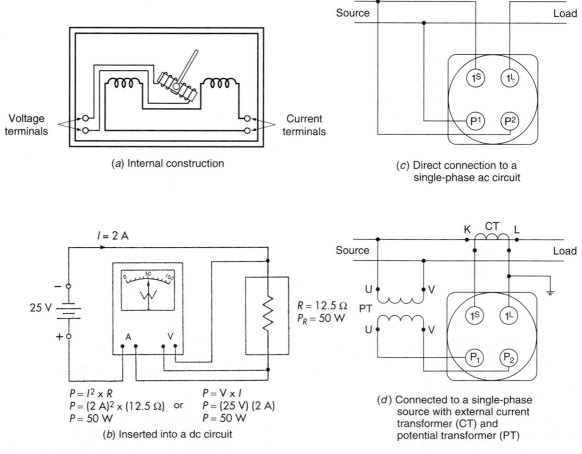

(a) Internal construction

(b) Inserted into a dc circuit

$P = I^2 \times R$

$P = (2\ A)^2 \times (12.5\ \Omega)$ or

$P = 50\ W$

$P = V \times I$

$P = (25\ V)\,(2\ A)$

$P = 50\ W$

(c) Direct connection to a single-phase ac circuit

(d) Connected to a single-phase source with external current transformer (CT) and potential transformer (PT)

Fig. 3-41 Wattmeter.

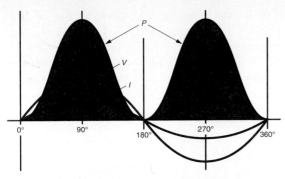

(a) AC power waveform for circuit with resistance

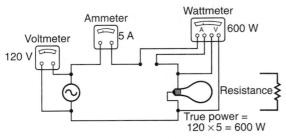

True power =
120 × 5 = 600 W

(b) Wattmeter measures real power

Fig. 3-42   Real power (watts).

tance, on the other hand, converts power to heat or a motor converts electric energy to mechanical energy. Thus, the power used by the resistance is not returned to the source, and there is a net consumption (conversion) of power. In the capacitor circuit of Fig. 3-43 we see that although the voltmeter and ammeter both register readings, the reading on the wattmeter is *zero,* indicating zero power con-

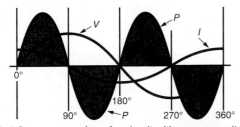

(a) AC power waveform for circuit with pure capacitance

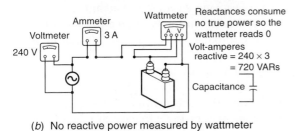

Reactances consume no true power so the wattmeter reads 0

Volt-amperes
reactive = 240 × 3
= 720 VARs

(b) No reactive power measured by wattmeter

Fig. 3-43   Reactive power (VAR).

sumption. To distinguish reactive power from real power, the watt power unit is not used. Instead, *volt-amperes reactive (VAR)* is used. The product of the capacitor voltage and current is used to calculate reactive power, with VAR as the unit of measurement.

Apparent power is the total power that the entire ac circuit is apparently using. It is the total power supplied to the circuit from the source, including any real and reactive power. Apparent power and real power for ac circuits are equal only when the circuits consist entirely of pure resistance (no reactance). In the motor circuit shown in Fig. 3-44 on page 70, of the total apparent power applied to the motor, some is being converted to heat and work and the rest is stored and returned to the circuit. Apparent power is measured in *volt-ampere (VA)* units to distinguish it from real power.

Power factor (PF) is the ratio of real power to apparent power:

$$PF = \frac{\text{watts}}{\text{volt-ampere}} = \frac{\text{real power}}{\text{apparent power}}$$

The power factor of a circuit is an indication of the portion of volt-amperes that are actually real power. In practice, the power factor of a circuit can be found by dividing a wattmeter reading by the product of a voltmeter reading that has been multiplied by the ammeter reading (Fig. 3-44).

Power factor is, therefore, real power divided by apparent power. It is commonly expressed as a percentage. When real power is equal to apparent power—such as in a circuit having only resistance or one in which the reactances exactly cancel one another—the power factor is equal to 1.00, or 100 percent.

A power factor of less than 100 percent indicates that not all the current is doing useful work. Yet the conductors must be sized to carry the total current and the capacity of the equipment in the system, such as generators, and transformers must be capable of handling this total current. The power factor is leading for a capacitance load, lagging for an inductive load, or in phase for a resistive load.

Most plant electrical systems have a lagging-power factor owing to transformers, motors, coils, fluorescent lighting ballasts, and the like. When the power factor of an entire system drops below 80 percent, measures must be taken to increase the power factor. A low power factor makes increased demands on the power company without consuming power (on which the utility bases its bill), so power com-

Volt-amperes
reactive (VAR)

Apparent power

Volt-ampere (VA)

Power factor (PF)

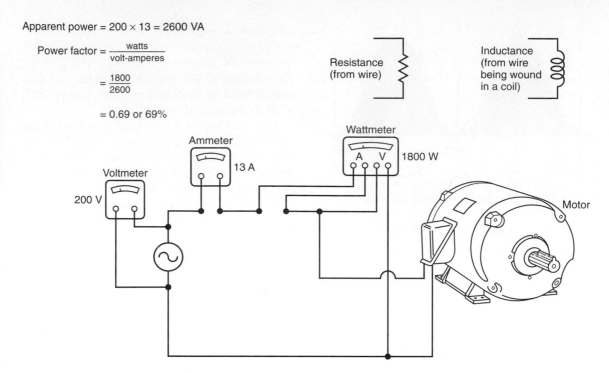

Apparent power = 200 × 13 = 2600 VA

$$\text{Power factor} = \frac{\text{watts}}{\text{volt-amperes}}$$

$$= \frac{1800}{2600}$$

= 0.69 or 69%

Fig. 3-44   Apparent power (VA) and power factor (PF).

panies often charge penalty fees for customers having excessively low power factors (i.e., below 70 percent lagging). In most cases, a low lagging-power factor is corrected by connecting capacitors in parallel with the system. Capacitors produce a leading-power factor, which tends to neutralize the lagging-power factor (Fig. 3-45).

▐▐▶ **YOU MAY RECALL**   that the practical unit for measuring electrical energy is the *kilowatthour* (*kWh*). To calculate the energy used in kilowatthours, multiply the power in kilowatts (kW) times the time in hours (h):

Kilowatts (kW) × hours (h) =
                    kilowatthour (kWh)

For example, if you turned on ten 100-W light bulbs and left them on for one hour, the energy used would be 1 kWh.

The *kilowatthour meter* is used to measure the amount of electric energy used to determine cost of electricity (Fig. 3-46). The meter records how many watts of power are used over a period of time. At regular intervals, a power company employee comes to read a customer's meter. The power company calculates the amount of electric energy used and charges the customer for it.

It is important to understand the difference between demand and energy. *Demand* is the

rate (kW) at which the customer uses electricity during the billing period. Energy is the quantity (kWh) of electricity used during a period of time. Figure 3-47 on page 72 shows a comparison between two factories consuming the same energy but having different demands. Both consumed a total of 720,000 kWh over the month-long time period. The demand for Factory A, however, was at a constant rate of 1000 kW, whereas the demand for Factory B fluctuated between 50 kW and 3000 kWh. The more regular the power flow, the less the cost. The electric utility, therefore, must invest more capital to service Factory B; consequently, it is reasonable that Factory B should pay more for its energy.

A *demand meter* is normally installed at the service entrance of most industries to monitor the power flowing into the plant. It automatically measures the average power during successive demand intervals of 10, 15, or 30 minutes. High power surges of shorter durations do not last long enough to warrant the installation of correspondingly large equipment by the utility company; hence, no additional charges are imposed.

We all tend to use more electricity during the period of 7 A.M. to 11 P.M. every day. Therefore, the utilities that serve us must have sufficient generating capacity in place to meet the demand. This period would be referred to as the *peak demand period* for the utility.

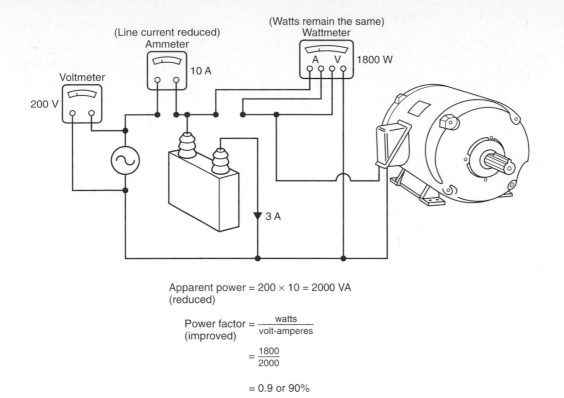

Apparent power = $200 \times 10 = 2000$ VA (reduced)

Power factor (improved) $= \dfrac{\text{watts}}{\text{volt-amperes}}$

$$= \dfrac{1800}{2000}$$

$$= 0.9 \text{ or } 90\%$$

Fig. 3-45 Power factor correction—capacitive current cancels inductive current.

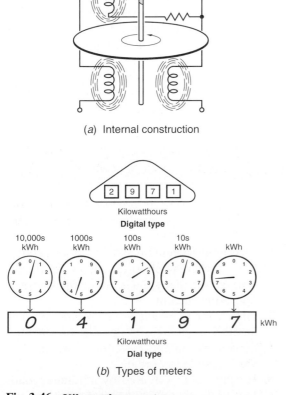

(a) Internal construction

Kilowatthours
Digital type

Kilowatthours
Dial type

(b) Types of meters

Fig. 3-46 Kilowatthour meter.

From the 11 P.M. to 7 A.M. period, most of us do not demand the same rate of electricity as during the day period. Therefore, we underutilize the utility's capacity. This period is referred to as the *off-peak period*, indicating that unused capacity is available for our use. The peak capacity or demand that the utilities must supply or meet affect the cost of electricity. One cost-effective way that utilities can meet new demands is by shifting some demand to the off-peak period, where there is excess capacity.

The use of electronic control devices to improve the energy efficiency of electrical systems is increasing. For example, substantial savings can be made by keeping the maximum demand as low as possible. Loads that are not absolutely necessary can be automatically switched off until the peak has passed.

Self-Test

Answer the following questions.

66. Electric energy is the energy carried by moving _____.

67 Electric power is defined as the _____ at which electric energy is converted.

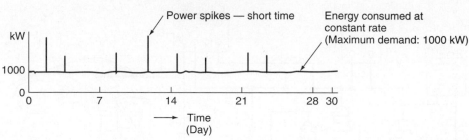

Factory A — 720,000 kWh consumed

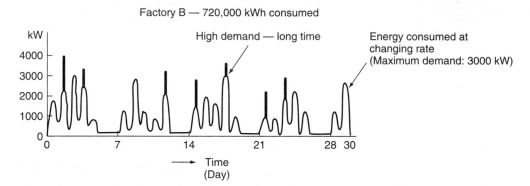

Factory B — 720,000 kWh consumed

Fig. 3-47 Comparison between two factories consuming the same energy but having different demands.

68. One kilowatt is equal to _____ watts.

69. A wattmeter has connections for measurement of both _____ and _____.

70. True or false? Real power (watts) in an ac circuit is the electric power that is actually converted into heat.

71. To distinguish it from real power, reactive power is measured in _____.

72. Apparent power and real power for ac circuits are equal only when the circuits consist entirely of pure _____.

73. True or false? A power factor of less than 100 percent indicates that not all current is doing useful work.

74. True or false? Power companies often charge lower rates for customers having very low power factors.

75. True or false? The more regular the power flow to a plant, the less the cost.

76. True or false? The hours from 11 P.M. to 7 A.M. would be referred to as the off-peak period for the utility.

SUMMARY

1. High voltages are used in transmission lines to reduce current and minimize line losses.

2. Substations contain step-down transformers that reduce the voltage to the final end-user voltage.

3. Single-phase power is supplied to residential customers; three-phase power is supplied to industrial customers.

4. Electrical distribution systems are grounded to protect human lives from the danger of electric shock and property from damage.

5. Lightning arresters are used to drain the energy from power lines struck by lightning.

6. Overload currents occur at rather modest levels, whereas short-circuit or fault current can be many times larger than normal operating current.

7. High-voltage circuit breakers may be designated as being air, oil, gas, or vacuum, depending on the medium in which they are designed to operate.

8. Harmonics are created by nonlinear loads that draw current in abrupt pulses.

9. Standby power supply systems are switched into emergency use, whereas uninterruptible power supply systems are on-line or continuous.
10. Transformers are used to transfer energy from one ac circuit to another.
11. In an ideal transformer, power output is equal to the power input, the voltage ratio is equal to the turns ratio, and the current ratio is inversely proportional to the turns ratio.
12. Potential and current transformers are used to step down voltage and current for metering purposes.
13. Transformer output is rated in kilovolt-amperes rather than watts.
14. Transformers are broadly classified as wet or dry.
15. Three-phase transformation is accomplished by using wye or delta connections.
16. The phases of a three-phase system are always 120 electrical degrees apart; therefore, the number 1.73 ($\sqrt{3}$) is used in three-phase voltage, current, and power calculations.
17. A unit substation consists of the high-voltage primary switch-gear section, a transformer section, and a low-voltage distribution section.
18. The maximum line voltage drop for a branch circuit should not exceed 3 percent.
19. A switchboard takes a large block of electric power and breaks it down into smaller blocks for distribution throughout the building.
20. A panelboard contains protective devices for lighting, convenience receptacles, or power distribution branch circuits.
21. A motor control center is a modular structure designed specifically for plug-in-type control units and power control.
22. Power in an ac circuit may consist of real power (watts), reactive power (VAR), and apparent power (VA).
23. Power factor (PF) is the ratio of real power to apparent power.
24. Low power factor makes increased demands on the power company without consuming power.
25. Demand is the rate (kW) at which the customer uses electricity during the billing period.

CHAPTER REVIEW QUESTIONS

Answer the following questions.

3-1. **a.** Why are high voltages used when transmitting electric power over long distances?
   b. What limitation is there to the use of high-voltage transmission systems?
3-2. What causes the voltage on distribution systems to vary?
3-3. Explain the function of earth grounding and equipment grounding as part of an effective grounding system.
3-4. How does a lightning arrester work to protect power lines that are struck by lightning?
3-5. Why are short-circuit currents considered to be more destructive than overload currents?
3-6. Compare the operation of a fuse and a circuit breaker.
3-7. List four ratings used in specifying high-voltage fuses.
3-8. What does selective coordination of protection devices ensure?
3-9. Describe four ways in which an electrical disturbance source can be coupled into other circuitry.
3-10. **a.** What is the fourth harmonic of a 60-Hz power source?
   b. What type of loads create harmonics?
3-11. What corrective actions can be taken to reduce harmonic current?
3-12. Which type of emergency power supply system is most reliable and effective? Why?
3-13. What is the function of the transformer core?
3-14. **a.** What are eddy currents?
   b. How is the core of a transformer constructed to reduce eddy currents?

3-15. In an ideal transformer, what is the relationship between
   a. The turns ratio and voltage ratio?
   b. The voltage ratio and current ratio?
   c. The primary power and secondary power?

3-16. A step-down transformer with a turns ratio of 10:1 has 120 V applied to its primary coil. A 3-Ω load resistor is connected across the secondary coil. Calculate the following:
   a. Secondary coil voltage.
   b. Secondary coil current.
   c. Primary coil current.

3-17. A step-up transformer has a primary coil current of 32 A and an applied voltage of 240 V. The secondary coil has a current of 2 A. Calculate the following:
   a. Power input of the primary coil.
   b. Power output of the secondary coil.
   c. Secondary coil voltage.
   d. Turns ratio.

3-18. **a.** What is the main purpose of an isolation transformer?
   b. What is the turns ratio for an isolation transformer?

3-19. For what purpose are autotransformers used on distribution lines?

3-20. Why are transformers rated in kilovolt-amperes or volt-amperes rather than kilowatts?

3-21. **a.** What are the two main types of losses associated with a transformer?
   b. In what form do these losses appear?

3-22. How are transformers broadly classified according to method of cooling?

3-23. How are three-phase transformers broadly classified according to type of connection?

3-24. If the phase-to-neutral voltage is 208 V in a wye transformer system, what would be the phase-to-phase voltage?

3-25. Why is it important to balance the load connected to a three-phase transformer bank?

3-26. How many electrical degrees apart are the phases of a three-phase system?

3-27. What is the function of the "taps" on a transformer?

3-28. What are the three main parts of a unit substation?

3-29. Why are dry-tape, air-cooled transformers universally used for unit substations?

3-30. List five factors that determine the ampacity rating of conductors installed in a raceway.

3-31. **a.** How is the voltage drop of a branch circuit most easily measured?
   b. What is the maximum voltage drop allowed for any one branch circuit?

3-32. List three pieces of electrical equipment designed to take a large block of electric power and break it down into smaller blocks for distribution purposes.

3-33. **a.** Compare electrical energy and electric power.
   b. State the basic unit used to measure each.

3-34. **a.** A wattmeter is a combination of what two types of meters?
   b. When connecting a wattmeter into a circuit, how is each meter section connected with regard to the load?

3-35. The voltage current and power of a capacitor connected to an ac source are metered. If the voltmeter reads 240 V and the ammeter reads 10 A, what should the wattmeter reading be?

3-36. **a.** Why do power utilities charge penalty fees for industrial customers that operate with low power factor?
   b. How can low power factor be corrected?

3-37. **a.** What is the difference between demand and energy?
   b. What type of demand is less costly to supply?

3-1. You have been assigned the task of checking a manufacturing plant for a harmonics problem. Suggest ways to determine if such a problem exists.

3-2. A single-phase transformer is required to step down a 600-V system to a 3-wire 120/240-V system for lighting and heating loads. The loads consist of 40-kVA resistive 120-V loads and 50-kVA resistive 240-V loads. Assuming that the 120-V loads will be balanced evenly between L_1 and L_2, find the following:

    a. Total load in kilovolt-amperes.
    b. Current in X_1 and X_4.
    c. Current in the neutral.
    d. Primary current (neglecting losses).
    e. The percentage load if a 100-kVA transformer is used.

3-3. Three single-phase transformers rated at 100 kVA, 4160/240 V are connected wye-to-wye. A three-phase 4160-V source is applied to the primaries. A balanced three-phase load of 200 kW at an 80 percent power factor is connected to the secondaries. Determine the following:

    a. The total three-phase kilovolt-ampere rating of the bank.
    b. The load on the transformer bank in kilovolt-amperes.
    c. The secondary line current.
    d. The secondary coil current.
    e. The primary line current (assume no losses).
    f. The primary coil current (assume no losses).
    g. The percentage load on the transformers bank.

3-4. a. What is the resistance of 500 ft of No. 10 AWG solid copper wire? (Refer to Table 3-3.)

    b. Calculate the line voltage drop if the *overall* length of wire in (**a**) is used in a circuit that draws 25 A of current.
    c. Calculate the line power loss of this circuit.

Answers to Self-Tests

1. false	19. true	37. true
2. current	20. false	38. resistance; inductive
3. transformers	21. false	reactance; capacitive
4. 60 Hz	22. true	reactance
5. false	23. secondary	39. false
6. true	24. mutual	40. true
7. true	25. exciting	41. false
8. earth; equipment	26. core	42. kilovolt-amperes
9. true	27. losses	43. 240/120 V; 3
10. false	28. true	44. wye; delta
11. true	29. true	45. voltage
12. true	30. false	46. current
13. false	31. false	47. neutral
14. false	32. 1:1	48. false
15. coordination	33. X_1; X_2	49. true
16. magnitude; frequency	34. true	50. tap switching
17. nonlinear	35. false	51. false
18. false	36. true	52. true

53. (1) high voltage primary switch gear section; (2) transformer section; (3) low-voltage distribution section
54. false
55. true
56. true
57. three
58. true
59. true
60. false
61. current
62. true
63. false
64. true
65. starters
66. electrons
67. rate
68. 1000
69. voltage; current
70. true
71. volt-amperes reactive (VAR)
72. resistance
73. true
74. false
75. true
76. true

CHAPTER 4

Industrial Control Devices

∎

CHAPTER OBJECTIVES

This chapter will help you to:

1. *Identify* manually operated switches commonly found in industry.
2. *Identify* mechanically operated switches commonly found in industry.
3. *Explain* how different sensors detect and measure the presence of something.
4. *Describe* the operating characteristics of a relay, solenoid, solenoid valve, stepper motor, and brushless dc motor.

Control devices are components that govern the power delivered to an electrical load. Every industrial system makes use of a wide variety of control devices. The industrial control devices introduced in this chapter range from simple pushbutton switches to more complex solid-state sensors. The terms and practical applications presented here illustrate how selection of a control device depends upon the specific application.

∎

4-1 PRIMARY AND PILOT CONTROL DEVICES

A *control device* is a component that governs the power delivered to an electrical load. All components used in motor control circuits may be classed as either primary control devices or pilot control devices. A *primary control device,* such as a motor contactor, starter, or controller, connects the load to the line. A *pilot control device,* such as a relay or contactor that activates a power circuit, directs the operation of another device. Pilot devices include pushbuttons, flow switches, pressure switches, and thermostats (Fig. 4-1 on page 78). Pilot-duty devices should not be used to switch horsepower loads unless they are specifically rated to do so. Contacts selected for both primary and pilot control devices must be capable of handling the voltage and current to be switched. In Fig. 4-1, closing the flow switch contact completes the circuit to energize the *coil* of the magnetic contactor which then causes the *power contacts* of the contactor to close and complete the main power circuit to the motor.

4-2 MANUALLY OPERATED SWITCHES

A *manually operated switch* is one that is controlled by hand. A *toggle switch* (Fig. 4-2 on page 78) is an example of a manually operated switch. This type of switching or contact arrangement is specified by the appropriate abbreviation. Electrical ratings are expressed in terms of maximum interrupting voltage and current; they should not be exceeded. A switch rated at 5 A will not last very long in a circuit that must break 10 A. Also, ac and dc ratings are not the same for a given switch. The dc current rating of a switch would have a *lower* magnitude than the ac rating.

The *slide switch* (Fig. 4-3 on page 78) uses a simple slide action to produce the same connections as a toggle switch. Except for the different type of operator action, the switched poles accomplish the same results. The slide switch is often used as a mode switch to select a certain mode of operation such as HIGH and LOW. The *rocker switch* is often a modified slide switch. Pressing one side of the rocker-

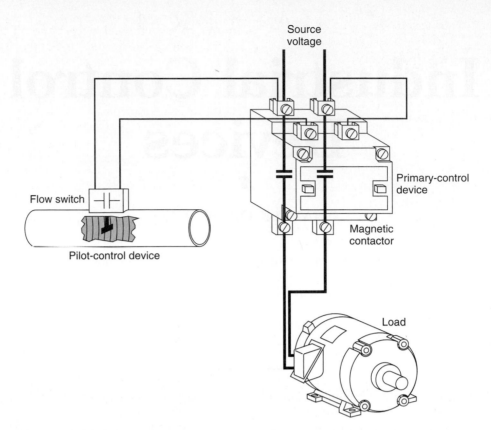

Fig. 4-1   Primary and pilot control devices.

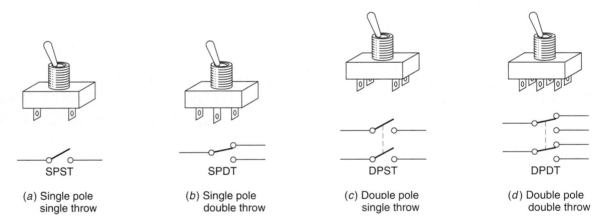

SPST	SPDT	DPST	DPDT
(*a*) Single pole single throw	(*b*) Single pole double throw	(*c*) Double pole single throw	(*d*) Double pole double throw

Fig. 4-2   Toggle switches.

Fig. 4-3   Slide switch.

arm mechanism causes the slide to be forced in the other direction.

DIP (*dual in-line package*) *switches* are small switch assemblies designed for mounting on printed circuit boards (Fig. 4-4). The pins or terminals on the bottom of the DIP switch are the same size and spacing as an integrated circuit (IC) chip. The individual switches may be of the toggle, rocker, or slide kind. Switch settings are seldom changed, and the changes occur mainly during installation, testing, and troubleshooting.

The *rotary switch* is often used for more complex switching operations, such as those found on oscilloscopes and multimeters. This type of switch is sometimes called a wafer switch because the main shaft passes through the center of one or more ceramic, fiberglass,

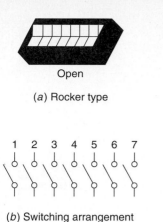

Open

(a) Rocker type

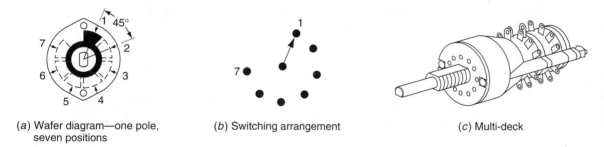

(b) Switching arrangement

Fig. 4-4   DIP switch.

or phenolic wafers on which the terminals and contacts are mounted. The center section of each wafer can be rotated, passing a common contact to one of the many stationary contacts mounted around the wafer's perimeter (Fig. 4-5). The rotary switch can also have several

banks of switch sections on one shaft. This allows contacts to change simultaneously and in sequence.

Thumbwheel switches are used on numerical and computer-controlled equipment to input information from the operator to the computer. Their specially made decks output binary coded decimal (BCD), decimal, or hexadecimal codes necessary to communicate with digital computers. Figure 4-6 shows a four-ganged thumbwheel switch set to input the decimal number 5670.

The *selector switch* is another common manually operated switch. Switch positions are made by turning the operator knob right or left (Fig. 4-7 on page 80). Selector switches may have two or more selector positions, with either maintained contact position or spring return to give momentary contact operation.

Pushbutton switches are the most common form of manual control found in industry. The NO (normally open) pushbutton makes a cir-

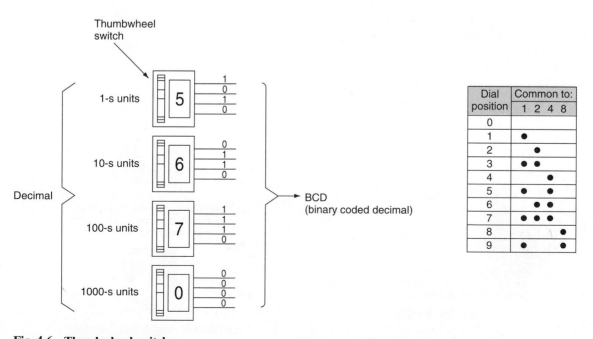

(a) Wafer diagram—one pole, seven positions

(b) Switching arrangement

(c) Multi-deck

Fig. 4-5   Rotary switch.

Thumbwheel switch

Decimal {

1-s units   5

10-s units   6

100-s units   7

1000-s units   0

BCD (binary coded decimal)

Dial position	Common to:			
	1	2	4	8
0				
1	●			
2		●		
3	●	●		
4			●	
5	●		●	
6		●	●	
7	●	●	●	
8				●
9	●			●

Fig. 4-6   Thumbwheel switch.

(a) Selector switch operator

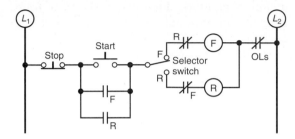

(b) Control circuit for selector switch used for reversing direction of rotation of a motor

Fig. 4-7   Selector switch.

cuit when it is pressed and returns to its open position when the button is released. The NC (normally closed) pushbutton opens the circuit when it is pressed and returns to the closed position when the button is released. The break-make pushbutton is used for *interlocking* controls. In this switch the top section is NC, whereas the bottom section is NO. When the button is pressed, the bottom contacts are closed after the top contacts open. When you have one or more pushbuttons in a common enclosure, it is referred to as a pushbutton station. The pushbutton may consist of one or more contact blocks, an operator device, and legend plate (Fig. 4-8).

A *drum switch* consists of a set of moving contacts mounted on and insulated from a

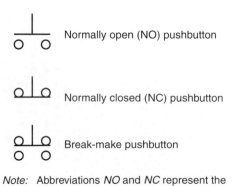

Normally open (NO) pushbutton

Normally closed (NC) pushbutton

Break-make pushbutton

Note: Abbreviations *NO* and *NC* represent the electrical state of the switch contacts when the switch is not actuated

(a) Pushbutton symbols

Fig. 4-8   Various types of pushbutton symbols and switches.

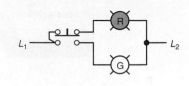

(b) Control circuit using a combination break-make pushbutton

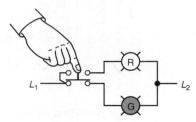

(b) Control circuit using a combination break-make pushbutton

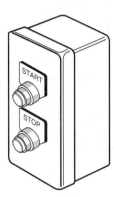

(c) Pushbutton station consists of one or more pushbuttons in a common enclosure

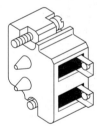

(d) Plastic contact block houses the switching contacts of the pushbutton

(e) Flush button operator (many types of operators may be attached to contact blocks depending on application)

(f) Legend plate denotes function

rotating shaft. The switch also has a set of stationary contacts that make and break contact with the moving contacts as the shaft is rotated. Drum switches (Fig. 4-9) are used for starting and reversing squirrel-cage motors, single-phase motors that are designed for reversing service, and dc shunt and compound-wound motors.

(Courtesy of Furnas Electric Company.)

(a) Switch

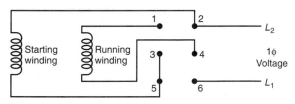

(b) Internal switching arrangement

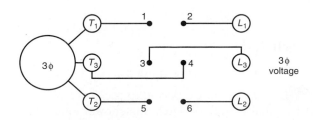

(c) Single-phase motor reversing—wiring diagram

(d) Three-phase motor reversing—wiring diagram

Fig. 4-9   Reversing drum switch.

4-3 MECHANICALLY OPERATED SWITCHES

A *mechanically operated switch* is one that is controlled automatically by factors such as pressure, position, and temperature. The *limit switch* (Fig. 4-10) is a very common industrial control device. Limit switches are designed to operate only when a predetermined limit is reached, and they are usually actuated by contact with an object such as a cam. These devices take the place of human operators. They are often used in the control circuits of machine processes to govern the starting, stopping, or reversal of motors.

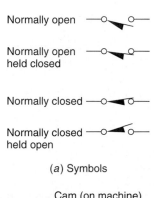

(a) Symbols

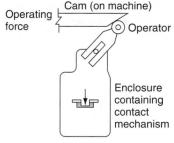

(b) Operation

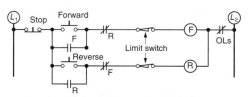

(c) Control circuit for starting and stopping a motor in forward and reverse with limit switches providing overtravel protection

Fig. 4-10   Limit switch.

A *microswitch* is a snap-acting switch housed in a small enclosure (Fig. 4-11 on page 82). In a snap-acting switch, as in a toggle switch, the actual switching of the circuit takes place at a fixed speed no matter how quickly

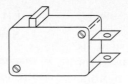

(a) Typical enclosure

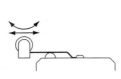

Roller leaf—low-force large movement actuation

Panel-mount roller plunger—actuation by cams

Lever—very low force, slow cams and slides

Panel-mount plunger—heavy duty in-line applications or slow cams. Cam rise should not exceed 30°

Leaf—low-force slow moving cams or slides

Pin plunger—in-line motion

Roller lever—very low force, fast moving cams

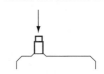

Overtravel plunger—in-line applications requiring additional overtravel

(b) Switch actuators

Fig. 4-11   Microswitch.

or slowly the activating mechanism moves. The small size and variety of operating levers make microswitches very useful as limit switches. They can operate with very small pressures on the operating levers, which allows a great deal of sensitivity.

Temperature switches, or *thermostats* (Fig. 4-12), are used to sense temperature changes. Although there are many types available, they are all actuated by some specific environmental temperature change. Temperature switches open or close when a designated temperature is reached. Industrial applications for these devices include maintaining the desired temperature range of air, gases, liquids, and solids.

The temperature switch is similar to a pressure switch but differs in that a closed, chemically filled bellows system is used. The pressure in the system changes in proportion to the temperature of the bulb. The temperature-responsive medium in the system is a volatile liquid whose vapor pressure increases as the temperature of the bulb rises. Conversely, as the temperature of the bulb falls, the vapor pressure decreases. The pressure change is transmitted to the bellows through a capillary tube operating a precision switch at a predetermined setting.

Pressure switches (Fig. 4-13) are used to control the pressure of liquids and gases. Again, although many types are available, they are all basically designed to actuate (open or close) their contacts when a specified pressure is reached. Pressure switches are pneumatically (air) operated switches. Generally a bellows or diaphragm presses up against a small microswitch and causes it to open or close.

Level switches (Fig. 4-14 on page 84) are used to sense the height of a liquid. The raising or lowering of a float, which is mechanically attached to the level switch, trips the level switch. The level switch itself is used to control motor-driven pumps that empty or fill tanks. Level switches are also used to open or close piping solenoid valves to control fluids.

Self-Test

Answer the following questions.

1. True or false? Pilot-duty devices are normally used to switch horsepower loads.

2. _____ switches are small assemblies designed for mounting on printed circuit boards.

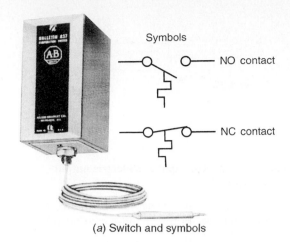

(a) Switch and symbols

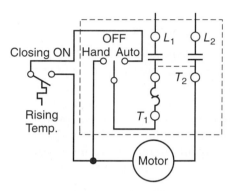

(b) Temperature switch used to automatically control a motor

Fig. 4-12   Temperature switch.

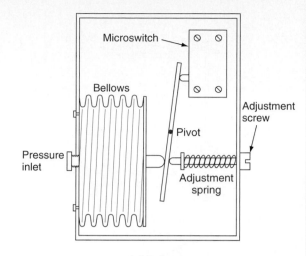

(a) Bellows

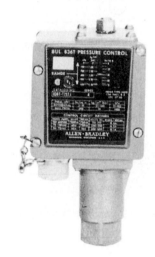

Symbols

NO contact

NC contact

(b) Switch and symbols

(c) Starter operated by pressure switch

Fig. 4-13   Pressure switch.

3. An _____ switch is often used for switching operations on multimeters.

4. _____ switches output binary codes necessary to communicate with digital computers.

5. True or false? All selector switches operate with spring return to provide momentary contact operation.

6. True or false? An NC (normally closed) pushbutton makes a circuit when it is pressed and returns to its open position when the button is released.

7. Drum switches are used for starting and _____ motors.

8. True or false? Limit switches installed on a machine are normally actuated by the operator.

9. True or false? A microswitch is capable of operating with extremely small pressure on its operating lever.

10. True or false? Temperature switches open or close when a designated temperature is reached.

11. True or false? All pressure switches use only NO (normally open) contacts.

12. _____ switches are used to sense the height of a liquid.

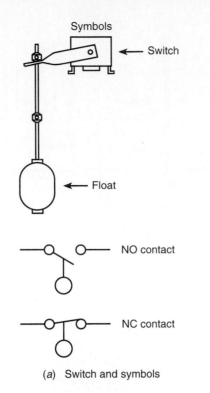

NO contact

NC contact

(a) Switch and symbols

(b) Two-wire level switch control of starter

Fig. 4-14   Level switch.

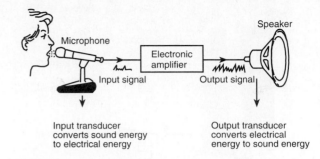

Fig. 4-15   Electric-input and electric-output transducers.

4-4 TRANSDUCERS AND SENSORS

IIII▶ **YOU MAY RECALL**   that a *transducer* is any device that converts energy from one form to another.

Transducers may be divided into two classes: *input transducers* and *output transducers* (Fig. 4-15). Electric-input transducers convert nonelectric energy, such as sound or light, into electric energy. Electric-output transducers work in the reverse order. They convert electric energy to forms of nonelectric energy.

Sensors are devices that are used to *detect,* and often to *measure,* the magnitude of something. They are a type of transducer used to convert mechanical, magnetic, thermal, optical, and chemical variations into electric voltages and currents. Sensors are usually categorized by what they measure and play an important role in modern manufacturing process control. They provide the equivalents

of our eyes, ears, nose, and tongue to the microprocessor brain of industrial automation systems (Fig. 4-16).

PROXIMITY SENSOR

Proximity sensors or switches are pilot devices that detect the presence of an object (usually called the target) *without physical* contact. They are solid-state electronic devices that are completely encapsulated to protect against excessive vibration, liquids, chemicals, and corrosive agents found in the industrial environment. Proximity sensors are used when:

- The object being detected is too small, too lightweight, or too soft to operate a mechanical switch.
- Rapid response and high switching rates are required, as in counting or ejection control applications.
- An object has to be sensed through nonmetallic barriers such as glass, plastic, and paper cartons.
- Hostile environments demand improved sealing properties, preventing proper operation of mechanical switches.
- Long life and reliable service are required.
- A fast electronic control system requires a bounce-free input signal.

An *inductive proximity sensor* is a sensing device that is actuated by a metal object. A typical application is shown in Fig. 4-17(*c*). Proximity sensors (A′ and B′) detect targets A and B moving in the directions indicated by the arrows. When A reaches A′, the machine reverses its motion; the machine reverses again when B reaches B′. In principle, an *inductive sensor* consists of a coil, an oscillator, a detector circuit, and a solid-state output (Fig. 4-17).

IIII▶ **YOU MAY RECALL**   that an oscillator is an electronic circuit for generating ac waveforms and frequencies from a dc energy source.

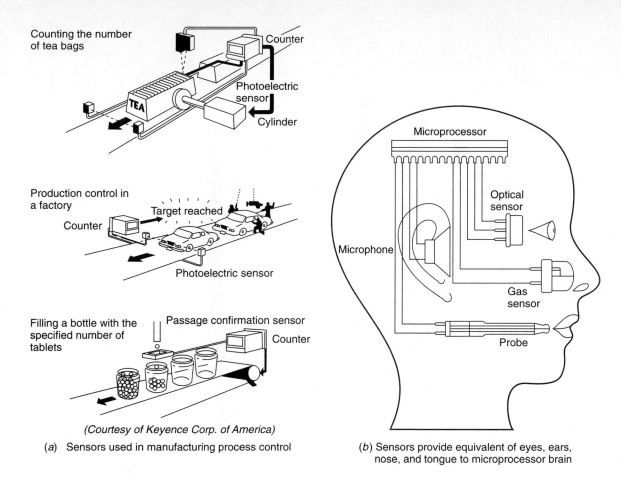

Counting the number of tea bags

Counter

Photoelectric sensor

Cylinder

TEA

Production control in a factory

Target reached

Counter

Photoelectric sensor

Filling a bottle with the specified number of tablets

Passage confirmation sensor

Counter

(Courtesy of Keyence Corp. of America)

(*a*) Sensors used in manufacturing process control

Microprocessor

Optical sensor

Microphone

Gas sensor

Probe

(*b*) Sensors provide equivalent of eyes, ears, nose, and tongue to microprocessor brain

Fig. 4-16 Sensors.

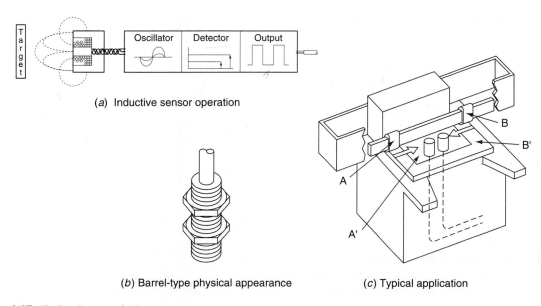

Target

Oscillator   Detector   Output

(*a*) Inductive sensor operation

(*b*) Barrel-type physical appearance

(*c*) Typical application

B
B'
A
A'

Fig. 4-17 Inductive proximity sensor.

When energy is supplied, the oscillator operates to generate a high-frequency field. At this moment, there must not be any conductive material in the high-frequency field. When a metal object enters the high-frequency field, eddy currents are induced in the surface of the target. This results in a loss of energy in the oscillator circuit; consequently, this causes a smaller amplitude of oscillation. The detector circuit recognizes a specific change in amplitude and generates a signal that will turn the solid-state output ON or OFF. When the metal object leaves the sensing area, the oscillator regenerates, allowing the sensor to return to its normal state.

The method of connecting and exciting a proximity sensor will vary with the type of sensor and its application (Fig. 4-18). With a *current-sourcing output* (*PNP*) the load is connected between the sensor and ground. Current flows from the sensor through the load to ground (open emitter). With a *current-sinking output* (*NPN*) the load is connected between positive supply and sensor. Current flows from the load through the sensor to ground (open collector). Remember, these sensors are used as pilot devices for loads such as starters, contactors, solenoids, and so on. Never directly operate a motor with a proximity sensor.

As a result of solid-state switching of the output, a leakage current flows through the sensor even when the output is turned OFF. Similarly, when the sensor is ON, a small voltage drop is lost across its output terminals. In order to operate properly, a proximity sensor should be powered continuously. The difference between the "operate" and "release" points of the sensor is called *hysteresis*, or *differential travel*. Inductive sensors can be actuated in an axial or radial approach (Fig. 4-19). It is important to maintain a minimum air gap between the target and the sensing face to prevent physically damaging the sensors.

A minimum amount of current must be allowed to continuously flow through the sensor in order to maintain operation. When load current is less than this minimum, a bleeder resistor is connected parallel to the load. See Fig. 4-19(*a*). In order to operate properly, a proximity sensor should be powered continuously. A bypass can be added across the mechanical contact to keep the sensor ready for instantaneous operation. See Fig. 4-19(*b*). With the sensor ON, a small voltage is lost across its output. See Fig. 4-19(*c*).

Hysteresis is the distance between the operating point when the target approaches the proximity sensor face and the release point when the target is moving away from the sensor face. It is given as a percentage of the nominal sensing range. Hysteresis is needed to keep proximity sensors from chattering when subjected to shock and vibration, slow-moving targets, or minor disturbances such as electrical noise and temperature drift. See Fig. 4-19(*d*).

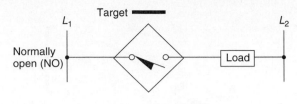

(*a*) 2-wire sensor connection

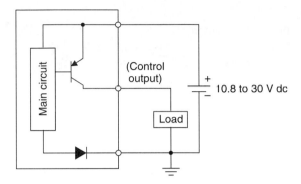

(*b*) Current sourcing output (PNP)

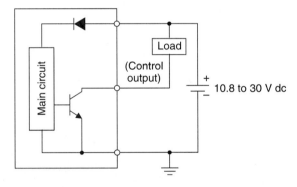

(*c*) Current sinking output (NPN)

DO NOT: Directly operate a motor with a proximity sensor. The in-rush current can cause an overload condition. Always use a motor starter, relay, or other appropriate device.

(*d*) Warning

Fig. 4-18   Proximity sensor connections.

A capacitive proximity sensor is a sensing device that is actuated by conductive and nonconductive materials. The operation of *capacitive sensors* is also based on the principle of an oscillator. Instead of a coil, however, the active face of a capacitive sensor is formed by two metallic electrodes—rather like an "opened" capacitor. See Fig. 4-20(*a*). The electrodes are

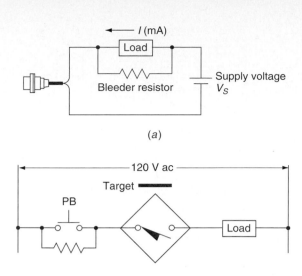

(a)

(b)

(c)

V_2 = Supply voltage $-V_1$

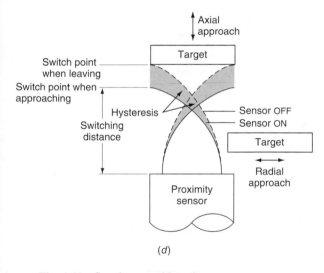

(d)

Fig. 4-19 Sensing considerations.

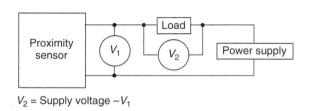

(a)

Fig. 4-20 Capacitive proximity sensor.

placed in the feedback loop of a high-frequency oscillator that is inactive with "*no target present.*" As the target approaches the face of the sensor, it enters the electrostatic field that is formed by the electrodes. This causes an increase in the coupling capacitance, and the circuit begins to oscillate. The amplitude of these oscillations is measured by an evaluating circuit that generates a signal to turn the solid-state output ON or OFF.

A typical application is shown in Fig. 4-20(*b*). Liquid filling a glass or plastic container can be monitored from the outside of the container with a capacitive proximity sensor. In some applications, the empty container is detected by a second sensor, which starts the flow of liquid. The flow is shut off when the level reaches the upper sensor.

In order to actuate inductive sensors, we need a conductive material. Capacitive sensors may be actuated by both conductive materials and by nonconductive materials such as wood, plastics, liquids, sugar, flour, and wheat. Along with this advantage of the capacitive sensor (compared to the inductive sensor) comes some disadvantages. For example, inductive proximity switches may be actuated only by a metal and are insensitive to humidity, dust, dirt, and the like. Capacitive proximity switches, however, can be actuated by any dirt in their environment. For general applications, the capacitive proximity switches are not really an alternative but a *supplement* to the inductive proximity switches. They are a supplement when there is no metal available for the actuation (e.g., for woodworking machines and for determining the exact level of liquids or powders).

MAGNETIC SWITCHES

A *magnetic switch* (also called a *reed relay*) contact is composed of two flat contact tabs that are hermetically sealed (airtight) in a glass

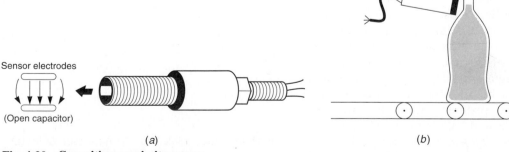

(b)

tube filled with protective gas. As a permanent magnet approaches, the ends of the over-lapped contact tab attract one another and come into contact. As the permanent magnet is moved further away, the contact tab ends are demagnetized and return to their original positions (Fig. 4-21). The magnetic switch switches virtually inertia free. Because the contacts are sealed, they are unaffected by dust, humidity, and fumes; thus, their life expectancy is quite high.

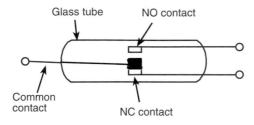

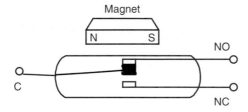

Fig. 4-21 Magnetic switch (reed relay).

A permanent magnet is the most common actuator for a reed relay. Permanent magnet actuation can be arranged in several ways, dependent upon the switching requirement. The most commonly used arrangements are proximity motion, rotation, and the shielding method (Fig. 4-22). The device can also be actuated by a dc electromagnet. Reed relays are faster, more reliable, and produce less arcing than conventional electromechanical switches. The current-handling capabilities of the reed relay, however, are limited.

LIGHT SENSORS

The *photovoltaic,* or *solar cell,* is a common light-sensor device that converts light energy directly into electric energy (Fig. 4-23). Modern silicon solar cells are basically PN junctions with a transparent P-layer. Shining light on the transparent P-layer causes a movement of electrons between P- and N-sections, thus producing a small dc *voltage.* Typical output voltage is about 0.5 V per cell in full sunlight.

The *photoconductive cell* (also called *photoresistive cell*) is another popular type of light

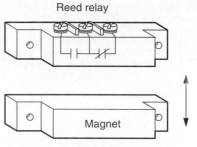

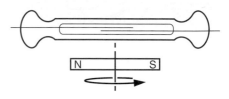

(a) Proximity motion—movement of the relay or magnet will activate the relay

(b) Rotary motion—relay is activated twice for every complete revolution

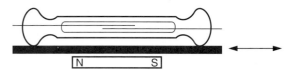

(c) Shielding—ferromagnetic (iron-based) shield short-circuits the magnetic field holding the contacts; relay is activated by removal of the shield

Fig. 4-22 Reed-relay activation.

transducer (Fig. 4-24). Light energy falling on a photoconductive cell will cause a change in the *resistance* of the cell. One of the more popular types is the *cadmium sulfide photocell.* When the surface of this device is dark, the resistance of the device is high. When brightly lit, its resistance drops to a very low value.

There are two main types of *photoelectric sensors* that are used to sense position. Each emits a light beam (visible, infrared, or laser) from its light-emitting element (Fig. 4-25). A *reflective-type photoelectric sensor* is used to detect the light beam reflected from the target. A *through-beam photoelectric sensor* is used to measure the change in light quantity caused by the target's crossing the optical axis.

Figure 4-26 on page 90 shows typical photoelectric sensor applications. Features of this type of sensor include:

■ *Noncontact detection.* Noncontact detection eliminates damage either to the target or

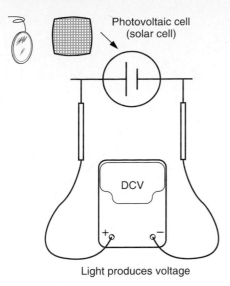

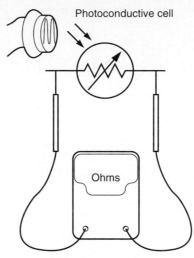

Fig. 4-23 Photovoltaic or solar cell.

Fig. 4-24 Photoconductive cell.

sensor head, ensuring long service life and maintenance-free operation.

- *Detection of targets of virtually any material.* Detection is based on the quantity of light received, or the change in the quantity of reflected light. This method allows detection of targets of diverse materials such as glass, metal, plastics, wood, and liquid.

- *Long detecting distance.* The reflective-type photoelectric sensor has a detecting distance

of 1 m, and the through-beam type has a detecting distance of 10 m.

- *High response speed.* The photoelectric sensor is capable of a response speed as high as 50 μs (1/20,000 s).

- *Color discrimination.* The sensor has the ability to detect light from an object based on the reflectance and absorptance of its color, thus permitting color detection and discrimination.

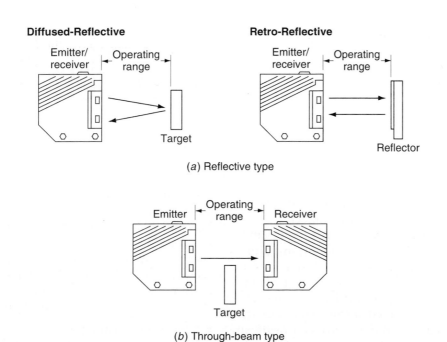

Fig. 4-25 Photoelectric sensor.

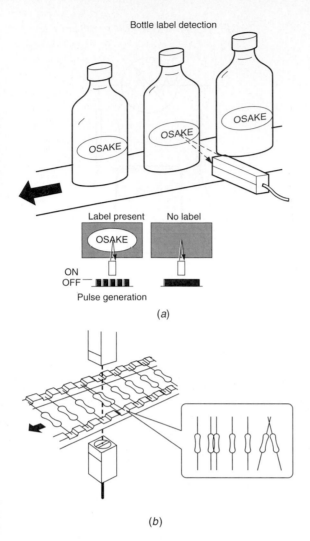

Bottle label detection

Label present   No label

(a)

(b)

Fig. 4-26   Photoelectric sensor applications.

- *Highly accurate detection.* A unique optical system and a precision electronic circuit allows highly accurate positioning and detection of minute objects.

For example, a reflective-type photoelectric sensor used for the detection of the presence or absence of a label is shown in Fig. 4-26(*a*). In Fig. 4-26(*b*), through-beam photoelectric sensor heads are positioned above and below the resistors traveling on a production line. A variation on the line changes the quantity of the laser beam, thus signaling a defect.

In most photoelectric sensors, a light-emitting diode (LED) is the light-transmitting source, and a phototransistor is the receiving source (Fig. 4-27). In operation, light from the LED falls on the input of the phototransistor, and the amount of conduction through the transistor changes. *Analog* outputs provide an output proportional to the quantity of light seen by the photodetector.

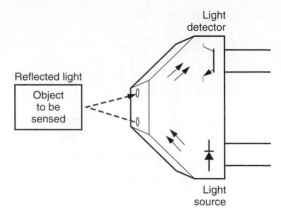

Fig. 4-27   Photoelectric sensor operation.

Bar code technology is widely implemented in industry and is rapidly gaining in a broad range of applications. It is easy to use, can be used to enter data much more quickly than manual methods, and is very accurate. A bar code system consists of three basic elements: the bar code symbol, a scanner, and a decoder.

The *bar code symbol* contains up to 30 characters that are encoded in a machine-readable form. The characters are usually printed above or below the bar code, so data can be entered manually if a symbol cannot be read by the machine. The blank space on either side of the bar code symbol, called the quiet zone, along with the start and stop characters, lets the scanner know where data begins and ends (Fig. 4-28).

Bar code   Quiet zone

→ 100 74983 12345 9
Human readable characters

Fig. 4-28   Bar code symbol.

There are several different kinds of bar codes. In each one, a number, letter, or other character is formed by a certain number of bars and spaces. In the United States, the Universal Product Code (UPC) is the standard bar code symbol for retail food packaging. The UPC symbol (Fig. 4-29) contains all of the encoded information in one symbol. It is strictly a numeric code containing the:

- UPC type (1 character).
- UPC manufacturer, or vendor, ID number (5 characters).

Fig. 4-29 UPC bar code symbol.

- UPC item number (5 characters).
- Check digit (1 character) used to mathematically check the accuracy of the read.

Bar code scanners are the eyes of the data collection system. A light source within the scanner illuminates the bar code symbol; those bars absorb light, and spaces reflect light. A photo detector collects this light in the form of an electronic-signal pattern representing the printed symbol. The *decoder* receives the signal from the scanner and converts this data into the character data representation of the symbol's code. Although the scanner and decoder operate as a team, they can be integrated or separate, depending on the application (Fig. 4-30).

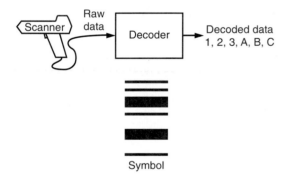

Fig. 4-30 Bar code scanner and decoder.

HALL-EFFECT SENSORS

The *Hall-effect sensor* is designed to sense the presence of a magnetic object, usually a *permanent magnet*. It is used to signal the *position* of a component. Because of its *accuracy* in sensing position, the *Hall-effect sensor* is a popular type of sensing device. Figure 4-31 on page 92 illustrates how a Hall element operates. The Hall element is a small, thin, flat slab of semiconductor material. When a current is

passed through this slab and *no* magnetic field is present, zero output voltage is produced. When a magnet is brought close to the semiconductor material, the current path is distorted. This distortion causes electrons to be forced to the right side of the material, which produces a voltage across the sides of the device. Hall-effect devices use two terminals for excitation and two for output voltage.

Digital Hall-effect integrated circuits (*ICs*), used in proximity switches, can be envisioned as a mechanical switch that allows current to flow when turned ON and blocks current when turned OFF. The digital Hall-effect sensor can be used to measure *speed*. When the magnet passes the sensor, the Hall switch activates, and a pulse is issued. By measuring the frequency of the pulses, the shaft speed can be determined. See Fig. 4-32(*a*).

The *analog Hall-effect transducer* puts out a continuous signal proportional to the sensed magnetic field. The analog Hall-effect sensor can be used to sense *position*. As the magnet moves back and forth, the field seen by the sensor becomes negative as it approaches the north pole and positive as it approaches the south pole. See Fig. 4-32(*b*) on page 93.

ULTRASONIC SENSORS

An *ultrasonic sensor* operates by sending sound waves toward the target and measuring the time it takes for the pulses to bounce back. The time taken for this echo to return to the sensor is directly proportional to the distance or height of the object because sound has a constant velocity. In Fig. 4-33 on page 93, the returning echo signal is electronically converted to a 4- to 20-mA output which supplies the monitored flow rate to external control devices. Solids, fluids, granular objects, and textiles can be detected by an ultrasonic sensor. The sonic reflectivity of liquid surfaces is the same as solid objects. Textiles and foams absorb the sonic waves and reduce sensing range.

PRESSURE SENSORS

A *strain wire gauge transducer* converts a mechanical strain into an electric signal. Strain gauges are based on the principle that the resistance of a conductor varies with length and cross-sectional area. Figure 4-34 on page 94 illustrates the operation of a typical strain gauge. The force applied to the gauge causes the gauge to bend. This bending action also

Bar code scanner

Hall-effect sensor

Digital Hall-effect integrated circuit

Analog Hall-effect transducer

Ultrasonic sensor

Strain wire gauge transducer

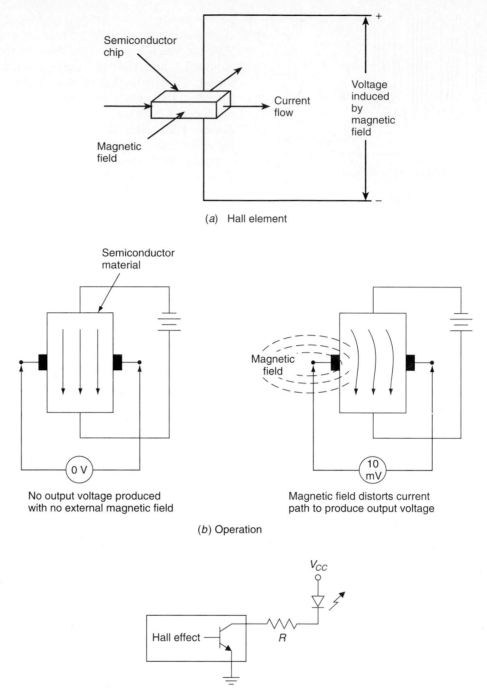

(a) Hall element

Semiconductor material

No output voltage produced with no external magnetic field

Magnetic field

10 mV

Magnetic field distorts current path to produce output voltage

0 V

(b) Operation

V_{CC}

Hall effect

R

(c) Interfaced for driving an LED indicator

Fig. 4-31   Hall-effect sensor.

distorts the physical size of the gauge, which, in turn, changes its *resistance*. This resistance change is fed to a bridge circuit that detects small changes in the gauge's resistance. *Strain gauge load cells* [see Fig. 4-34(*d*)] are usually made with steel and sensitive strain gauges. As the load cell is loaded, the metal elongates or compresses very slightly. The strain gauge detects this movement and translates it to an mV signal. Many sizes and shapes of load cells are available that range in sensitivity from grams through to millions of pounds.

The *semiconductor strain gauge* uses a piezo-electric *crystal* as its sensing element. As a force is applied to the crystal, the shape of the crystal changes. This change of shape develops a *voltage* at the output terminals of the crystal (Fig. 4-35 on page 94). The semiconductor

Linear
displacement
transducer

Angular
displacement
transducer

Linear variable
differential
transformer (LVDT)

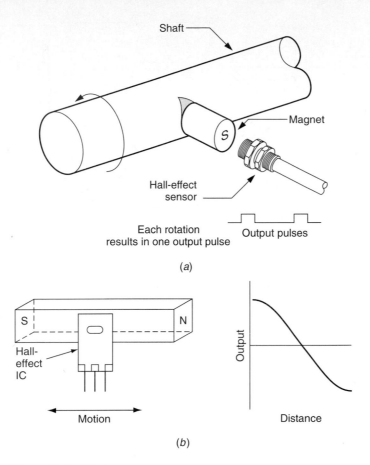

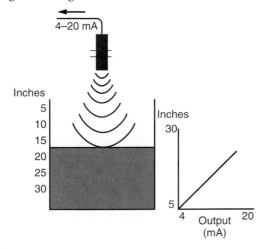

Fig. 4-32 Digital and linear Hall-effect sensors.

Fig. 4-33 Ultrasonic sensor.

strain gauge reacts more quickly to changes taking place on the measuring device and has a higher sensitivity than the strain wire gauge.

DISPLACEMENT TRANSDUCERS

Displacement refers to an object's physical position with respect to a reference point. Displacement transducers can be *linear* (straight-line) or *angular* (rotary). Potentiometers can be used to measure both linear and angular displacement, as illustrated in Fig. 4-36 on page 95. In Fig. 4-36(*a*) the input shaft turns, and the wiper contact moves with it. As the wiper moves, the resistance across the wiper contact and either end contact changes. In Fig. 4-36(*b*), an excitation voltage is contacted across the resistance element. The output voltage is a function of shaft position or shaft displacement. With either type of transducer, there is a direct relationship between the wiper shaft position and the output voltage.

The most common displacement transducer used in industry is the *linear variable differential transformer* (*LVDT*) (Fig. 4-37 on page 95), which is basically a transformer with a movable core and two secondary windings. The movable core is connected to the input shaft. The primary is excited by an ac source. When the core is in its exact center location, the amplitude of the voltage induced into secondary 1 is the same as the voltage induced into secondary 2. The secondaries are connected series opposing, so the output voltage will be zero at that point. If the core is moved off center, the mutual inductance of the pri-

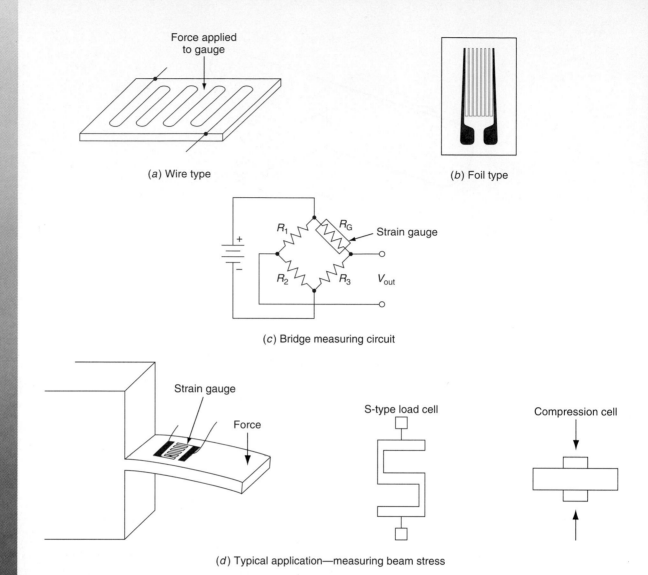

(a) Wire type

(b) Foil type

(c) Bridge measuring circuit

(d) Typical application—measuring beam stress

Fig. 4-34 Strain wire gauge.

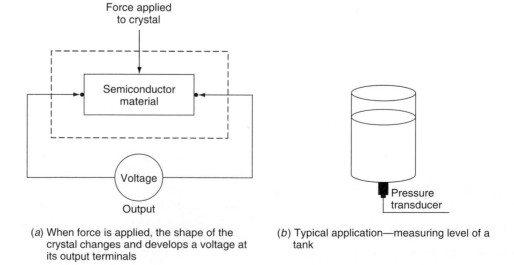

(a) When force is applied, the shape of the crystal changes and develops a voltage at its output terminals

(b) Typical application—measuring level of a tank

Fig. 4-35 Semiconductor strain gauge.

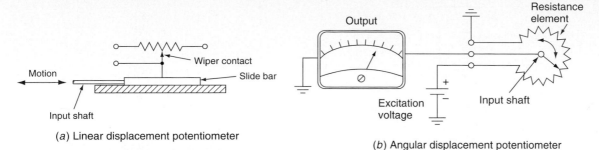

(a) Linear displacement potentiometer

(b) Angular displacement potentiometer

Fig. 4-36 Potentiometer transducers.

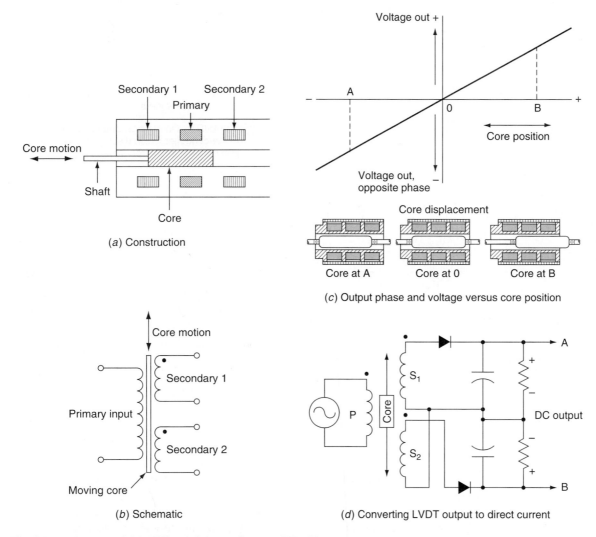

(a) Construction

(c) Output phase and voltage versus core position

(b) Schematic

(d) Converting LVDT output to direct current

Fig. 4-37 Linear variable differential transformer (LVDT).

mary with one secondary will be greater than with the other, and a differential voltage will appear across the secondaries in series. When a dc output is required, the ac output is rectified and filtered.

The excitation frequency for an LVDT varies. Typical values range from 50 Hz to 30 kHz. If the transducer must accurately track rapidly changing displacement, the higher fre-

quencies are advantageous. The voltage applied to the primary is usually around 10 V. Displacements of as little as 50 μin. are detected by LVDTs. In addition to displacement, LVDTs are used to measure weight, pressure, and force (Fig. 4-38 on page 96).

In Fig. 4-38(a) an LVDT sensor is used to control the water level in a tank. When the water level is low, the core moves into the

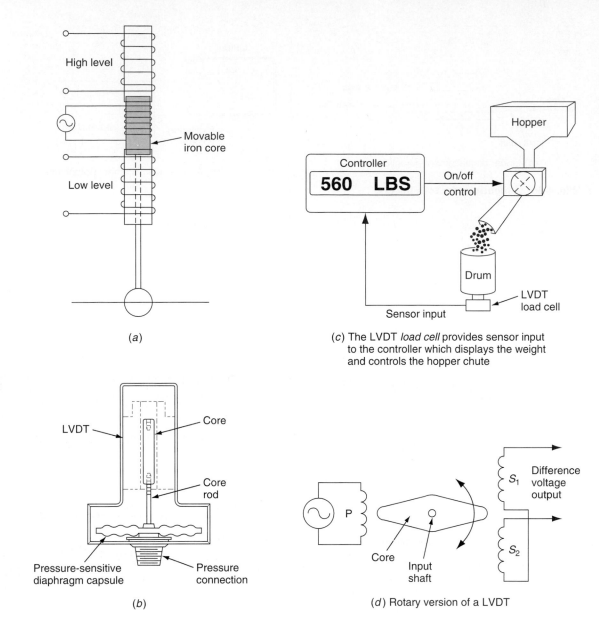

(a)

(c) The LVDT *load cell* provides sensor input to the controller which displays the weight and controls the hopper chute

(b)

(d) Rotary version of a LVDT

Fig. 4-38 LVDT applications.

lower coil, and its output will be greater than the upper coil's. By applying this signal to the proper control circuit, a valve may be opened or closed, or a pump turned off. Figure 4-38(*b*) shows an LVDT-type pressure sensor used for very low pressures, typically in millimeters of water column. The deflection of the capsule when pressurized is measured by the LVDT.

TEMPERATURE SENSORS

There are four basic types of temperature sensors commonly used today: *thermocouple, resistance temperature detector (RTD), thermistor,* and *IC sensor.* Figure 4-39 compares the important features of these devices.

A *thermocouple (T/C)* consists essentially of a pair of dissimilar conductors welded or fused together at one end to form the "hot" or measuring junction, with the free ends available for connection to the "cold" or reference junction. A temperature difference between the measuring and reference junctions must exist for this device to function as a thermocouple. When this occurs, a small dc voltage is generated. Thermocouples, because of their ruggedness and wide temperature range, are used in industry to monitor and control oven and furnace temperatures (Fig. 4-40 on page 98).

A reference junction is required for measuring temperature with thermocouples. The output voltage of a thermocouple is approxi-

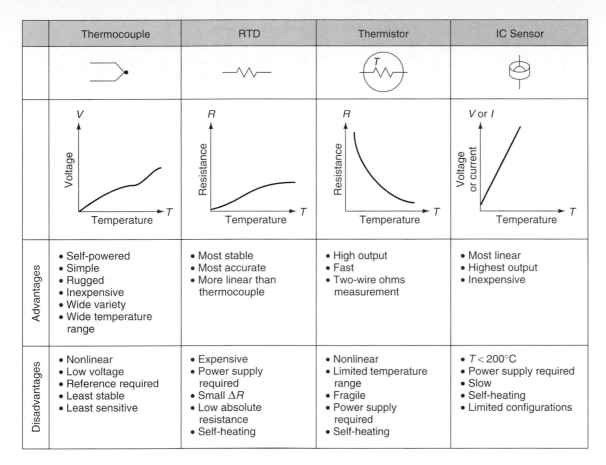

	Thermocouple	RTD	Thermistor	IC Sensor
	V vs Temperature (Voltage increasing curve)	*R* vs Temperature (Resistance increasing curve)	*R* vs Temperature (Resistance decreasing curve)	*V* or *I* vs Temperature (Voltage or current linear increasing)
Advantages	• Self-powered • Simple • Rugged • Inexpensive • Wide variety • Wide temperature range	• Most stable • Most accurate • More linear than thermocouple	• High output • Fast • Two-wire ohms measurement	• Most linear • Highest output • Inexpensive
Disadvantages	• Nonlinear • Low voltage • Reference required • Least stable • Least sensitive	• Expensive • Power supply required • Small ΔR • Low absolute resistance • Self-heating	• Nonlinear • Limited temperature range • Fragile • Power supply required • Self-heating	• $T < 200°C$ • Power supply required • Slow • Self-heating • Limited configurations

Fig. 4-39   Common temperature sensors.

mately proportional to the temperature difference in voltage between the measuring (hot) junction and the reference (cold) junction. This constant of proportionality is known as the *Seebeck coefficient* and ranges from 5 to 50 V per °C for commonly used thermocouples. The best way to know the temperature at the reference junction is to keep this junction in an ice bath; this will result in zero output voltage of 0°C (32°F). A more convenient approach used in electronic instruments is known as *cold junction compensation*. This technique adds a compensating voltage to the thermocouple's output so that the reference junction appears to be at 0°C, independent of the actual temperature. If this compensating voltage is proportional to temperature with the same constant of proportionality that occurs with the thermocouple, changes in ambient temperature will have no effect on output voltage.

The basic concept underlying the measurement of temperature by *resistance temperature detectors* (*RTDs*) is that the electrical resistance of metals varies proportionally with temperature. This proportional variation is precise and repeatable, therefore allowing the consis-

tent measurement of temperature through electrical resistance detection. Platinum is the material most often used in RTDs because of its superiority regarding temperature limit, linearity, stability, and reproducibility (Fig. 4-41 on page 99). Some *airflow* sensors use heated RTDs that sense either a reduction or increase of air flow through the cooling effects of the air passing across the sensing element.

A *thermistor* is a thermally sensitive resistor that usually has a negative temperature coefficient. As the temperature increases, the thermistor's resistance decreases, and vice versa. Thermistors are very sensitive (as much as 5 percent resistance change per °C); therefore, they are capable of detecting minute changes in temperature. Their sensing area is small, and their low mass allows a fairly fast response time of measurement (Fig. 4-42 on page 100). They are available in a variety of sizes and styles [Fig. 4-42(*a*)]. Because of the large voltage output produced by a typical thermistor bridge [see Fig. 4-42(*b*)], amplification is normally unnecessary. As shown in Fig. 4-42(*b*), the thermistor sensing unit is inserted into the system to be monitored. When the coolant is

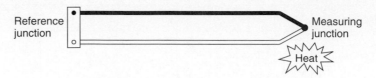

(a) Measuring and reference junctions

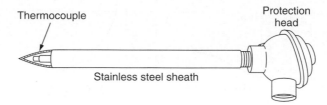

(b) Industrial thermocouple physically mounted in protective sheath

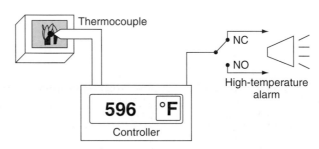

(c) Controller uses signal from thermocouple to monitor oven temperature and provide relay control to an external alarm

Fig. 4-40   Thermocouple.

flowing, the heat generated in the probe is conducted away. When the coolant is at rest, the sensor heats up by a few degrees and produces a change in resistance.

Integrated circuit (IC) temperature sensors use a silicon chip for the sensing element. They are available in both voltage- and current-output configurations. Although limited in temperature range (below 200°C), they produce a very linear output over their operating range (Fig. 4-43 on page 101).

VELOCITY/RPM SENSORS

||||➡ **YOU MAY RECALL**   that the output voltage of a generator varies with the speed at which the generator is driven.

A *tachometer* normally refers to a small permanent magnet dc generator. When the generator is rotated, it produces a dc voltage directly proportional to speed. Tachometers coupled to motors are commonly used in motor speed control applications to provide a feedback voltage to the controller that is proportional to motor speed (Fig. 4-44 on page 101).

The rotating speed of a shaft is often measured using a *magnetic (inductive) pickup sensor*. A magnet is attached to the shaft. A small coil of wire held near the magnet receives a pulse each time the magnet passes. By measuring the frequency of the pulses, the shaft speed can be determined. The voltage output of a typical pickup coil is quite small and requires amplification to be measured (Fig. 4-45 on page 101).

ENCODER SENSORS

An encoder sensor is used to convert linear or rotary motion into a digital signal (Fig. 4-46 on page 102). A *rotary encoder* monitors the rotary motion of a device. There are two types:

- The *incremental encoder*, which transmits a specific quantity of pulses for each revolution of a device.

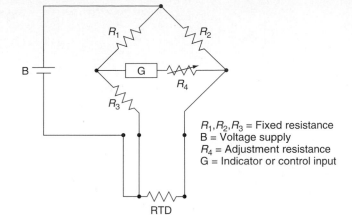

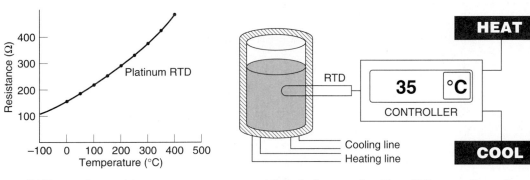

(*a*) Glass-encapsulated-type RTD wound with platinum wire

R_1, R_2, R_3 = Fixed resistance
B = Voltage supply
R_4 = Adjustment resistance
G = Indicator or control input

RTD

(*c*) 3-wire RTD bridge configuration; typical RTD will increase its resistance by 0.385Ω/°C (this small change in resistance demands the accuracy and sensitivity of a bridge circuit)

Platinum RTD

(*b*) Temperature-resistance curve

(*d*) Controller uses signal from RTD sensor to monitor temperature of liquid in the tank and control heating and cooling lines

Fig. 4-41 Resistance temperature detector (RTD).

- The *absolute encoder*, which provides a specific binary code for each angular position of the device.

The optical-type incremental rotary encoder, shown in Fig. 4-46(*a*), creates a series of square waves as its shaft is rotated. The encoder disk interrupts the light as the encoder shaft is rotated to produce the square wave output waveform. It is very sensitive and accurate (normally from 100 to 4000 pulses per revolution). The optical-type absolute rotary encoder, shown in Fig. 4-46(*b*), operates in the same manner with the exception that more traces have been applied to the encoder disk. When scanned in a parallel fashion, these traces provide angle information in the form of a code. The number of traces corresponds to the number of steps per rotation. The traces are arranged in Gray code. The Gray coding has the advantage over other coding in that only one bit (trace) is changed per step. This helps avoid misreadings.

Referring to Fig. 4-46(*c*), the number of square waves obtained from the output of the encoder can be made to correspond to the mechanical movement required. For example, to divide a shaft revolution into 100 parts, an encoder could be selected to supply 100 square wave cycles per revolution. By using a counter to count those cycles, it is possible to tell how far the shaft has rotated.

FLOW MEASUREMENT

Many industrial processes depend on accurate measurement of fluid flow. Although there are a variety of ways to measure fluid flow, the usual approach is to convert the kinetic energy that the fluid has into some other measurable form. This can be as simple as connecting a paddle to a potentiometer or as complex as connecting rotating vanes to a pulse-sensing system or tachometer.

Figure 4-47 on page 103 shows different types of flowmeters. The principle of the dif-

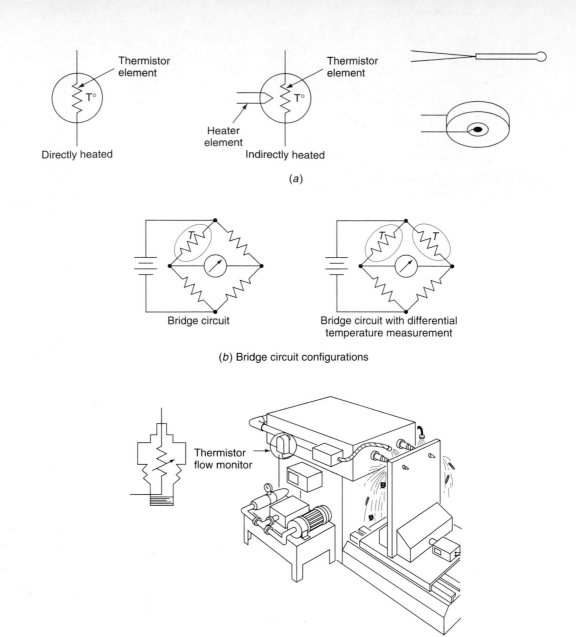

(a)

(b) Bridge circuit configurations

(c) Thermistor-type flowmeter

Fig. 4-42   Thermistors.

ferential pressure flowmeter in Fig. 4-47(a) is called the *Bernoulli effect*. When fluid is flowing, pressure P_1 will be greater than P_2, and the pressure difference is proportional to the flow. The bellows extend in proportion to pressure. When P_1 is greater than P_2, the core in the LVDT will move to the right. If the flow stops, there will be no Bernoulli effect and the core will be centered. With the turbine flowmeter shown in Fig. 4-47(b), the turbine blades turn at a rate proportional to the fluid velocity and are magnetized to induce voltage pulses in the coil. In the *target flowmeter* shown in Fig. 4-47(c), the fluid exerts a pres-

sure on the target that is proportional to fluid velocity. The resultant force on the target is sensed by a strain gauge. Figure 4-47(d) shows an electronic magnetic flowmeter, which can be used with electrically conducting fluids and offers *no* restriction to flow. A coil in the unit sets up a magnetic field. If a conductive liquid flows through this magnetic field a voltage is induced, which is sensed by two electrodes.

SIGNAL CONDITIONING

The input electrical signal produced by a sensor often is not in a form that can be used

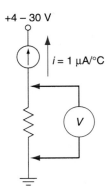

(a) Typical case style

+4 – 30 V

$i = 1\ \mu A/°C$

V

(b) Produces output current that varies with
temperature (typical sensitivity value:
1 μA / °C)

+4 – 30 V

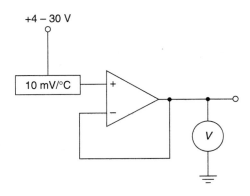

10 mV/°C

V

(c) Produces output voltage that varies with
temperature (typical sensitivity value:
10 mV/°C)

Fig. 4-43   IC temperature sensor.

directly. A *signal conditioner* changes a signal in a desired manner to make it easier to measure or to make it more stable.

Amplification and *attenuation* are common signal conditioning techniques. Amplification is needed when the output of the sensor is too small to be directly useful in a measurement or control system; attenuation is used to reduce a voltage before measurement [Fig. 4-48(a) and (b) on page 104]. The current to voltage converter shown in Fig. 4-48(c) uses the voltage drop produced across the resistor to convert the standard 4- to 20-mA current loop to a 1- to 5-V source. *Filtering* a signal to change its

frequency content is another common technique [see Fig. 4-48(d)] The major types of filters are the high pass, which passes only higher frequencies; the bandpass, which passes frequencies in a certain band; and the low pass, which passes only lower frequencies.

In certain instances, the signal may need to be changed from analog to digital, or vice versa, by means of an *A/D converter* (analog to digital) or *D/A converter* (digital to analog). For example, Fig. 4-48(e) shows an A/D converter used to convert an analog signal generated by a sensor into a digital signal that can be used by a microprocessor. The *smart signal conditioner* shown in Fig. 4-48(f) is microprocessor-based and capable of converting the majority of electrical sensor signals.

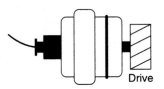

Drive

(a) Tachometer is normally a small
permanent magnet dc generator

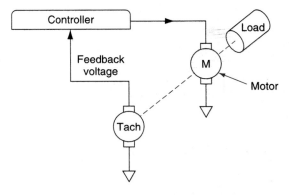

Controller

Feedback
voltage

Load

M

Motor

Tach

(b) Tachometer provides feedback voltage to the
controller that is proportional to motor speed

Fig. 4-44   Tachometer.

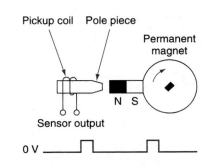

Pickup coil   Pole piece

Permanent
magnet

N   S

Sensor output

0 V

Fig. 4-45   Magnetic pickup sensor.

Signal conditioner

Amplification

Attenuation

Filtering

A/D converter

D/A converter

Smart signal
conditioner

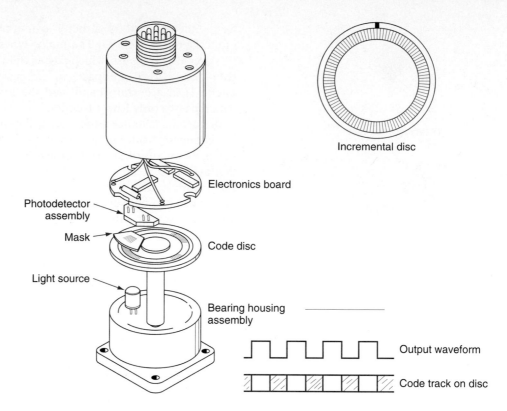

Incremental disc

Electronics board

Photodetector assembly

Mask

Code disc

Light source

Bearing housing assembly

Output waveform

Code track on disc

Courtesy of BEI Motion Systems Company

(*a*)

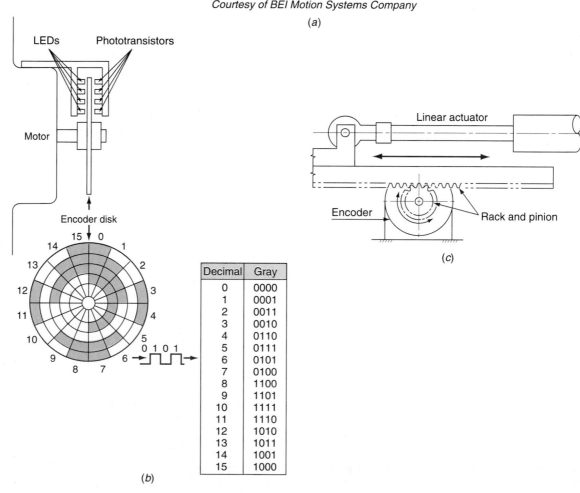

LEDs   Phototransistors

Motor

Encoder disk

15 0
14   1
13   2
12   3
11   4
10   5
9   0 1 0 1
8 7 6

Decimal	Gray
0	0000
1	0001
2	0011
3	0010
4	0110
5	0111
6	0101
7	0100
8	1100
9	1101
10	1111
11	1110
12	1010
13	1011
14	1001
15	1000

(*b*)

Linear actuator

Encoder   Rack and pinion

(*c*)

Fig. 4-46   Encoder sensors.

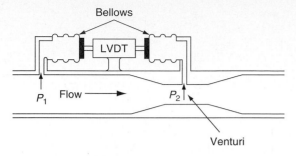

(a) Differential pressure flowmeter

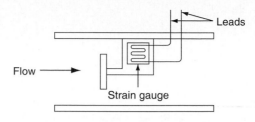

(c) Target flowmeter

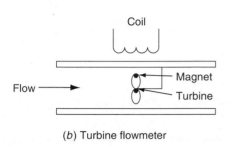

(b) Turbine flowmeter

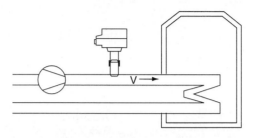

(d) Electronic magnetic flowmeter

Fig. 4-47 Flow measurement.

Self-Test

Answer the following questions.

13. A transducer is any device that converts _____ from one form to another.
14. Sensors are devices that are used to _____ and often to _____ the presence of something.
15. True or false? A proximity sensor can detect the presence of an object without physical contact.
16. True or false? An inductive proximity sensor can be actuated by both conductive and nonconductive materials.
17. True or false? With a current-sourcing output, the load is connected between the sensor and ground.
18. True or false? Unlike mechanical switches, solid-state switches produce a leakage current when the output is turned OFF.
19. True or false? A permanent magnet is the most common actuator for a reed relay.
20. True or false? Photovoltaic (solar) cells produce ac voltages.
21. True or false? Light shining on a photoconductive (photoresistive) cell will cause its resistance to drop.
22. The two main types of photoelectric sensors are the _____ type and the _____ type.
23. True or false? Color detection and discrimination are not possible with photoelectric sensors.

24. The three basic elements of a bar code system are: _____, _____, and _____.
25. The Hall-effect sensor is designed to sense the presence of a(n) _____ object.
26. An ultrasonic sensor operates by measuring the _____ taken for the echo to return to the sensor.
27. Strain wire gauges are operated on the principle that the _____ of a conductor varies with length and cross-sectional area.
28. The semiconductor strain gauge uses a piezoelectric _____ as its sensing element.
29. True or false? A potentiometer can be used to measure both linear and angular displacement.
30. The LVDT (linear variable differential transformer) is basically a transformer with a _____ core and _____ secondary windings.
31. True or false? The basic operating principle of thermocouples is that the electrical resistance of metals varies with temperature.
32. True or false? Resistance temperature detectors (RTDs) require cold junction compensation.
33. True or false? As the temperature of a thermistor with a negative temperature coefficient increases, its resistance decreases.
34. A tachometer produces a dc voltage proportional to _____.

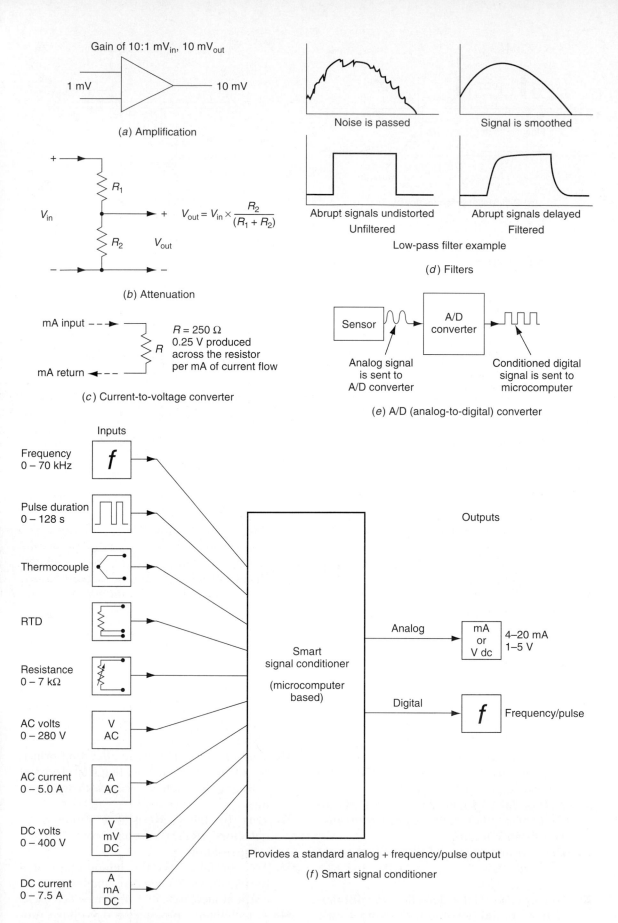

Gain of 10:1 mV$_{in}$, 10 mV$_{out}$

1 mV — 10 mV

(a) Amplification

V_{in}

R_1

R_2 V_{out}

$V_{out} = V_{in} \times \dfrac{R_2}{(R_1 + R_2)}$

(b) Attenuation

mA input

mA return

$R = 250\ \Omega$
0.25 V produced
across the resistor
per mA of current flow

(c) Current-to-voltage converter

Noise is passed

Signal is smoothed

Abrupt signals undistorted
Unfiltered

Abrupt signals delayed
Filtered

Low-pass filter example

(d) Filters

Sensor — A/D converter

Analog signal
is sent to
A/D converter

Conditioned digital
signal is sent to
microcomputer

(e) A/D (analog-to-digital) converter

Inputs

Frequency
0 – 70 kHz f

Pulse duration
0 – 128 s

Thermocouple

RTD

Resistance
0 – 7 kΩ

AC volts
0 – 280 V   V
AC

AC current
0 – 5.0 A   A
AC

DC volts
0 – 400 V   V
mV
DC

DC current
0 – 7.5 A   A
mA
DC

Smart
signal conditioner

(microcomputer
based)

Outputs

Analog   mA
or
V dc   4–20 mA
1–5 V

Digital f Frequency/pulse

Provides a standard analog + frequency/pulse output

(f) Smart signal conditioner

Fig. 4-48   Signal conditioning.

35. True or false? A magnetic pickup sensor generates pulses that are counted to determine shaft speed.

36. True or false? An absolute encoder provides a specific binary code for each angular position.

37. The usual approach taken to measure fluid flow is to convert the _____ energy that the fluid has into some other measurable form.

38. True or false? Amplification is a common signal conditioning technique.

4-5 ACTUATORS

An *actuator,* in the electrical sense, is any device that converts an electrical signal into mechanical movement. The principal types of actuators are relays, solenoids, and motors.

RELAYS

A *relay* is an electrically operated device that mechanically switches electric circuits. The relay is an important part of many control systems because it is useful for remote control and for controlling high-voltage and current devices with a low-voltage and current control signal. When current flows through the electromagnet in an electromechanical control relay (Fig. 4-49 on page 106), a magnetic field that attracts the iron arm of the armature to the core of the magnet is set up. As a result, the contacts on the armature and relay frame are switched. Relays may have NO contacts or NC contacts or combinations of both.

SOLENOIDS

A *solenoid* is a device used to convert an electrical signal or an electrical current into linear mechanical motion. As shown in Fig. 4-50 on page 107, the solenoid is made up of a coil with a movable iron core. When the coil is energized, the core, or armature as it is sometimes called, is pulled inside the coil. The amount of pulling or pushing force produced by the solenoid is determined by the number of turns of copper wire and the amount of current flowing through the coil.

The coil current for *dc solenoids* is limited by coil *resistance only.* With the plunger seated, pull is greater than needed, so on smaller solenoids a partial voltage coil is often used. On larger solenoids a two-section coil can be used where a cut-off switch operates

when the plunger is just about seated and opens up the circuit for part of the coil. The iron core is made of soft steel with low reluctance. For dc solenoids, a solid core is acceptable because the current is in one direction and continuous. The dc solenoid has a time constant; because coil inductance slows down magnetizing, the action is slower than that of the ac solenoid.

The frame and plunger of *ac solenoids* consists of thin silicon-steel that has been laminated and varnished to reduce eddy currents. The head is carefully ground so it will seal accurately. An air gap or spacer of nonmagnetic material must be in the magnetic path so that the plunger will not stick closed because of residual magnetism. The resistance of the ac coil is *very low,* so current flow is primarily limited by the inductive reactance (X_L) of the coil.

||||➤ YOU MAY RECALL That the inductive reactance (X_L) of a coil can be calculated using the formula:

$$X_L = 2\,\pi f L$$

where X_L = the inductive reactance in ohms
f = the frequency of the alternating current in hertz
L = the inductance in henries

The current is determined by Ohm's law, with X_L replacing R, as follows:

$$I = \frac{V}{X_L}$$

The plunger must seal fully against the seat. If it doesn't, the air gap in the magnetic path is increased and will reduce the inductance of the coil. As a result, the coil will overheat.

The length of the solenoid *stroke* is very important. Shorter strokes result in faster operating rates; they also require less power. More force is available at shorter strokes, which allows a smaller size solenoid to be used. Shorter strokes have less impact force, which helps to reduce solenoid wear. The maximum stroke should not exceed one-half of the plunger length. In ac solenoids, a relatively high in-rush current occurs, which decreases as the plunger moves toward the seated position. This high in-rush current generally provides higher forces at longer strokes than the dc versions (Fig. 4-51 on page 108). The reason for

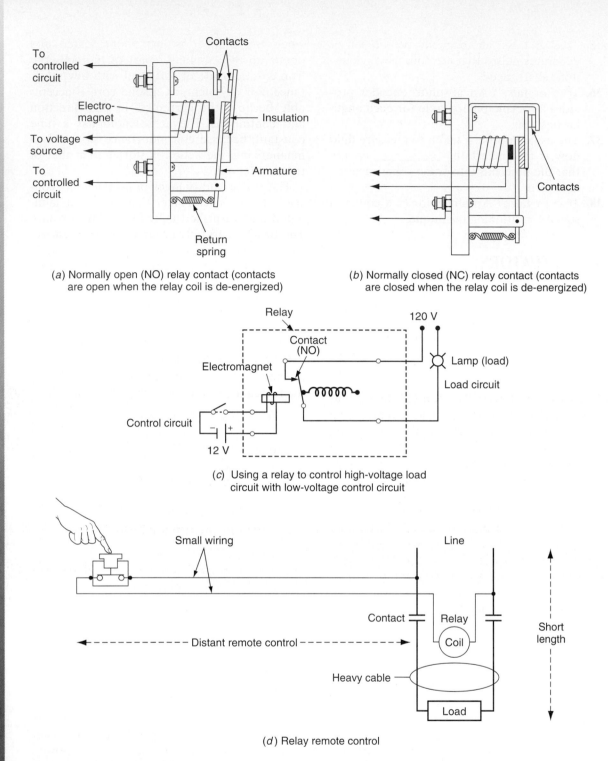

(a) Normally open (NO) relay contact (contacts are open when the relay coil is de-energized)

(b) Normally closed (NC) relay contact (contacts are closed when the relay coil is de-energized)

(c) Using a relay to control high-voltage load circuit with low-voltage control circuit

(d) Relay remote control

Fig. 4-49   Electromechanical control relay.

such a high in-rush current is the fact that the basic opposition to current flow when a solenoid is energized is only the resistance of the copper coil. Upon energizing, however, the armature begins to move iron into the core of the coil. This large amount of iron in the magnetic circuit greatly increases the inductance of the coil and thus decreases the current through the coil. The heating effect of the coil further reduces current flow because copper wire, when heated, increases in resistance, limiting some current flow.

Basic solenoid ratings include force, stroke, duty cycle, temperature, and power. Other considerations are size, mounting, electrical connections, life expectancy, and environment.

The solenoid selected for a particular application must be one that produces the force

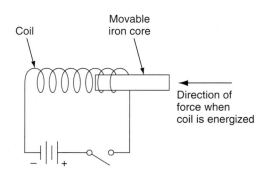

(a) Symbol

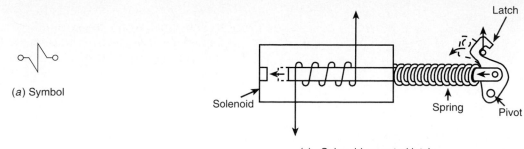

(c) Solenoid-operated latch

Coil

Movable iron core

Direction of force when coil is energized

(b) When energized, the solenoid produces a straight-line mechanical force

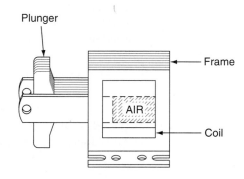

Plunger

Frame

AIR

Coil

(d) Heavy-duty solenoid

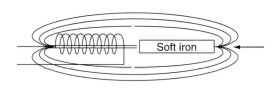

Soft iron

In pulling-type solenoid, magnetic field pulls a core into a coil

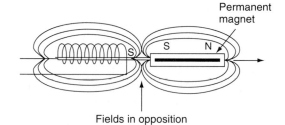

Permanent magnet

S

S   N

Fields in opposition

In push-pull solenoid, a permanent magnet is used for the core, and the core is *pulled in* or *pushed out* by reversing the direction of current flow through the coil

(e) Solenoid may *push* or *pull*, depending on design and application

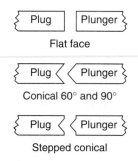

Plug   Plunger

Flat face

Plug   Plunger

Conical 60° and 90°

Plug   Plunger

Stepped conical

(f) Common dc coil plug and plunger configurations (flat-faced configuration is best for short strokes and high holding force; 60° conical is best for longer strokes; 90° concial and stepped conical are better for medium strokes)

Fig. 4-50   Solenoids.

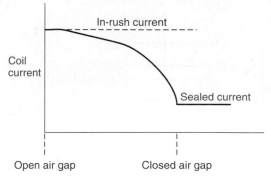

In-rush current to solenoid is six to ten times the sealed current

Fig. 4-51   AC coil in-rush and sealed currents.

the load so that the plunger is free to move in a straight line. When the solenoid is energized, the plunger must be free to center itself in the coil and seat firmly. If this is not done, the unit may buzz and is likely to overheat. If the possibility exists, in application, that the load on the unit could ever "stall" the plunger, some means of protection against overheating should be employed. This could be a fuse in the circuit, a built-in thermal cut-out, or the application of the load through a spring (Fig. 4-52).

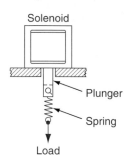

Spring provides overload protection by allowing the complete seating of the plunger when load is stalled

Fig. 4-52   AC solenoid burnout protection.

SOLENOID VALVES

A *solenoid valve* is a combination of two basic functional units:

- A solenoid (electromagnet) with its core or plunger.
- A valve body containing an orifice in which a disc or plug is positioned to restrict or allow flow.

Flow through an orifice is OFF or allowed by the movement of the core and depends on whether the solenoid is energized or de-energized. When the coil is energized, the core is drawn into the solenoid coil to *open* the valve. The spring returns the valve to its original *closed* position when the current ceases. Solenoid valves are available to control *hydraulics* (oil fluid), *pneumatics* (air), or *water* flow (Fig. 4-53).

Standard applications of solenoid valves generally require that the valve be mounted directly in line in the piping with the inlet and outlet connections directly opposite each other (Fig. 4-54). The valve body is usually a brass forging. It is recommended that *strainers* be used to prevent grit or dirt from lodging in the orifice and causing leakage. A valve must be installed with direction of flow in accordance with the arrow cast on the side of the valve

required throughout its entire stroke and operating temperature range. The load must never exceed the force developed at the stroke. If the load is too great, the plunger will not pull in or seat. On the other hand, a highly overrated solenoid that develops substantially more force than required by the load should not be used unless speed of operation is the determining factor. Excessive energy imparted to the solenoid must be dissipated by some other means. If it is not dissipated, the plunger and field piece assembly must absorb the energy of impact, causing premature failure.

Duty cycle, expressed in the form of a percent, is the ratio of on time to total time of a cycle of operation. Continuous duty is a 100 percent duty cycle. If a solenoid is energized for 100 s and de-energized for 300 s, the duty cycle is as follows:

$$\text{Duty cycle in } \% = \frac{\text{on time}}{\text{on time} - \text{off time}} \times 100$$

$$= \frac{100}{100 - 300} \times 100 = 25\%$$

For best performance, an ac solenoid with a coil designed for the exact supply frequency should be used. A 50-Hz coil will not develop the rated force on 60 Hz. A 60-Hz coil will tend to overheat when operated at 50 Hz. For some applications, a dual frequency 50-60-Hz coil can be used. An ac solenoid will never be completely silent, but any hum can be held to a minimum by using a shading coil and ensuring that the plunger seats firmly. Any wear or dirt on the plunger will cause excessive noise as well as overheating.

Alignment of load and solenoid on ac units is even more important than it is on dc units. The solenoid should be mounted and linked to

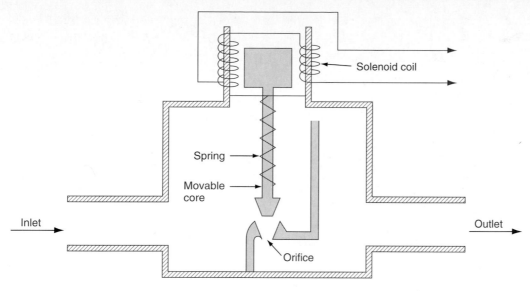

Fig. 4-53 Solenoid valve.

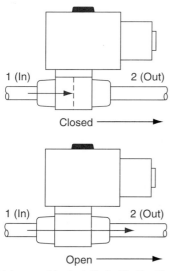

Valve must be installed with direction
of flow in accordance with markings

Fig. 4-54 Solenoid valve installation.

body, or the markings "IN" and "OUT" on the pipe connections. A solenoid valve is suitable for handling flow *in one direction only,* with pressure applied to the top of the valve disc.

Directional valves start, stop, and control the direction of the flow path. These valves direct the flow by opening and closing flow paths in definite valve positions (Fig. 4-55 on pages 110 and 111). They are classified by their number of connections and valve positions. For instance, a valve with four connections and three possible switching positions is referred to as a *4/3-way valve.*

The graphic symbol for a directional valve contains a separate square for each finite posi-

tion and shows the flow in that position. Valve circuits are normally shown in their de-actuated state. The lines are shown connected to the *de-actuated position.* When the valve is actuated, you should mentally shift the lines to the other position. Two-way valves are considered to be either normally open (NO) or normally closed (NC). The NC and NO symbols are shown in Fig. 4-55(*b*). A normally closed solenoid valve blocks flow when the solenoid is de-energized and allows flow when the solenoid is energized. The reverse is true of a normally open solenoid valve. Figure 4-55(*c*) is a solenoid operated 2/2-way valve. When the solenoid coil is energized, it forms a magnetic field that pulls the rod or spool inside the valve and causes it to shift to the actuated side. As a result, the air is switched or redirected as per the valve's function. When the solenoid is de-energized, the spring pressure returns the valve spool to its normal, de-energized position.

Often directional solenoid valves are used to operate double-acting cylinders (Fig. 4-56 on page 112). In one valve position, pressure is applied to one side of the cylinder; the other side is connected to exhaust. In the other position, pressure and exhaust are reversed. The electrohydraulic circuit shown in Fig. 4-55(*a*) uses a two-position single-solenoid valve to operate a double acting cylinder. When the pushbutton is depressed, the valve spool in the solenoid valve is shifted so that the fluid is directed to the back side of the cylinder piston, causing it to move forward. When the pushbutton is released, the valve spool returns to its original de-energized position so that the

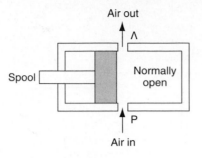

(a) Spool-type air valve

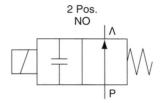

(b) Normally closed (NC) and normally open (NO)
two-way valve graphic symbols

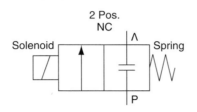

(c) Solenoid operated 2/2-way valve

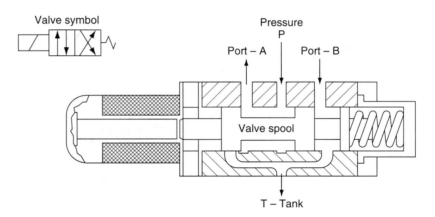

The valve in the de-energized state (when solenoid
is de-energized, pressure is free to flow to port A,
port B is open to tank)

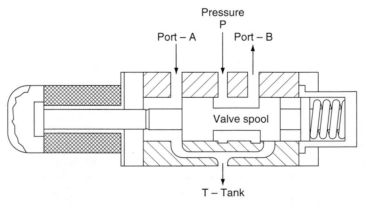

The valve in the energized state (when solenoid is
energized, spring is loaded, or depressed, pressure
is free to flow to port B, and port A is open to tank)

(d) 4/2-way single-solenoid, spring-return hydraulic valve

Fig. 4-55 **Directional control valves.**

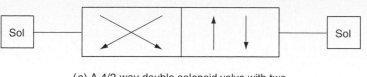

(e) A 4/2-way double solenoid valve with two
positions maintained (will maintain the
position in which it was last energized)

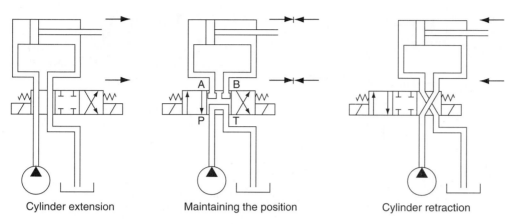

Cylinder extension　　　Maintaining the position　　　Cylinder retraction

(f) A 4/3-way valve, double-solenoid actuated and
with spring centering for positioning cylinder in
the three possible switched positions

Fig. 4-55　Directional control valves (continued).

fluid is directed to the front side of the cylinder piston, causing it to move in reverse.

Both on-off and analog valves are available. *On-off* valves will usually be based on a solenoid arrangement. Analog valves may use a motor or throttling control to gradually increase or decrease the amount of flow.

STEPPER MOTORS

A *stepper motor* converts electrical pulses applied to it into discrete rotor movements called *steps*. A one-degree-per-step motor will require 360 pulses to move through one revolution. Microstep motors, with thousands of steps per revolution, are also available. The rating of the stepper motor is generally given in steps per revolution of the motor. They are generally low speed and low torque, and they provide precise position control of movement.

Figure 4-57 on page 113 illustrates the basic operation of a dc stepper motor. It consists of an electromagnetically excited stator and a permanent magnet rotor. When the polarity of the exciting winding is suitably reversed, the rotor turns in the chosen direction by exactly one step to a new rest position. The number of steps per revolution is determined by the number of pole pairs on the rotor and stator. The greater the number of poles on both parts, the greater the steps per revolution of the motor.

The operation of a stepper motor greatly depends upon the power supply that drives it. The power supply generates the pulses, which, in turn, are usually initiated by a microcomputer. The computer initiates a series of pulses in order to move the controlled device to whatever position is desired. In this way, the stepper motor provides precise position control of movement. By keeping count of the pulses applied, the computer knows exactly what position the motor is in; therefore, it is unnecessary to use a feedback signal.

A typical stepper motor control system consists of a stepper motor and a drive package that contains control electronics and a power supply [Fig. 4-58(a) on page 113]. The driver is the interface between the computer and the stepper motor. It contains the logic to convert or "translate" digital information into motor shaft rotation. The motor will move one step for each pulse received by the driver. The computer provides the desired number of pulses at a specified or programmed rate which translates into distance and speed.

**Stepper motor
holding torque**

Detent torque

**Variable
reluctance motors**

Hybrid motors

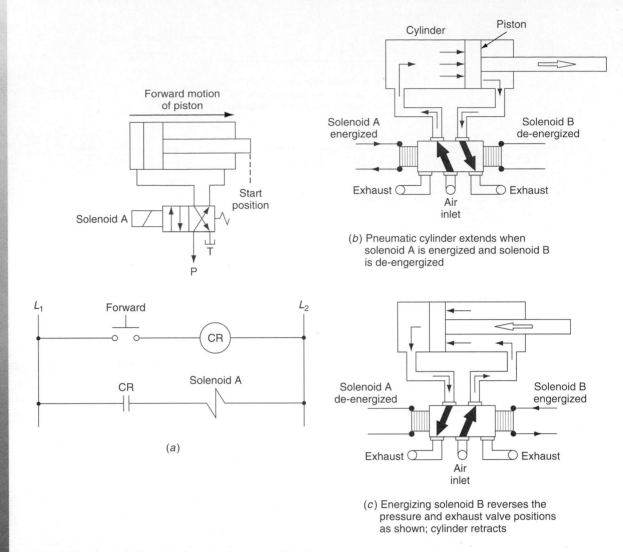

Fig. 4-56   Typical directional control valve applications.

(b) Pneumatic cylinder extends when
solenoid A is energized and solenoid B
is de-engergized

(c) Energizing solenoid B reverses the
pressure and exhaust valve positions
as shown; cylinder retracts

(a)

The number of steps per revolution is determined by the number of pole pairs on the rotor and stator. Once voltage is applied to the windings, the permanent magnet rotor of a stepper motor assumes its unloaded holding position. This means that the permanent magnet poles of the rotor are aligned according to the electromagnetic poles of the stator. The maximum torque with which this excited motor can now be loaded without causing a continuous rotation is termed the *stepper motor holding torque.* A torque can also be perceived with a nonexcited motor. This is because of the pole induction of the permanent magnet on the stator. This effect (cogging), together with the motor internal friction, produces *detent torque,* which is the torque with which a nonexcited motor can be statically loaded [see Fig. 4-58(*b*)].

There are three main types of stepper motors: permanent magnet motors, variable reluctance motors, and hybrid motors. In the permanent magnet type, shown in Fig. 4-59(*a*) on page 114, the motor construction results in relatively large step angles. This type is suited to applications in fields such as computer peripherals. *Variable reluctance motors* have *no* permanent magnet, so they require a different driving arrangement from the other types [see Fig. 4-59(*b*)]. The rotor spins freely without "detent" torque. Torque output for a given frame size is restricted. This type is used in small sizes for applications such as micropositioning tables. As the name implies, *hybrid motors* [see Fig. 4-59(*c*)] combine the operating principle of the other two types. The hybrid type is the most widely used stepper motor in industrial applications. The rotor consists of two pole pieces with three teeth on each. In between the pole pieces is a perma-

Control of the step sequence

Step	Switch position SA	SB	Counter-clockwise	Clock-wise
0	1	1		
1	2	1		
2	2	2		
3	1	2		
4	1	1		
etc.				

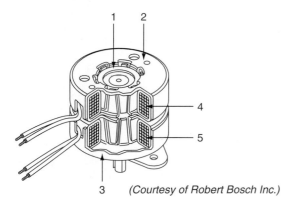

Claw pole stepper motor: 1 Rotor, 2 Stator segment A, 3 Stator segment B, 4 Winding, 5 Winding B

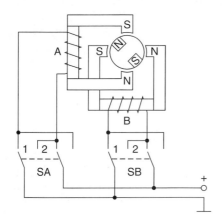

(Courtesy of Robert Bosch Inc.)

Fig. 4-57   Permanent magnet dc stepper motor.

nent magnet which is magnetized along the axis of the rotor, making one end a north pole and the other a south pole. The teeth are offset at the north and south ends as shown in the diagram. The stator consists of a shell having four teeth which run the full length of the rotor. Coils are wound on the stator teeth and are connected together in pairs.

The increasing trend toward digital control of machines and process functions has generated a demand for stepper motors. Industrial uses include a wide variety of control and positioning applications. A stepper motor can replace devices such as brakes, clutches, and

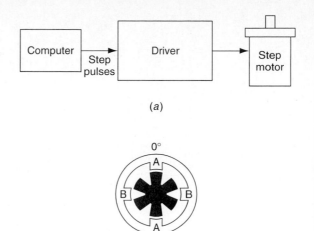

Fig. 4-58   Stepper motor control.

gears with an overall improvement in life expectancy and accuracy.

BRUSHLESS DC MOTOR

As the name implies, *dc brushless motors* have no brush or commutation mechanisms. Instead, solid-state electronic devices are used to switch the armature current.

▐▶ **YOU MAY RECALL**   that in a dc motor, the commutator and brushes change the direction of current in the armature coil so that the direction of the torque is always the same.

Figure 4-60 on page 114 illustrates the operation of a permanent magnet dc brushless motor. The permanent magnets are put on the rotating part, and the windings are put on the stator. Unlike the dc brush motor, the brushless version cannot be driven by connecting it up to a source of direct current. The current in the stator circuit must be reversed at defined rotor positions, so the motor is in fact being driven by alternating current. The stator field windings are energized in sequence to produce a revolving magnetic field. This current is supplied by the commutation encoder in response to signals from an optical or Hall-effect sensor.

Brushless dc motors are used in servo and robotic systems. They have high efficiency, long life, low electrical noise, and low power consumption. The brushless dc motor is *not* a stepper motor. It has smooth continuous shaft rotation like a conventional dc permanent magnet motor, not the fixed-steps detents found in the stepper motor.

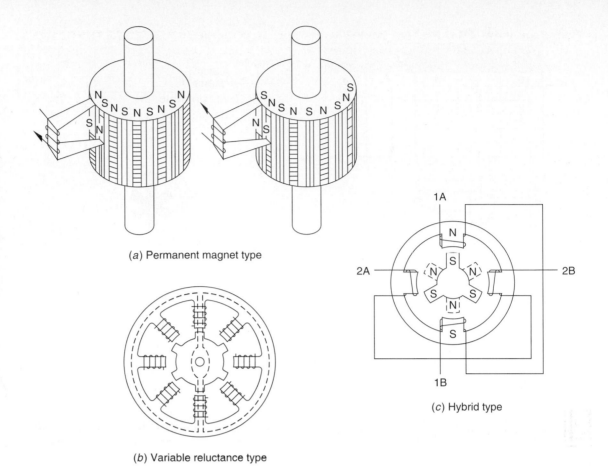

(a) Permanent magnet type

(b) Variable reluctance type

(c) Hybrid type

Fig. 4-59 Types of stepper motors.

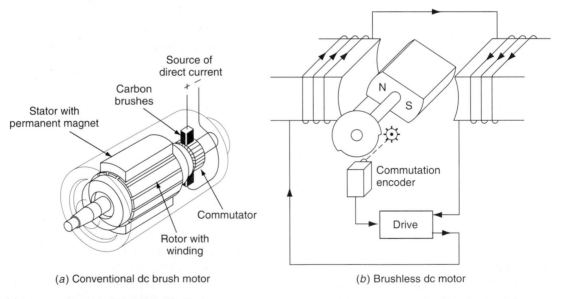

(a) Conventional dc brush motor

(b) Brushless dc motor

Fig. 4-60 DC permanent magnet motor.

Self-Test

Answer the following questions.

39. An actuator is any device that converts an electrical signal into _____.

40. The contacts of an electromechanical relay are switched by energizing and de-energizing a(n)_____.

41. The solenoid is made up of a coil with a(n) _____ iron core.

4-24. List two advantages of the semiconductor strain gauge over the strain wire gauge.
4-25. In what way are potentiometers used to measure linear and angular displacement?
4-26. Explain how the movement of the core of an LVDT senses displacement.
4-27. List the four basic types of temperature sensors and the principle of operation of each.
4-28. Compare the operation of an incremental and an absolute rotary encoder.
4-29. What is the usual approach taken to measure fluid flow?
4-30. List four common signal conditioning techniques.
4-31. Define the term *actuator* as it applies to an electric circuit.
4-32. State two useful applications for relays.
4-33. Describe the construction and function of a solenoid.
4-34. List four reasons why the length of stroke of a solenoid is kept as short as possible.
4-35. State two undesirable effects of an ac solenoid plunger not seating firmly.
4-36. What are the two basic functional units of a solenoid valve?
4-37. Why is it recommended that strainers be used in conjunction with solenoid valve-operated systems?
4-38. State three functions of a directional valve.
4-39. How are directional valves classified?
4-40. In what state are valve circuits normally shown on diagrams?
4-41. Explain how a directional valve is used to operate a double-acting cylinder.
4-42. What is the basic operating principle of a stepper motor?
4-43. How is the rating of a stepper motor generally expressed?
4-44. State the three main types of stepper motors.
4-45. In a brushless dc motor, what system is used to change the direction of current so that the direction of the torque is always the same?

CRITICAL THINKING QUESTIONS

4-1. A catalog lists the interrupting current rating of a switch contact as 10 A ac and 5 A dc. Explain why the dc current rating is much lower.
4-2. The coil resistance of two identical solenoids—one rated for 120 V dc and the other rated for 120 V ac—are measured using an ohmmeter. How should the resistance of the two compare? Why?
4-3. An inductive proximity sensor has the following specifications:
   ■ Hysteresis, 5 percent.
   ■ Voltage drop across conducting sensor, 0.7 V at 150 mA.
   ■ Continuous load current, 150 mA.
   ■ Leakage current, 10 μA.
   ■ No-load current, 10 mA.

   a. What does the 5 percent hysteresis mean?
   b. What is the largest size load that can be directly operated by the sensor?
   c. For what is the 10-mA, no-load current used?
   d. What does the 10 μA leakage current represent?
   e. How much voltage is lost across the sensor's output terminals when the sensor is ON?
   f. Assume that this sensor is required to operate a 5-A load. Suggest a way this could be accomplished without overloading the sensor.

42. True or false? Normally, an ac solenoid that is rated for 120-V alternating current can be safely used in a 120-V direct current circuit.
43. True or false? For most effective and efficient operation, the length of a solenoid stroke is kept as short as possible.
44. True or false? In ac solenoids, coil current decreases as the plunger moves toward the seated position.
45. The duty cycle for a solenoid that is energized for 10 s and de-energized for 40 s of a cycle of operation is _____.
46. True or false? An ac solenoid is energized and fails to seat firmly. This can cause the coil to burn out.
47. True or false? A normally open solenoid valve allows flow when the solenoid is de-energized and blocks flow when the solenoid is energized.
48. True or false? A solenoid valve must be installed in accordance to the direction of flow.
49. A directional valve with three connections and two possible switching positions is referred to as a(n) _____ valve.
50. Directional valves are used to _____, _____, and _____ the direction of the flow path.
51. _____ valves may use a motor or throttling control to gradually increase or decrease the amount of flow.
52. It would take _____ pulses to move a 200-step stepper motor 18°.
53. True or false? Stepper motor systems can provide precise position control of movement without the use of a feedback signal.
54. True or false? A brushless dc motor, like the brush type, can be driven by connecting it to a source of direct current.

SUMMARY

1. A primary control device connects the load to the line.
2. A pilot control device directs the operation of another device that activates a power circuit.
3. Manually operated switches are controlled by hand and include toggle switches, slide switches, DIP switches, rotary switches, thumbwheel switches, selector switches, pushbutton switches, and drum switches.
4. The dc current rating of a switch would have a lower magnitude than the ac rating of a switch.
5. The slide switch is often used as a mode switch to select a certain mode of operation, such as HIGH and LOW.
6. DIP switches are designed for mounting on printed circuit boards. Switch settings are seldom changed and occur mainly during installation.
7. The rotary switch is used for more complex switching operations.
8. Thumbwheel switches output special codes necessary to communicate with computers.
9. Selector switch positions are made by turning the operator knob right or left.
10. Pushbutton switches may consist of one or more NO or NC contact blocks, an operator device, and legend plate.
11. Drum switches are used for starting and reversing motors.
12. Mechanically operated switches are controlled automatically. Examples include limit switches, microswitches, temperature switches, pressure switches, and level switches.
13. Limit switches are actuated by contact with an object only when a predetermined limit is reached.
14. Microswitches operate with very small pressures, and therefore permit a great deal of sensitivity.
15. Temperature switches are all actuated by some specific environmental temperature change.
16. Pressure switches, used to control the pressure of liquids and gases, are designed to open or close when a specified pressure is reached.
17. Level switches are used to sense the height of a liquid by raising or lowering a float that is mechanically attached to the switch.
18. A transducer is any device that converts energy from one form into another.
19. Sensors are used to detect and often to measure the presence of something.
20. Sensors are usually categorized by what they measure.

21. Proximity sensors detect the presence of an object without physical contact.
22. An inductive proximity sensor is actuated by a metal object and uses an oscillator and detector circuit for the basis of its operation.
23. Both current sourcing (load connected between the sensor and ground) outputs and current sinking (load connected between the positive supply and sensor) outputs of proximity sensors are available.
24. With solid-state switching, a leakage current flows through the sensor when the output is turned OFF. A small voltage is dropped across the sensor output when it is turned ON.
25. A capacitive proximity sensor can be actuated by both conductive and nonconductive materials.
26. A magnetic switch, or reed relay, consists of flat contact tabs that are hermetically sealed and actuated by a permanent magnet.
27. The photovoltaic, or solar cell, light sensor converts light energy directly into electric energy.
28. The photoconductive or photoresistive cell increases in resistance when the surface changes from brightly lit to dark.
29. A reflective-type photoelectric sensor is used to detect the light beam reflected from the target.
30. A through-beam photoelectric sensor is used to measure the change in light quantity caused by the target's crossing the optical axis.
31. Photoelectric sensors often use an LED as the light-transmitting source and a phototransistor as the receiving source.
32. A bar code system consists of three basic elements: the bar code symbol, a scanner, and a decoder.
33. The Hall-effect sensor is a semiconductor designed to sense the presence of a magnetic object, usually a permanent magnet.
34. An ultrasonic sensor operates by sending sound waves toward the target and measuring the time it takes for the pulses to bounce back.
35. A strain wire gauge operates on the principle that the resistance of a conductor varies with length and cross-sectional area.
36. The semiconductor crystal strain gauge develops an output voltage when a force is applied that changes the shape of the crystal.

37. Potentiometers can be used to measure displacement because there is a direct relationship between the wiper shaft position and the output signal.
38. A linear variable differential transformer (LVDT) is designed to sense position through the movement of a transformer core and the resultant voltages induced in two secondary windings.
39. A thermocouple consists of a pair of dissimilar conductors welded at one end to form the measuring junction; the free ends are available for connection to the reference junction. A temperature difference between the measuring and reference junctions causes a small dc voltage to be generated.
40. Resistance temperature detectors (RTDs) operate on the principle that the electrical resistance of metals varies proportionally with temperature.
41. A thermistor is a thermally sensitive resistor whose resistance decreases with an increase in temperature (negative temperature coefficient type).
42. Integrated circuit (IC) temperature sensors use a silicon chip for the sensing element in a variable voltage or variable current configuration.
43. A tachometer is a small permanent magnet dc generator that produces a dc voltage directly proportional to the speed at which it is driven.
44. A magnetic (inductive) pickup sensor contains a small coil of wire that generates a pulse each time a magnet passes it.
45. Encoder sensors are used to convert linear or rotary motion into a digital signal.
46. The usual approach taken to fluid flow measurement is to convert the kinetic energy that the fluid has into some other measurable form.
47. A signal conditioner changes a signal in a desired manner to make it easier to measure or to make it more stable.
48. An actuator is any device that converts an electrical signal into mechanical movement.
49. An electromagnetic relay consists of a coil and contacts that mechanically switch electric circuits.
50. A solenoid is made up of a coil with a movable core and is used to convert an electrical signal to linear mechanical motion.
51. Basic solenoid ratings include force, stroke, duty cycle, temperature, and power.

52. A solenoid valve is a combination of a solenoid and a valve body.
53. Directional valves start, stop, and control the direction of the flow path.
54. Stepper motors convert electrical pulses applied to them into discrete rotor movements called steps.
55. The three main types of stepper motors are permanent magnet motors, variable reluctance motors, and hybrid motors.
56. DC brushless motors use a commutation encoder in place of brushes and commutator to switch current.

CHAPTER REVIEW QUESTIONS

Answer the following questions.

4-1. Define the term *control device*.
4-2. List three examples of primary motor control devices.
4-3. List three examples of pilot motor control devices.
4-4. Identify the type of switch referred to based on the description given:
   a. Used on computer-controlled equipment to input binary coded information from the operator to the computer.
   b. Used for complex switching operations such as those found on test instruments.
   c. A small switch assembly designed for mounting on printed circuit boards.
   d. Makes a circuit when it is pressed and returns to its open position when released.
   e. Used for starting and reversing the direction of rotation of a motor.
4-5. Define the term *mechanically operated switch*.
4-6. Describe the operation of a limit switch.
4-7. State a practical industrial application for each of the following:
   a. Temperature switch.
   b. Pressure switch.
   c. Level switch.
4-8. Define the term *transducer*.
4-9. What are sensors used for?
4-10. List four instances in which the use of a proximity sensor would be preferable to that of a mechanically operated switch.
4-11. In principle, what does an inductive proximity sensor consist of?
4-12. Compare the load connection in current sourcing and current sinking sensor outputs.
4-13. Define the term *hysteresis* as it applies to a proximity sensor.
4-14. In what instance would a capacitive proximity sensor be selected over the inductive type?
4-15. State an advantage and limitation of the magnetic reed relay switch.
4-16. Explain how light is sensed in:
   a. A photovoltaic cell.
   b. A photoconductive cell.
4-17. What are the two main types of photoelectric sensors?
4-18. List four important features of photoelectric sensors.
4-19. Explain the function of each of the basic components of a bar code system.
4-20. Explain the principle of operation of a Hall-effect sensing element.
4-21. Compare the operation of the digital- and analog-type Hall-effect sensor.
4-22. Explain the principle of operation of an ultrasonic sensor.
4-23. Explain how a strain wire gauge is able to sense force.

Answers to Self-Tests

1. false
2. DIP
3. rotary
4. thumbwheel
5. false
6. false
7. reversing
8. false
9. true
10. true
11. false
12. level
13. energy
14. detect; measure
15. true
16. false
17. true
18. true
19. true
20. false
21. true
22. reflective; through-beam
23. false
24. symbol; scanner; decoder
25. magnetic
26. time
27. resistance
28. crystal
29. true
30. movable; two
31. false
32. false
33. true
34. speed
35. true
36. true
37. kinetic
38. true
30. mechanical movement
40. electromagnet
41. movable
42. false
43. true
44. true
45. 20 percent
46. true
47. true
48. true
49. 3/2-way
50. start; stop; control
51. analog
52. 10
53. true
54. false

CHAPTER 5

Power Electronics

■

CHAPTER OBJECTIVES

This chapter will help you to:

1. *Explain* multimeter specifications and special features.
2. *Operate* an oscilloscope to observe and measure signals.
3. *Describe* the Wheatstone bridge method of measurement.
4. *Discuss* various practical applications for diodes.
5. *Explain* the operation and application of the different types of transistors.
6. *Describe* the operation of an SCR and a triac as current control devices.
7. *Compare* the operation of digital and analog ICs.

Electronic systems and controls have gained wide acceptance in industry; consequently, it has become essential to be familiar with power electronics. In this chapter, which is a broad review of general semiconductor types, the fundamentals of diodes, transistors, thyristors, and integrated circuits (ICs) will be covered.

■

5-1 ELECTRONIC TEST INSTRUMENTS

MULTIMETERS

The *multimeter* is a combination voltmeter, ammeter, and ohmmeter in a single instrument. Technicians have come to prefer digital multimeters over analog multimeters because the digital type is easier to read and more accurate.

A typical *digital multimeter* (*DMM*) is shown in Fig. 5-1. The digital readout is the main meter output, indicating the numerical value of the measurement. A key component for the operation of the multimeter is its front-panel switching arrangement. A *function switch* selects the type of measurement: voltage, current, or resistance. A *range switch* selects the full-scale range of the measurement. *Jacks* accept the plug-in test leads. This permits proper selection of the internal circuits to ensure that only one range, of one type of measurement, is selected at any one time. Because digital multimeters use electronic circuits to produce their measurements, they need internal batteries to supply power for *all* measurements. The ON/OFF power switch connects the power supply to the electronic circuits.

⫸ YOU MAY RECALL THAT

- The *voltmeter* has high resistance value and is connected in parallel or across the load or circuit.
- The *ammeter* has low resistance value and is connected in series with the load or circuit.
- The *ohmmeter* is self-powered and can be damaged if connected to a live circuit.

When selecting a multimeter it is important to be sure that the meter's capabilities will cover the types of test procedures that you usually do. Following are some important specifications and features to consider.

Input Impedance Input impedance refers to the combined resistance to current created by the resistance, capacitance, and inductance of the voltage-measuring circuits. A meter with low impedance can draw enough current to cause an inaccurate measurement of voltage drop. A high-impedance meter will draw little current, ensuring accurate readings. Most digital voltmeters have a 10-MΩ input impedance, which means there is little or no loading effect on a circuit while measurements are being taken.

Accuracy and Resolution Simple accuracy specifications are given as a plus/minus percentage of full scale. Resolution refers to the smallest numerical value that can be read on the display of a digital meter. Factors that determine resolution are the number of digits displayed and the number of ranges available for each function. Digital meter error is less than ±1 percent of the actual count, plus one or more counts of the least-significant digit. Analog meter error is in the range of ±1 to 3 percent of the actual count.

Battery Life Whereas the analog multimeter draws power from the circuit being tested for voltage and current measurements, the digital multimeter requires a battery to operate. *Battery life* (rated in hours) is a major consideration in selecting a digital multimeter. Either a light-emitting diode (LED) or liquid-crystal display (LCD) can be used in a digital multimeter. The LCD-type digital meter requires less power and, therefore, has a much longer battery-life rating.

Protection Meter-protection circuits prevent damage to the meter in cases of accidental overloading. The resistance and current measuring circuitry is usually protected by a fuse connected in series with the input lead. Safety precautions, involving the possible misuse of meters, have lead to the use of a second *high-energy fuse* within the meter circuit. This fuse is a high-voltage-rated element (around 600 V)

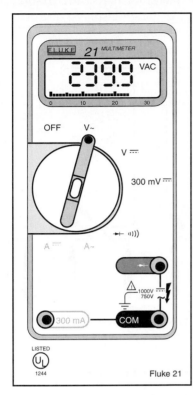

(*a*) Typical front-panel switching arrangement

From page 120:

Multimeter

Digital multimeter

Function switch

Range switch

Jack

Voltmeter

Ammeter

Ohmmeter

Input impedance

On this page:

Battery life

High-energy fuse

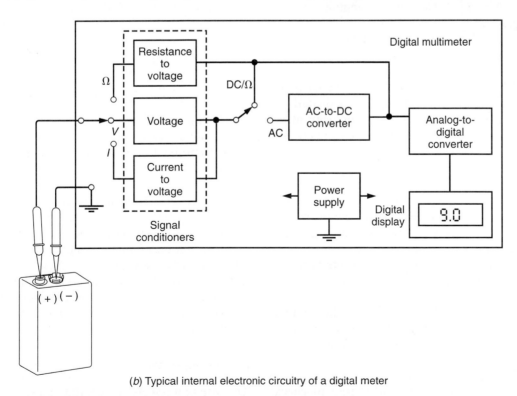

(*b*) Typical internal electronic circuitry of a digital meter

Fig. 5-1   Digital multimeter (DMM).

and is usually two or three times higher in fusing current than the standard user-replaceable fuse.

Combination Digital and Analog Display
Analog meters are particularly suited for trend observation, as in slowly changing voltage levels. Some digital multimeters use a combination display that includes a bar graph. The graph provides a simulation of an analog needle (Fig. 5-2) for watching changing signals or for adjusting circuits.

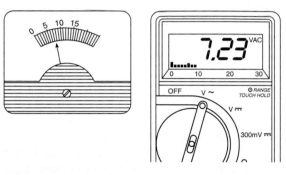

(Reproduced with permission from the John Fluke Mfg. Co., Inc. © John Fluke Mfg. Co., Inc.)

Fig. 5-2   Bar graph display used to simulate an analog needle.

Auto Ranging   The auto ranging feature automatically adjusts the meter's measuring circuits to the correct voltage, current, or resistance ranges.

Auto Polarity   With the automatic polarity feature, a plus (+) or minus (–) activated on the digital display indicates the polarity of dc measurements and eliminates the need for reversing leads.

Hold Feature   Many digital multimeters have a HOLD button that captures a reading and displays it from memory even after the probe has been removed from the circuit. This is particularly useful when making measurements in a confined area where you cannot quite read the meter.

Response Time   Response time is the number of seconds a digital multimeter requires for its electronic circuits to settle to their rated accuracy.

Diode Test   The diode test is used to check the forward and reverse bias of a semiconduc-

tor junction. Typically, when the diode is connected in *forward bias*, the meter displays the *forward voltage drop* and beeps briefly (Fig. 5-3). When connected in *reverse bias* or open circuit, the meter displays *OL*. If the diode is shorted, the meter displays zero and emits a continuous tone.

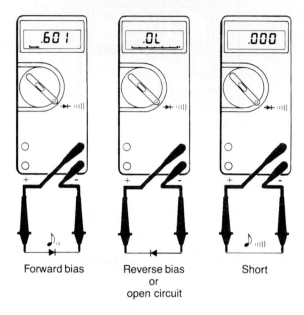

Forward bias   Reverse bias
                              or
                       open circuit

Short

Fig. 5-3   Diode test function.

Averaging or True RMS   A *true rms* (root mean square) *meter* responds to the effective heating value of an ac waveform. Meters that have rectifier-type circuits have scales that are calibrated in rms values for ac measurements, but, these meters actually are measuring the *average* value of the input voltage and are depending on the voltage to be a sine wave. When the ac signal approximates a pure sine wave, there is little or no difference in the two readings (Fig. 5-4).

Multifunction Digital Meters   The *integrated-circuit* (IC) chip revolution has helped to combine the capabilities of other test instruments into a multifunction digital meter. Volts, ohms, and amps are the most often used functions; however, the multifunction digital meter allows the reading of dBm (decibels above 1 mW) frequency, capacitance, logic level, and temperature measurements, as well.

OSCILLOSCOPES

The *oscilloscope* is a versatile piece of test equipment primarily used to *measure and dis-*

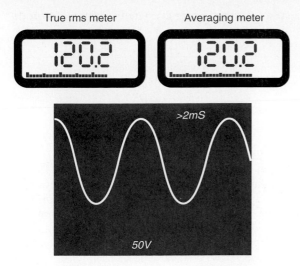

(a) Sine wave: True rms meter and averaging meter give same reading

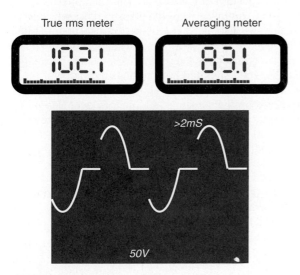

(b) Non-sine wave: true rms meter reads higher than the averaging meter

Fig. 5-4   Averaging or true rms.

play voltages. It can be used to measure ac and dc voltages like a voltmeter. In addition, it can provide information about the shape, time period, and frequency of voltage waveforms. The main advantage of the oscilloscope is its ability to stop a rapidly oscillating input signal to examine its waveform or shape.

The heart of an oscilloscope is the *cathode-ray tube (CRT)*, which is a special type of electron tube (Fig. 5-5). Electrons emitted by the heated cathode of the tube are focused and accelerated. They form a narrow beam of high-velocity electrons. The direction in which this beam travels is then controlled, and it is allowed to strike a fluorescent screen. Light is emitted at the spot of impact with the screen and produces a visual indication of the beam position.

The basic quantity that the oscilloscope measures is *voltage.* The test leads that are hooked up to a circuit are the same as the voltage test leads used with a multimeter. The screen on the oscilloscope ·measures voltage with the vertical lines on the screen. A number of control circuits are required in order for the oscilloscope to convert the voltage signal fed to it into a visual display on the CRT screen (Fig. 5-6 on page 125). An understanding of the front-panel controls (see Table 5-1 on page 124) is essential in order to use the oscilloscope properly. Your instructor can demonstrate how these controls apply to the oscilloscope you will be using.

Oscilloscopes are normally powered from a standard 120-V grounded receptacle. The test probe of the scope contains two leads. One is connected to the input circuit of the scope and the other *directly to ground.* When the scope is used to test a circuit that has one side

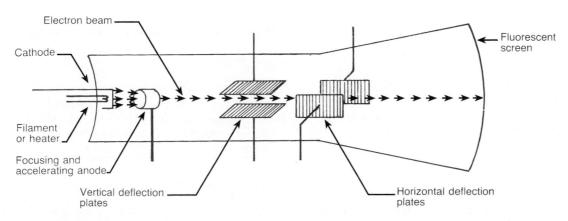

Fig. 5-5   Cathode-ray tube.

Table 5-1 · Oscilloscope Controls and Their Functions

Control	Function
POWER ON	A switch for the instrument's main voltage supply ON/OFF control.
INTENSITY	A rotary control used to vary the brightness of the displayed waveform or trace.
FOCUS	A rotary control used to vary the sharpness or definition of the trace.
Y SHIFT (Vertical position)	A rotary control used to adjust the position of the displayed waveform vertically up or down.
X SHIFT (Horizontal position)	A rotary control used to adjust the position of the displayed waveform horizontally to the left or right.
VOLTS/DIV	A rotary switch used to select the height or amplitude of the display. It works similarly to the range switch of a voltmeter. The switch is calibrated in volts per division of the vertical axis. These calibration markings permit you to determine the magnitude of the voltage display. The small control knob at the center of the VOLTS/DIV knob adjusts the amplitude between the calibrated settings. When the VOLTS/DIV selection is to be used for measuring the amplitude of a waveform, the small center knob must be turned to its extreme clockwise (calibrate) position.
TIME/DIV	A rotary switch used to select the width of the display. The calibration markings on this control allow you to measure the elapsed time between any two horizontal points on the display. This time period measurement is used in calculating the frequency of the waveform. The small control knob at the center of the TIME/DIV knob adjusts the width of the display between the calibrated time settings. The calibrated time settings only apply if the center control knob is turned to the extreme clockwise (calibrate) position.
INPUT	The input signal to be displayed on the screen is fed into this jack. Normally a coaxial cable equipped with a connector and probes is used to make the electrical connection from the instrument to the circuit. The use of coaxial cable helps prevent unwanted, stray signals being picked up by the oscilloscope input.
AC/DC	A switch that permits the input signal to be directly connected (DC position) or connected via a capacitor (AC position). In the DC position, both the ac and dc components of the waveform would be displayed. In the AC position, the capacitor blocks the direct current, so that only the ac component of the waveform would be displayed.
TRIGGER CONTROLS	Controls used to provide a stable (no-jitter) CRT display. Triggering defines the starting point of every successive sweep of the displayed trace. The functions of the four trigger controls are:
AUTO NORMAL SWITCH	Determines whether the time base will be triggered automatically or if it is to be operated in a free-running mode. If it is operated in the normal setting, the trigger signal is taken from the line to which the probe is connected. The scope is generally operated with the trigger set in the automatic position.
TRIGGER LEVEL KNOB	Adjusts the instant in time at which the waveform display commences. The display may be made to commence exactly when the signal goes through its zero position or at some time shortly before or after this instant.
SLOPE CONTROL	Permits the scope to trigger on the positive or negative half of the waveform.
INT-LINE-EXT SWITCH	Selects the source of trigger signal to be used. The scope is generally operated in the internal (INT) mode in which the trigger signal is provided by the scope. In the line mode, the trigger signal is provided from a sample of the line. The external (EXT) mode permits an external trigger signal to be applied.

grounded and the other side ungrounded, care must be taken to ensure that the *grounded conductor of the probe is connected to the* *grounded side of the circuit* (Fig. 5-7). If the grounded conductor of the probe is connected to the ungrounded side of the circuit, a direct

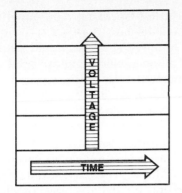

(a) Oscilloscope measures voltage versus time

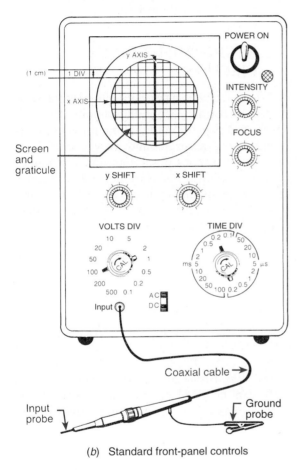

Coaxial cable →

Input probe

Ground probe

(b) Standard front-panel controls

Fig. 5-6   Oscilloscope.

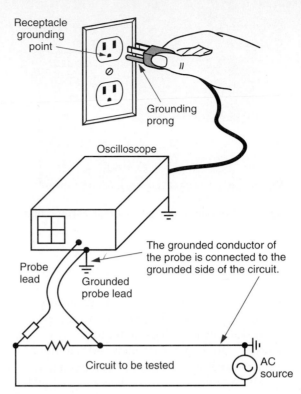

Fig. 5-7   Oscilloscope probe connection.

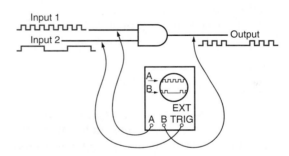

Fig. 5-8   Dual-trace oscilloscope connected to compare waveforms.

short to ground through the probe lead and the case of the scope will be created.

The *dual-trace oscilloscope* can display two different traces *at the same time.* It has two sets of controls, one for each trace. The dual-trace oscilloscope is particularly useful in *comparing* two waveforms to show amplitude differences, distortion, or time (phase) differences. Figure 5-8 illustrates a dual-trace scope connected to compare waveforms in a digital logic circuit, where, along with signal levels, timing is of

critical importance. In this example the "external" triggering input is used as an extra input to the scope.

The modern and powerful *digital-storage oscilloscopes* (*DSOs*) allow you to capture, store, and analyze a variety of repetitive or single-event signals (Fig. 5-9 on page 126). The benefits of DSOs over analog oscilloscopes are numerous. DSOs can include these features:

SAVE/RECALL. Different scope setups can be stored in the oscilloscope memory and recalled by pressing a button.

AUTOSET. The AUTOSET feature puts a meaningful waveform on display at the touch of a single button. AUTOSET

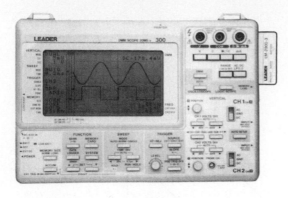

Fig. 5-9 Combination digital storage oscilloscope and DMM.

automatically sets voltage, time, triggering, and position.

STEP SEQUENCE. The STEP SEQUENCE allows different setups to be executed in a fixed sequence by pressing one button.

CURSORS. Direct measurement and readout of basic waveform parameters can be made without counting display divisions and scale-factor multiplication.

COUNTER/TIMER. The COUNTER/TIMER feature adds frequency, period, and event counting to the scope.

DMM. The DMM feature adds digital multimeter measurements to the oscilloscope.

Bandwidth, rise time, and sensitivity are the fundamental specifications for an oscilloscope; they determine what can be displayed on the screen. *Bandwidth* is the frequency range a scope can display. *Rise time* determines the ability of the instrument to deal with short events, such as rise times of glitches. *Sensitivity* is the amount of signal amplitude (volts per division) the scope can display.

Self-Test

Answer the following questions.

1. A multimeter is a combination _____, _____, and _____ in a single instrument.
2. True or false? The LED-type digital multimeter requires less operating power than the LCD-type.
3. The _____ feature of a digital multimeter automatically adjusts the meter's measuring circuits to the correct range.
4. True or false? For a non-sine wave, a true rms meter reads higher than the averaging meter.

5. The oscilloscope is primarily used to measure and display _____.
6. True or false? The X SHIFT and Y SHIFT controls on an oscilloscope are used to position the waveform on the screen.
7. The width of the display on an oscilloscope screen is adjusted by means of the _____ rotary switch.
8. The dual-trace oscilloscope is particularly useful in _____ two waveforms.
9. True or false? The SAVE/RECALL feature is found only on digital-storage oscilloscopes.
10. The _____ refers to the frequency range a scope can display.

5-2 BRIDGE MEASURING CIRCUITS

A *bridge circuit* consists of four sections connected to form two parallel circuits. The balanced-bridge method of making measurements is extremely accurate and very useful. It is often used in control circuits and test equipment for accurate measurement of resistance, capacitance, or inductance. The resistance and voltage characteristics of a balanced-bridge circuit (Fig. 5-10) are as follows:

$$\frac{R_1}{R_2} = \frac{R_3}{R_4}$$
$$V_1 = V_3$$
$$V_2 = V_4$$
$$V_A = V_B$$
$$V_{AB} = 0$$

The *Wheatstone bridge circuit*, shown in Fig. 5-10, can be used to measure resistance. It requires fixed, known resistance values for R_1

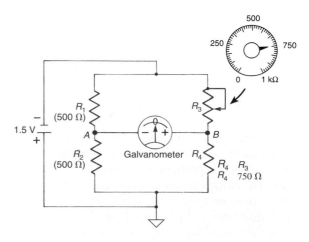

Fig. 5-10 Wheatstone bridge circuit.

and R_2 and a variable resistor for R_3. Variable resistor R_3 is attached to a calibrated dial that shows its adjusted resistance value. Resistor R_4 represents the resistance value to be measured. To determine R_4, the variable resistance R_3 is adjusted until the galvanometer indicates zero current. Consequently, the galvanometer is usually termed a *null detector*. The zero galvanometer reading indicates that the bridge is balanced. When the galvanometer indicates a null condition, the voltage on each side of the galvanometer is equal. If R_1 and R_2 are the same value, then the calibrated dial reading of R_3 indicates the resistance value of R_4.

If the values of R_1 and R_2 are different, the unknown resistance value for a balanced bridge can be calculated. The equation used is:

$$R_4 = \frac{R_2}{R_1} \times R_3$$

EXAMPLE 5-1

Assume that a balanced bridge (see Fig. 5-11), has the following resistance values:

$$R_1 = 400 \ \Omega$$
$$R_2 = 800 \ \Omega$$
$$R_3 = 250 \ \Omega$$

The value of the unknown resistor R_4 is:

$$R_4 = \frac{R_2}{R_1} \times R_3$$

$$R_4 = \frac{800 \ \Omega \times 250 \ \Omega}{400 \ \Omega}$$

$$R_4 = 500 \ \Omega$$

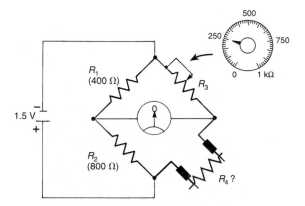

Fig. 5-11   Resistance measurement with the Wheatstone bridge.

More precise control can be obtained with a bridge circuit, than with a straight series or parallel connection. For example, a photoconductive cell light sensor is placed in one arm of the bridge, as shown in Fig. 5-12, so that it can detect changes in light. The bridge current is then balanced by adjustment of potentiometer R_3. Any changes in light will change the resistance of the photoconductive cell, and the bridge becomes unbalanced. With the bridge unbalanced, current flows through the galvanometer, which acts as a very sensitive light detector.

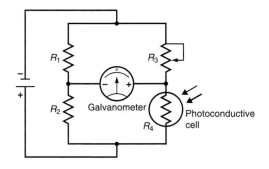

Fig. 5-12   Light-sensor bridge circuit.

Figure 5-13 shows a temperature-sensor bridge circuit that contains two thermistor sensors. Here, two matched thermistors are used to form the bridge measuring circuit. In normal operation, one thermistor is placed where its temperature is kept at a fixed value, and the other thermistor is used as a monitor. The bridge is then balanced by the adjustment of R_3. As long as the two thermistors have equal temperature, their resistance values will be the same, and the bridge circuit will remain balanced. Should the monitoring thermistor detect an increase or decrease in temperature, however, its resistance will change and the bridge will become unbalanced. With the bridge unbalanced, current flows through the

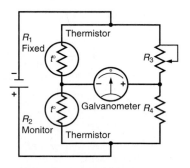

Fig. 5-13   Heat-sensor bridge circuit.

galvanometer, which acts as a very sensitive temperature sensor.

Self-Test

Answer the following questions.

11. A bridge circuit consists of _____ sections connected to form two parallel circuits.
12. Refer to Fig. 5-10. Assume the resistance value of R_2 is 1 kΩ. What resistance setting of R_3 will result in a balanced-bridge condition?
13. Refer to Fig. 5-11. Assume that the bridge is balanced with R_3 set to 750 Ω. What is the value of the unknown resistor R_4?
14. Refer to Fig. 5-12. Assume that the bridge is balanced with the galvanometer reading zero. How does the value of the voltage drop across R_2 compare with that across the photoconductive cell R_4?

5-3 SEMICONDUCTOR DIODES

▌▌▌▶ **YOU MAY RECALL** that a *PN-junction diode* allows the current to pass in one direction (positive source terminal connected to the anode) and blocks current in the other direction (negative source terminal connected to anode) (Fig. 5-14).

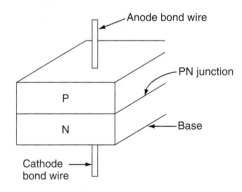

(a) PN-junction diode

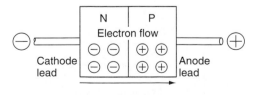

(b) PN-junction with forward bias applied

Fig. 5-14   Semiconductor diode.

RECTIFIER DIODE

Rectification is the process of changing alternating current to direct current. Because diodes allow current to flow in only one direction, they are used as rectifiers. Figure 5-15 shows the schematic for a *single-phase, half-wave rectifier circuit*. During the positive half-cycle of the ac input wave, the anode side of the diode is positive. The diode is then forward biased, allowing it to conduct a current flow to the load. Because the diode acts as a closed switch during this time, the positive half-cycle is developed across the load. During the negative half-cycle of the ac input wave, the anode side of the diode is negative. The diode is now reverse biased; as a result, no current can flow through it. The diode acts as an open switch during this time, so no voltage is produced across the load. Thus, applying a constant ac voltage produces a pulsating dc voltage across the load.

One type of single-phase, half-wave rectifier is the *Schottky diode*. It is constructed with an N-type chip of silicon bonded to platinum. This combination provides diode action and turns OFF much more quickly than a PN-junction diode. The Schottky diode is used in high-speed switch circuits.

Because the half-wave rectifier makes use of only half of the ac input wave, its use is limited to low-power applications. A less pulsating and a greater average direct current can be produced by rectifying both half-cycles of the ac input wave. Such a rectifier circuit is known as a *full-wave rectifier*.

The schematic for a *single-phase, full-wave bridge rectifier circuit* is shown in Fig. 5-16. During the positive half-cycle, the anodes of D_1 and D_2 are positive (forward biased), whereas the anodes of D_3 and D_4 are negative (reverse biased). Electron flow is from the negative side of the line, through D_1, to the load, then through D_2, and back to the other side of the line.

During the next half-cycle, the polarity of the ac line voltage reverses. As a result, diodes D_3 and D_4 become forward biased. Electron flow is now from the negative side of the line through D_3, to the load, then through D_4, and back to the other side of the line. Note that during this half-cycle the current flows through the load in the same direction, producing a full-wave pulsating direct current.

For heavier load requirements, such as those required for industrial applications, it is necessary to change three-phase alternating current

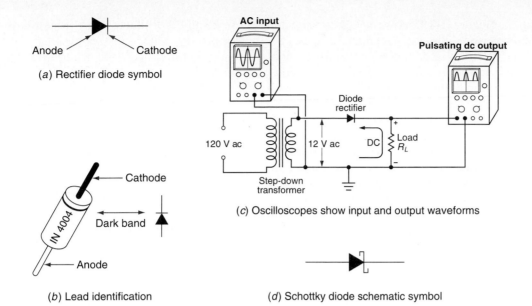

(a) Rectifier diode symbol

(b) Lead identification

(c) Oscilloscopes show input and output waveforms

(d) Schottky diode schematic symbol

Fig. 5-15 Single-phase, half-wave rectifier.

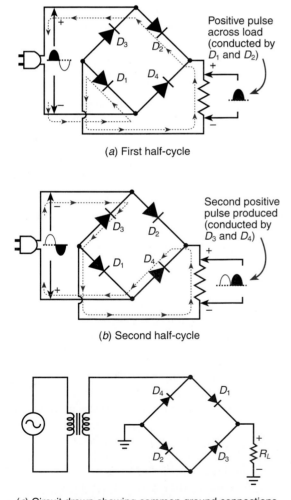

(a) First half-cycle

Positive pulse across load (conducted by D_1 and D_2)

(b) Second half-cycle

Second positive pulse produced (conducted by D_3 and D_4)

(c) Circuit drawn showing common ground connections

Fig. 5-16 Single-phase, full-wave rectifier circuit.

into direct current. The simplest configuration is the *three-phase, half-wave rectifier* shown in Fig. 5-17. A diode is connected to each leg of the wye-secondary of the three-phase power transformer. The diodes are forward biased when the voltage of each line becomes positive and reverse biased when the voltage becomes negative. As the voltage of each of the three-phase lines goes positive, current flows through the load to the center tap of the transformer to complete the circuit. This rectifier has a higher average voltage output and *less ripple* than a single-phase, full-wave rectifier.

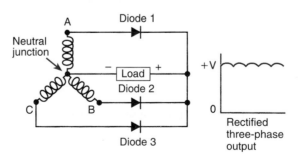

Fig. 5-17 Three-phase, half-wave rectifier.

The *three-phase, full-wave rectifier* has even less ripple than the three-phase, half-wave rectifier. It does not require a center-tapped, wye-connected, three-phase transformer for operation. It needs only to be connected to three-phase power for operation. Therefore, power can be supplied by either a wye or delta

CHAPTER 5 POWER ELECTRONICS **129**

system. Figure 5-18 shows a delta-connected three-phase, full-wave rectifier circuit. Six diodes are required. *Two* diodes must be conducting at all times in order to provide a complete path for the dc output. The diode with the anode voltage that is most positive will conduct. The diode with the cathode voltage that is most negative will also conduct. This three-phase, bridge-type rectifier also has a higher average dc voltage and less ripple than the three-phase, half-wave rectifier because the bridge rectifier changes both the positive and negative halves of the ac voltage into direct current.

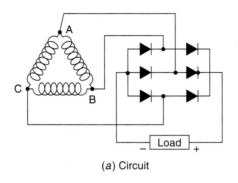

(a) Circuit

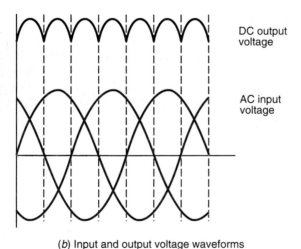

(b) Input and output voltage waveforms

Fig. 5-18   Three-phase, full-wave rectifier.

The pulsating dc output of rectifier circuits is not smooth enough to properly operate most electronic devices. They do not produce pure direct current. The power-supply output still has an ac component, which is called *ripple* (Fig. 5-19 shows peak-to-peak ripple). The ripple produced by changing ac voltage into dc voltage can be made much smoother by the use of filters. The amount of filtering required is partly determined by the amount of ripple

the rectifier produces. The three-phase rectifier produces much less ripple than the single-phase rectifier; therefore, it requires much less filtering.

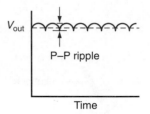

Fig. 5-19   Ripple: ac component in a dc power supply.

A *filter* is used to reduce the amount of ac ripple, thus providing a relatively pure form of direct current. The most common filter device is a *capacitor* connected in *parallel* with the output of the rectifier circuit. The filter capacitor is a large value capacitor. It makes an excellent filter because of its ability to store electric charges. The schematic diagram for a simple capacitive filter is shown in Fig. 5-20. It works by charging the capacitor when the diode conducts and discharging it when the diode does not conduct. When the rectifier circuit is conducting, the capacitor charges *rapidly* to approximately the *peak voltage* of the input wave. As the voltage from the rectifier drops—between the pulsations in the wave—the capacitor then discharges through the load. The capacitor, in effect, acts like a storage tank that accepts electrons at peak voltage and supplies them to the load when the rectifier output is low. The larger the load current, the larger the capacitor needed to provide adequate filtering. The capacitance value is determined from the value of the peak ripple voltage permitted.

Large power semiconductors must be *heat sinked* in order to operate at their listed power ratings. The cathode end of the high-current power rectifier diode in Fig. 5-21(a) forms a bolt for mounting into a heat sink. Heat sinks like the one in Fig. 5-21(b) vary in size and shape, but they have only one purpose. That purpose is to increase the surface area of the device connected to it. This permits air to contact a greater area and remove heat at a faster rate. When a component is mounted to a heat sink, *thermal compound* is generally used to ensure a good thermal contact between the device and the heat sink. Forcing a component to dissipate more heat than it was designed for can only shorten its life.

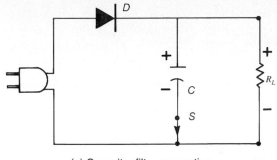

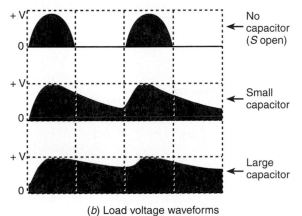

(a) Capacitor filter connection

No capacitor (S open)

Small capacitor

Large capacitor

(b) Load voltage waveforms

Fig. 5-20 Capacitor filter.

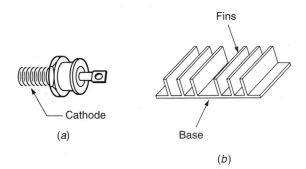

Fins

Cathode

(a)

Base

(b)

Fig. 5-21 Heat sinks.

CLAMPING OR DESPIKING DIODES

A *diode-clamping* or *despiking circuit* is a circuit that holds the voltage or current to a specific level. In certain inductive circuits, it is necessary to limit the amount of voltage present in the circuit. For example, if the dc voltage to a coil is switched OFF, there will be a noticeable arc at the switch contacts. As the current falls to a zero value, as shown in Fig. 5-22(*a*) on page 132, the collapsing magnetic field causes a high voltage to be induced in the coil. The arc produced at the contacts is a result of the induced voltage's attempting to maintain the current in the circuit.

Of special concern are the coils controlled by electronic control modules that contain transistors and integrated circuits. If the high-voltage spike generated by switching an inductive device OFF is allowed to reach the transistor, it can destroy it. As shown in Fig. 5-22(*b*), the clamping, or despiking, diode protects the circuit from surge currents developed when the current flowing through an electromagnetic device is abruptly interrupted. It is connected in *reverse bias* to block the battery's positive voltage; therefore, it allows the current to pass through the coil. When the circuit is switched off by the transistor, the magnetic field of the coil collapses, inducing current flow in the coil. This current then passes through the *diode.* Thus, the induced current flow is allowed to collapse upon itself and *not* through the transistor. A spark is not produced, and the sensitive electronic circuit controlling the coil is protected against damage. It is important to note that the clamping diode *must* be connected in reverse bias. Operating the circuit with the diode connected in forward bias creates a high-current *short circuit* through the diode and solid-state control module, which can damage both the diode and module (Fig. 5-22).

ZENER DIODES

The *zener diode* is like a rectifying diode in that it allows current to flow in the forward direction. It differs from a rectifying diode, however, in that its reverse-direction breakdown voltage is much lower than that of an ordinary rectifying diode. It is very heavily doped during manufacture. The large number of extra current carriers allows the zener diode to *conduct current in the reverse direction.* This reverse-bias current would destroy a normal diode, but the zener is made to operate this way. The specified zener voltage rating of a zener diode indicates the voltage at which the diode begins to conduct when reverse biased.

Zener diodes are often used as part of voltage-regulator circuits. A simple zener diode voltage-regulator circuit is shown in Fig. 5-23 on page 133. A 5-V zener diode is connected in series with a resistor R_1 to the variable dc input voltage. This input voltage is connected so that the zener diode is *reverse biased.* The series (R_1) is required to drop all the input voltage not dropped across the zener diode. As the input voltage is increased from 0 V, the voltage across the zener diode increases until the 5-V zener voltage is reached. At this point the zener diode conducts and *maintains a con-*

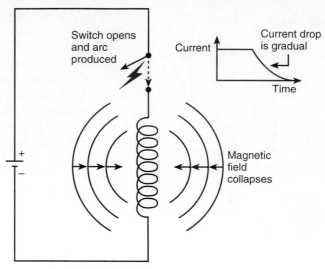

(*a*) With no clamping or despiking diode connected

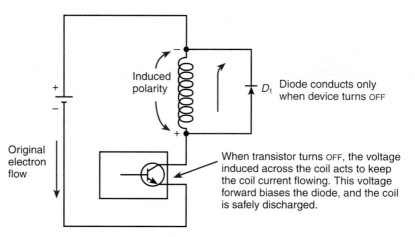

(*b*) With clamping or despiking diode connected

Fig. 5-22   Clamping or despiking diode.

stant 5-V output as the input voltage varies from 5 to 9 V.

The zener diode can be used to *shape*, or *condition*, signals from sensors for use by a digital computer. For example, the waveform-clipper circuit of Fig. 5-23 can be used to convert an incoming sine-wave signal to a near-square-wave signal that can be used by the computer. It clips both halves of the ac input wave equally when two zener diodes of the same voltage rating are used. Two back-to-back zener diodes can also limit ac voltage or *suppress* voltage transients on an ac line.

LIGHT-EMITTING DIODES

A *light-emitting diode* (*LED*) is a special semiconductor diode designed specifically to emit light when current flows through it (Fig. 5-24).

When forward biased, the energy of the electrons flowing through the resistance of the junction is converted directly to light energy. Because the LED is a diode, current will flow only when the LED is connected in forward bias. The LED must be operated within its specified voltage and current ratings to prevent irreversible damage. Most LEDs require approximately 1.5 to 2.2 V to forward bias them and can safely handle 20 to 30 mA of current. An LED is usually connected in series with a resistor that limits the voltage and current to the desired value.

The main advantages of using an LED as a light source rather than an ordinary light bulb are a much lower power consumption, a much higher life expectancy, and high speed of operation. Conventional silicon diodes convert energy to heat. *Gallium arsenide diodes* con-

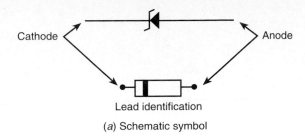

Cathode — Anode

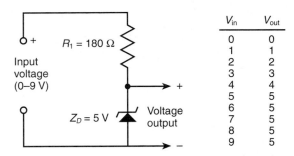

(a) Schematic symbol

Lead identification

$R_1 = 180\ \Omega$

Input voltage (0–9 V)

$Z_D = 5\ V$ Voltage output

+

−

V_{in}	V_{out}
0	0
1	1
2	2
3	3
4	4
5	5
6	5
7	5
8	5
9	5

(b) A 5-V zener diode voltage-regulator circuit

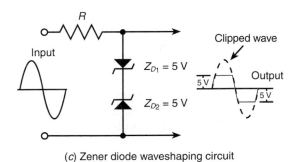

R

Input

$Z_{D_1} = 5\ V$

$Z_{D_2} = 5\ V$

Clipped wave

Output

5 V

5 V

(c) Zener diode waveshaping circuit

Fig. 5-23 Zener diode.

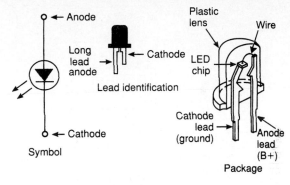

← Anode

Long lead anode

Cathode

← Cathode

Plastic lens

Wire

LED chip

Cathode lead (ground)

Anode lead (B+)

Package

Lead identification

Symbol

(a) Symbol and construction

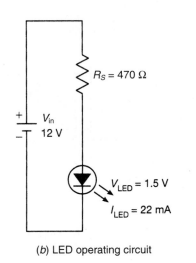

$R_S = 470\ \Omega$

V_{in} 12 V

$V_{LED} = 1.5\ V$

$I_{LED} = 22\ mA$

(b) LED operating circuit

Fig. 5-24 Light-emitting diode (LED).

vert energy to heat and infrared light. This type of diode is called an *infrared-emitting diode (IRED)*. Infrared light is not visible to the human eye. By doping gallium arsenide with various materials, LEDs with visible outputs of red, green, yellow, or blue light can be manufactured. Light-emitting diodes are used for alphabetical and digital displays and as signaling lamps.

LASER DIODES

Laser diodes are specially made LEDs that are capable of operating as lasers. *Laser* stands for "light amplification by stimulated emission of radiation." Unlike the LED, it has an *optical cavity,* which is required for laser production. The optical cavity is formed by coating opposite sides of a chip to create two highly reflective surfaces. Like the LED, it is a PN-junction diode, which at some current level will emit light. This emitted light is bounced back and forth internally between two reflecting surfaces. The bouncing back and forth of the light waves causes their intensity to reinforce and build up. The result is a high brilliance, single-frequency light beam that is emitted from the junction. Laser diodes are used in applications such as fiber-optic communications and bar code scanning systems.

PHOTODIODES

All PN-junction diodes are light sensitive. *Photodiodes* are PN-junction diodes specifically designed for light detection. The basic construction of a photodiode is shown in Fig. 5-25 on page 134. Light energy passes through the lens that exposes the junction. The photodiode is designed to operate in the *reverse-bias mode.* In this device, the reverse-bias leakage current increases with an increase in the light level. Typical current values are in the microampere range. The photodiode has a fast response time to variations in light levels.

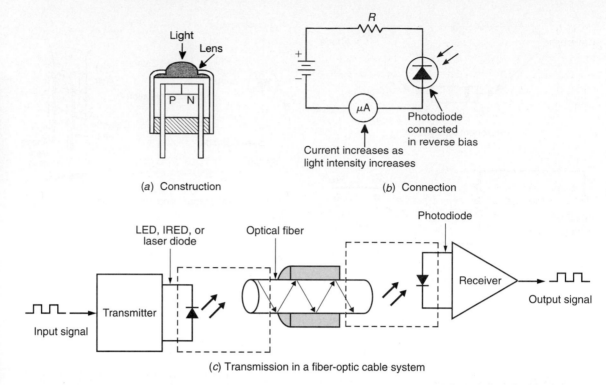

Fig. 5-25 Photodiode.

Photodiodes and LEDs are often used in conjunction with *fiber-optic cable* for the purpose of data transmission. Within the *transmitter*, the input signal (normally digital pulses) is converted from electric to light energy by flashing a light source within the transmitter OFF and ON very rapidly. The light-beam pulses are directed into a length of *light-conducting fiber*, where they reflect through the fiber core. At the other end of the fiber is a *receiver* (which contains a photodiode) that accepts the light-beam pulses and converts them back to their original electrical form.

Self-Test

Answer the following questions.

15. Rectification is the process of changing _____ to _____.
16. True or false? A PN-junction diode allows current to pass when its anode is connected negative with respect to its cathode.
17. True or false? A single-phase, half-wave rectifier rectifies both half-cycles of the ac input wave.
18. True or false? A single-phase, full-wave bridge rectifier requires four diodes.
19. True or false? Refer to Fig. 5-17. Assume

diode 1 becomes open-circuited. As a result, the dc output voltage drops to zero.
20. A(n) _____ is used to reduce the ac ripple in a dc power supply.
21. Large power diodes must be _____ to operate at their listed power ratings.
22. A clamping or despiking diode is normally connected in _____ bias.
23. True or false? A zener diode will only conduct current in the reverse-bias direction.
24. An LED is designed to emit _____ when current flows through it.
25. True or false? Laser diodes are used for digital displays.
26. True or false? Photodiodes are designed to operate in the reverse-bias mode.

5-4 TRANSISTORS

⏵ **YOU MAY RECALL** that *transistors* are semiconductor components which have three or more pins and with which power can be amplified. They function as *amplifiers* of electric signals, *variable resistors,* or as *switches*.

BIPOLAR-JUNCTION TRANSISTORS

Many electronic systems rely heavily on the transistor's ability to act as a switch. Used as a

NPN bipolar-
junction transistor
(BJT)

Base current

Collector current

Current gain

switch, the transistor has the advantages of having no moving parts, of being able to operate ON and OFF at a very high rate of speed, and of requiring very low driving voltages and currents in order to trigger the switching action.

Figure 5-26 illustrates the switching action of an *NPN bipolar-junction transistor* (*BJT*). The *emitter* is very heavily doped and acts as the main *source* of the electron current. The *base* is very lightly doped and acts to *control* the flow of the current. In an NPN transistor, the *collector* is moderately doped and *receives* most of the electrons from the emitter. Current in the base lead is called the *base current,* and the current in the collector lead is called the *collector current.* The amount of base current (I_B) determines the amount of collector current (I_C). With no base current there is no collector current (normally OFF). A small increase in I_B results in a large increase in I_C; thus, the base current acts to control the amount of collector current.

The *current gain* is the ratio of the collector current to the base current. Bipolar-junction transistors come in two variables: NPN and PNP. The action of each is the same, but the polarities are reversed.

Figure 5-27 on page 136 shows an experimental BJT timed-switching circuit. The length of the timing period is determined by the time it takes the capacitor to discharge after being charged. The operation of the circuit in Fig. 5-27(*a*) can be summarized as follows:

- With the pushbutton closed:
  - Sufficient base current flows to switch transistor ON.
  - The relay coil energizes.
  - Normally open relay contacts close to switch the lamp ON.
  - The capacitor charges to 9 V.
  - The circuit remains in this state with the lamp ON as long as the pushbutton is closed.
- When the pushbutton is released:
  - The timing action begins.
  - The base circuit to the 9-V dc source is opened.
  - The capacitor begins to discharge its stored energy through the base circuit.
  - The transistor continues to be switched ON until the charge on the capacitor is drained; then it switches OFF.
  - The relay coil de-energizes, and the relay contacts open to turn the lamp OFF.
  - Fixed resistor R_2 and variable resistor R_3

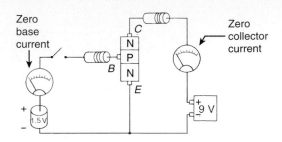

(*a*) No current passes from emitter to collector when base is not activated

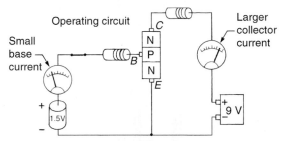

(*b*) The small base current controls a larger collector current

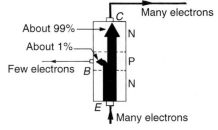

(*c*) Main and control current flows

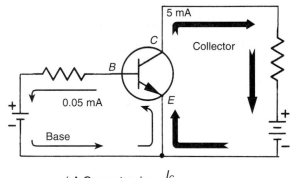

(*d*) Current gain $= \dfrac{I_C}{I_B}$

Current gain $= \dfrac{5\text{ mA}}{0.05\text{ mA}} = 100$

Fig. 5-26 Bipolar-junction transistor (BJT).

form a second parallel-current discharge path across the capacitor.
- The discharge rate and, thus, the timing period can be adjusted by varying the resistance of R_3.

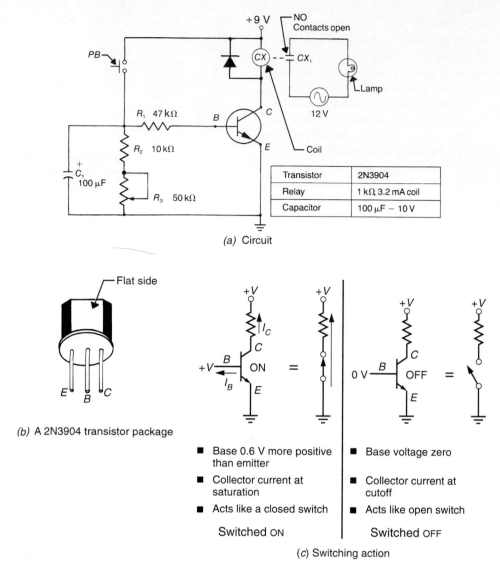

(a) Circuit

Transistor	2N3904
Relay	1 kΩ, 3.2 mA coil
Capacitor	100 μF − 10 V

(b) A 2N3904 transistor package

- Base 0.6 V more positive than emitter
- Collector current at saturation
- Acts like a closed switch

Switched ON

- Base voltage zero
- Collector current at cutoff
- Acts like open switch

Switched OFF

(c) Switching action

Fig. 5-27   BJT-timed switching.

FIELD-EFFECT TRANSISTORS

A BJT is a current-driven device. The current flow into or out of the base of the transistor controls the larger current that flows between the emitter and the collector. The *field-effect transistor* (*FET*) uses practically no input current. Instead, output current flow is controlled by a varying "electric field," which is created through the application of a voltage. This is the origin of the term *field-effect*.

Field-effect transistors are used in the same way as bipolar transistors. They have three connections: *source, gate,* and *drain*. These terminals correspond to the emitter, base, and collector of the bipolar transistor, respectively. Field-effect transistors are *unipolar;* their working current flows through only *one* type of

semiconductor material. This is in contrast to bipolar transistors, which have current flowing through both N-type and P-type regions (Fig. 5-28). In Fig. 5-28(a), the working current flows through the N-type semiconductor material. In Fig. 5-28(b), the working current flows through both the N-type and P-type regions. The FET was designed to get around the two major disadvantages of the bipolar transistor: low switching speed and high drive power, which are required because of base current.

Field-effect transistors can be divided into two main groups: the *junction field-effect transistor* (*JFET*) and the *metal-oxide semiconductor field-effect transistor* (*MOSFET*). Figure 5-29 on page 138 shows how current is conducted through an N-channel JFET. The N-channel of the JFET acts like a silicon resis-

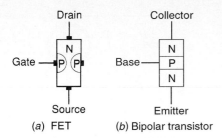

Fig. 5-28 **The FET and bipolar transistor compared.**

tor that conducts a current between the drain and source. The normal polarities for biasing the N-channel JFET are as indicated. A *positive* fixed V_{DS} supply voltage is connected between the source and drain. This sets up a current flow between the source and drain. A *negative* variable V_{GS} supply is connected between the gate and the source. This negative voltage at the gate increases the channel resistance and reduces the amount of current flow between the source and drain. Thus, the gate voltage controls the amount of drain current, and the control of the current is almost *powerless.* Junction field-effect transistors are *normally ON.* This means that JFETs allow full current to pass between the drain and the source when the gate circuit is *not activated.*

The JFET static charge detector circuit in Fig. 5-30 on page 139 demonstrates the almost powerless and highly sensitive simple gate-drive requirement of an FET. This circuit detects static electricity from a charged object (e.g., a plastic comb) over a foot away. When the charged object is brought near the antenna, the negative charge on the gate decreases the current flow and brightness of the LED.

The MOSFET is one of the most popular transistors in use today. The MOSFET is similar to the JFET in that current flows through a channel region where the effective diameter changes to alter the resistance; consequently, the current flow through the transistor is altered. Unlike the JFET, the gate of a MOSFET has *no electric contact* with the source and drain. A glasslike insulation of silicon dioxide separates the gate's metal contact from the rest of the transistor. As a result, gate current is extremely small whether the gate is *positive* or *negative.* The MOSFET is sometimes referred to as an *insulated-gate FET.*

Like JFETs, MOSFETs come in two varieties: N channel and P channel. The action of each is the same, but the polarities are reversed. There are also two different ways to make a metal-oxide semiconductor transistor: One is designed to operate in the *depletion mode,* where, with no gate voltage, the transistor is *normally* ON (Fig. 5-31 on page 139). The other is designed to operate in the *enhancement mode,* where, with no gate voltage, the transistor is *fully* OFF (Fig. 5-32 on page 140). Depletion-mode devices are often used in linear circuits, whereas enhancement devices are preferred for use in digital circuits.

Figure 5-33 on page 141 shows an experimental N-channel MOSFET timer circuit. The operation of the timer circuit in Fig. 5-33(*a*) can be summarized as follows:

- With the switch closed:
  - Positive voltage is applied to the gate.
  - The transistor switches ON to switch lamp ON.
  - The capacitor charges to 9 V.
  - The circuit remains in this state with the lamp ON as long as the switch is closed.
- When the switch is opened:
  - The timing action begins.
  - The positive gate circuit to the 12-V source is opened.
  - The positive charge on the capacitor keeps the transistor switched ON.
  - The capacitor begins to discharge its stored energy through R_1, and R_2 will still maintain a positive voltage at the gate.
  - The transistor and the lamp continue to conduct a current for as long as it takes the capacitor to discharge.
  - Because the gate current flow is negligible, very long time delays of from minutes to hours are possible.
  - The discharge rate and, thus, the timing period can be adjusted by varying the resistance of R_2.

Both *depletion-type* and *enhancement-type MOSFETs* use a thin layer of silicon dioxide to insulate the gate from the rest of the transistor. This insulating layer is kept as thin as possible to give the gate more control over drain current. Because the insulating layer is so thin, it is easily destroyed by too much gate voltage. These MOSFETs are *very sensitive to static charges* (which can be very high voltage) that would not harm other devices. For this reason, MOSFETS are kept either in special *static protection packages* or with their leads shorted together until they are placed into a circuit. If a MOSFET is removed from or inserted into a circuit while the power is still on, transient inductive voltages can destroy the MOSFET.

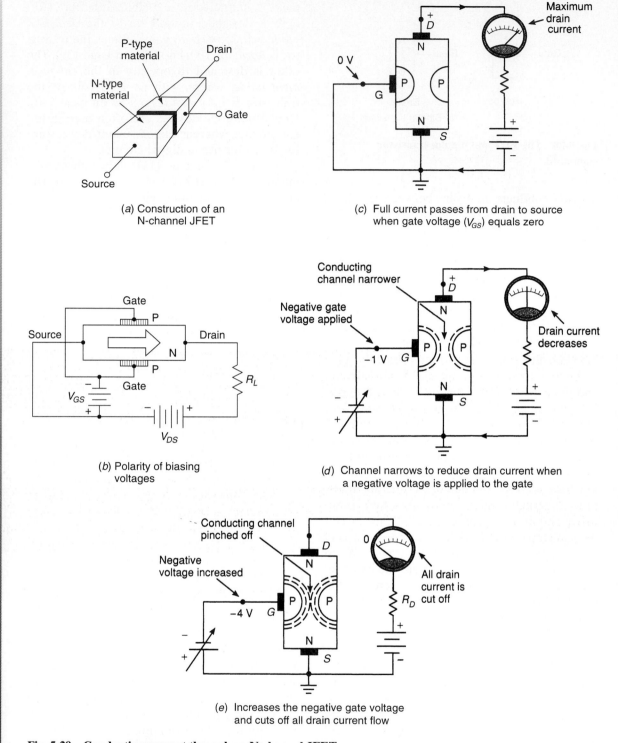

(a) Construction of an N-channel JFET

(b) Polarity of biasing voltages

(c) Full current passes from drain to source when gate voltage (V_{GS}) equals zero

(d) Channel narrows to reduce drain current when a negative voltage is applied to the gate

(e) Increases the negative gate voltage and cuts off all drain current flow

Fig. 5-29   Conducting current through an N-channel JFET.

UNIJUNCTION TRANSISTORS

Unlike a true transistor, a *unijunction transistor (UJT) does not amplify*. Instead, it works more like a voltage-controlled switch. It contains one emitter and two bases (Fig. 5-34 on page 142). When the variable emitter supply voltage is set to zero, a small current flows from base 1 to base 2. No current flows from base 1 to the emitter. When the positive emitter voltage reaches a certain *threshold* value (several volts), the UJT switches ON, and a high current flows from base 1 to the emitter. No current flows between base 1 and the emitter below this threshold voltage. Thus, the

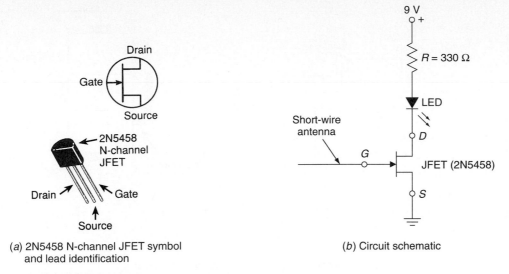

(a) 2N5458 N-channel JFET symbol
and lead identification

(b) Circuit schematic

Fig. 5-30   JFET static charge detector.

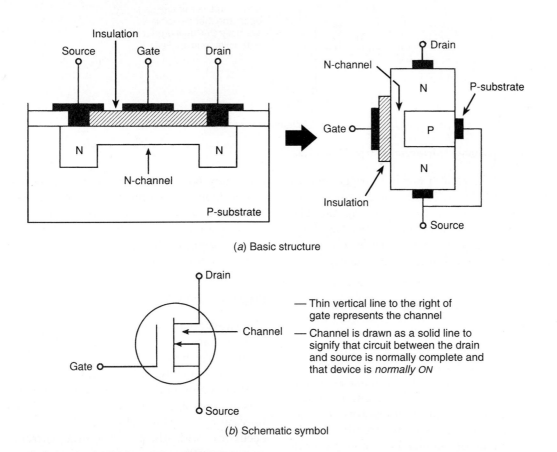

(a) Basic structure

— Thin vertical line to the right of
gate represents the channel

— Channel is drawn as a solid line to
signify that circuit between the drain
and source is normally complete and
that device is *normally ON*

(b) Schematic symbol

Fig. 5-31   N-channel depletion-type MOSFET.

UJT operates like a base 1-to-emitter switch that switches from OFF to ON once the threshold voltage level is reached.

Unijunction transistors are used as part of ramp circuits, oscillators, timing circuits, and trigger circuits. Figure 5-35 on page 143 shows how a UJT can be connected as a relaxation oscillator. Capacitor C_1 charges to the threshold trigger voltage level and then discharges through the emitter. As a result, three useful output voltage waveforms are produced. A sawtooth (or ramp) waveform is produced at the emitter. The gradual increase in this sawtooth voltage results from the charging of the

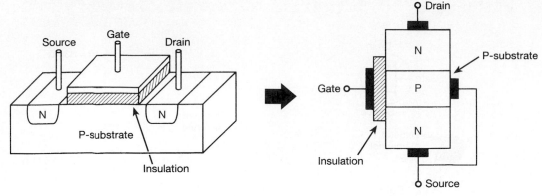

(a) Basic structure

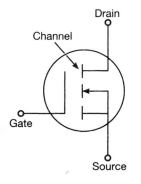

Channel is drawn as a broken line to signify that circuit between the drain and source is normally open and that device is *normally OFF* until proper gate voltage is applied

(b) Schematic symbol

Fig. 5-32 N-channel enhancement-type MOSFET.

capacitor. Positive and negative trigger pulses are produced at base 1 and base 2, respectively. These trigger pulses appear during the discharge of the capacitor because the UJT conducts heavily at this time. Variable resistor R_3 can be adjusted for different frequency rates.

DARLINGTON TRANSISTORS

The *Darlington pair* configuration is a way of hooking up two or more bipolar transistors so that a smaller transistor provides the base current for a larger transistor. This creates a transistor unit with a very high current-amplification factor. Two separate transistors may be used together, or—as is most often the case—the semiconductor manufacturer may put both transistors together in the same package (Fig. 5-36 on page 143).

INSULATED-GATE BIPOLAR TRANSISTORS

The *insulated-gate bipolar transistor (IGBT)* is a type of hybrid transistor that offers even higher current densities than bipolar Darlington transistors (Fig. 5-37 on page 143). In addi-

tion, they have the much simpler gate-driven requirements of a power MOSFET. They are used in a myriad of higher-voltage applications, including motor drives, uninterruptable power supplies (UPS), welders, industrial controls, and power supplies.

PHOTOTRANSISTORS

The most common *phototransistor* is an NPN bipolar transistor with a light-sensitive, collector-base PN junction (Fig. 5-38 on page 144). When this junction is exposed to light through a lens opening in the transistor package, it creates a control current flow that switches the transistor ON. This action is similar to that occurring with the base-emitter current of ordinary NPN transistors. A phototransistor can be either a two-lead or a three-lead device. The base lead may be brought out so that the device can be used as a conventional bipolar transistor, with or without the additional light-sensitivity feature. When there is no light entering the lens opening, only a very small leakage current flows between collector and emitter. This is called the *dark current*. When light strikes the collector-base PN junction, a

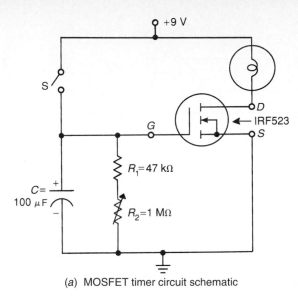

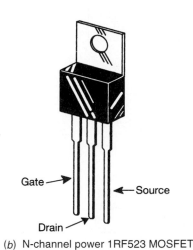

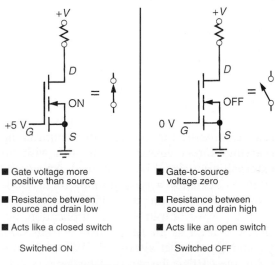

$+V$

D

ON $=$

$+5\,V_{\overline{G}}$

S

■ Gate voltage more positive than source

■ Resistance between source and drain low

■ Acts like a closed switch

Switched ON

$+V$

D

OFF $=$

$0\,V_{\overline{G}}$

S

■ Gate-to-source voltage zero

■ Resistance between source and drain high

■ Acts like an open switch

Switched OFF

(c) Switching action of a MOSFET

Fig. 5-33   MOSFET timer circuit.

base current is produced that is directly proportional to the light intensity. This action produces an *amplified* collector current. Except for the way base current is generated, the phototransistor behaves like a conventional bipolar transistor.

TRANSISTOR AMPLIFIERS

The transistor switching circuits are always operated at *saturation* or *cutoff*. *Transistor amplifiers* are operated in the linear region *between* these two points. In this region, the output signal varies in proportion to the input signal. When operated as an amplifier, a transistor can act like a rheostat, or variable resistor, to *vary* (rather than switch ON and OFF) the amount of current delivered to a load. Transistors have many advantages over the rheostats they replace. They are smaller and lighter, and they can control current much more efficiently than a rheostat.

An experimental *MOSFET lamp-dimmer circuit* is shown in Fig. 5-39 on page 144. Basically, it is a *dc amplifier*. Adjustment of R_2 varies the amount of positive dc voltage applied to the gate. Because the MOSFET operates in the enhancement, or normally OFF, mode, increasing the positive gate voltage lowers the channel resistance and increases drain current flow. Thus, changing the setting of R_2 changes the brightness of the lamp. This circuit illustrates how a transistor can be used as a variable resistor. The maximum load that can be safely connected into the circuit is determined by the power rating of the power MOSFET transistor.

An *ac amplifier* is used to increase the level or strength of small ac signals. Basically, an ac amplifier is supplied with a dc operating voltage and small ac signal input voltage. The circuit produces an ac output signal that is an amplified version of the applied ac input signal. Figure 5-40 on page 145 shows a complete ac common emitter voltage amplifier circuit using a BJT transistor. The ac *input* signal is applied through a coupling capacitor C_2 to the *emitter-base* circuit of the transistor. Use of a coupling capacitor blocks the dc path to the input devices so that its resistance will not affect the dc bias currents. Similarly, the ac *output* circuit is taken from the *collector to the emitter* through a second coupling capacitor C_3. Small base current changes created by the input signal are amplified by the collector circuit. Voltage gain of the amplifier depends on two factors. The first is the transistor

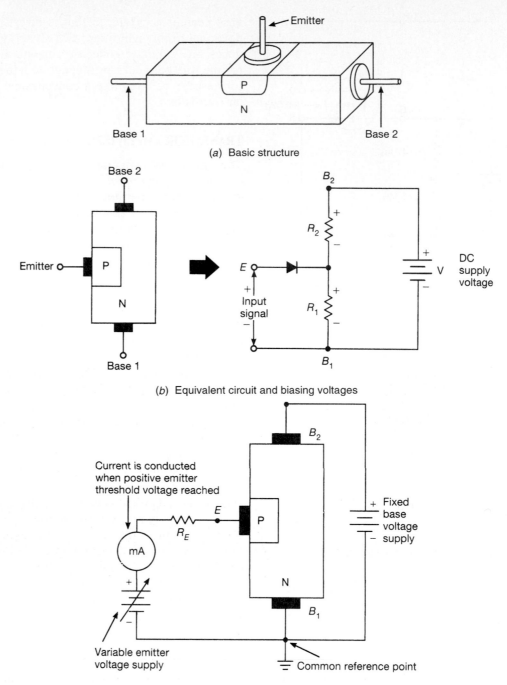

(a) Basic structure

(b) Equivalent circuit and biasing voltages

(c) Theory of operation

Fig. 5-34   Unijunction transistor (UJT).

current. In an amplifier circuit, the greater the current, the greater the voltage gain. Another important factor is the value of the collector load (R_3 and R_5). The higher the load resistance, the greater the voltage change across it, given a change in the current.

TRANSISTOR OSCILLATORS

A *transistor oscillator* is a special transistor amplifier circuit that converts pure direct cur-

rent to alternating current or to pulsating (varying) direct current on its own with no external input signal being applied. An experimental circuit for a *multivibrator oscillator* is shown in Fig. 5-41 on page 145. This circuit has a square-wave output. In this application, the circuit is used as an LED *flasher.* Transistors Q_1 and Q_2 conduct alternately. When Q_2 conducts, the LED is flashed ON. The rapid switching action is controlled by the values of the resistors R_2 and R_3 and capacitors C_1 and C_2.

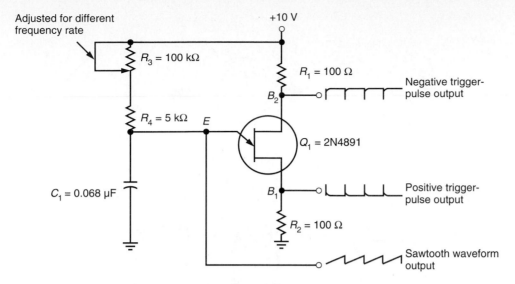

Fig. 5-35   UJT relaxation oscillator.

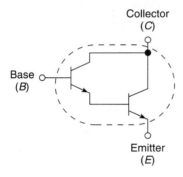

Fig. 5-36   NPN Darlington transistor.

(a) Schematic symbol for the
N-channel, enhancement type

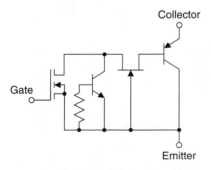

(b) Internal equivalent circuit

Fig. 5-37   Insulated-gate bipolar transistor (IGBT).

Increasing the values of C_1 and C_2 slows the flash rate.

Power inverters are oscillators used to convert dc power to ac power (Fig. 5-42 on page 145). An inverter is composed of electronic switches (transistors or thyristors) that switch the dc power ON and OFF to produce an ac power output at the desired frequency and voltage. A simple power inverter circuit is basically a free-running multivibrator with transistors Q_1 and Q_2 conducting alternately. When Q_1 conducts, current flows in the upper half of the primary winding. When Q_2 conducts, current flows in the lower half of the primary winding. This action produces an ac voltage at the secondary of the transformer.

Self-Test

Answer the following questions.

27. Transistors can function as _____, _____, or _____.
28. True or false? Bipolar-junction transistors are current driven, whereas FETs use practically no input current.
29. Current flow through a BJT is controlled by _____ current, whereas current flow through an FET is controlled by _____ voltage.
30. True or false? All JFETs are normally ON devices.
31. True or false? The gate of a MOSFET has no electric contact with the source and drain.
32. True or false? A depletion-type MOSFET is normally ON, whereas an enhancement-type MOSFET is normally OFF.

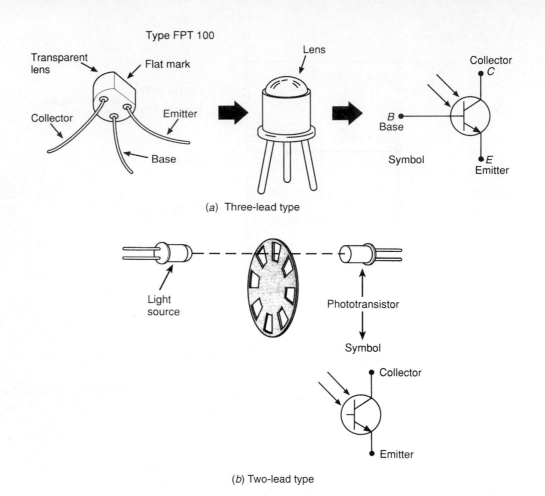

Type FPT 100

(a) Three-lead type

(b) Two-lead type

Fig. 5-38   NPN phototransistor.

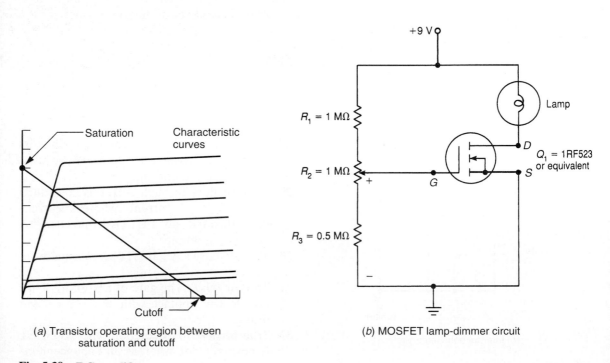

(a) Transistor operating region between saturation and cutoff

(b) MOSFET lamp-dimmer circuit

Fig. 5-39   DC amplifier.

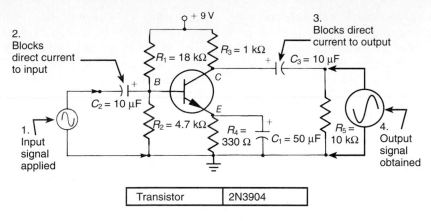

Fig. 5-40   AC amplifier.

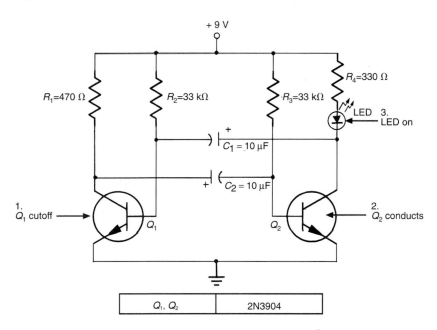

Fig. 5-41   Multivibrator oscillator.

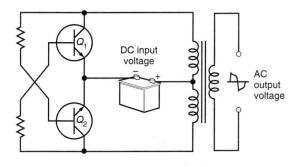

Fig. 5-42   DC to ac power inverter.

33. True or false? Metal-oxide semiconductor field-effect transistors are very sensitive to static charges.
34. Unlike a true transistor, a UJT does not _____.
35. A Darlington transistor is made up of two

bipolar transistors connected so that a smaller transistor provides the _____ for a larger transistor.
36. An IGBT is similar to a Darlington transistor except that the smaller bipolar transistor has been replaced with a(n) _____ transistor.
37. Transistor amplifiers are operated in the region between _____ and _____.
38. Transistor oscillators convert _____ to _____.

5-5 THYRISTORS

▶ **YOU MAY RECALL**   that *thyristors* are semiconductor components with at least three PN junctions. The word *thyristor* is used as the generic term for all kinds of components conforming to this definition. The operation of a thyristor is similar to that of a *switch*. Like a mechanical switch, it has two states: ON (conducting) and OFF (not conducting). There is *no* linear area between these two states as there is in a transistor.

The *silicon-controlled rectifier* and *triac* are the most frequently used thyristor devices. They are the workhorses of industrial electronics. The thyristor is used in power electronics for speed and frequency control, rectification, and power conversion. Common applications include control of motors, control and manipulation of robots, and control of heat and light.

SILICON-CONTROLLED RECTIFIERS

A *silicon-controlled rectifier,* commonly called an *SCR*, is a four-layer (PNPN) semiconductor device that makes use of three leads—the anode, the cathode, and the gate—for its operation (Fig. 5-43). Unlike transistors, SCRs *cannot amplify* signals. They are used strictly as solid-state switches and are categorized according to the amount of current they can switch. Low-current SCRs can operate with an anode current of less than 1 A. High-current SCRs can handle load currents in the thousands of amperes. Most SCRs have provisions for some type of heat sink to dissipate internal heat.

The schematic symbol for the SCR is very much like that of the diode rectifier. In fact, the SCR electrically resembles the diode because it conducts in only one direction. In other words, the SCR must be forward biased from anode to cathode for current conduction. It is unlike the diode because of the presence of a gate (*G*) lead, which is used to turn the device ON.

The operation of an SCR is similar to that of a standard diode except that it requires a *momentary positive* voltage applied to the gate to switch it ON. The schematic of an SCR switching circuit that is operated from a dc source is shown in Fig. 5-44. The anode is connected so that it is positive with respect to the cathode (forward biased). Momentarily closing pushbutton PB_1 applies a positive current-limited voltage to the gate of the SCR, which switches the anode-to-cathode circuit ON, or into conduction, thus turning the lamp ON. Once the SCR is ON, it stays ON, even after the gate voltage is removed. The only way to turn the SCR OFF is to reduce the anode-cathode current to zero by removing the source voltage from the anode-cathode circuit. This is accomplished by momentarily pressing pushbutton PB_2. It is important to note that the anode-cathode circuit will switch ON in *just one direction*. This occurs only when the anode is positive with respect to the cathode and a positive voltage is applied to the gate.

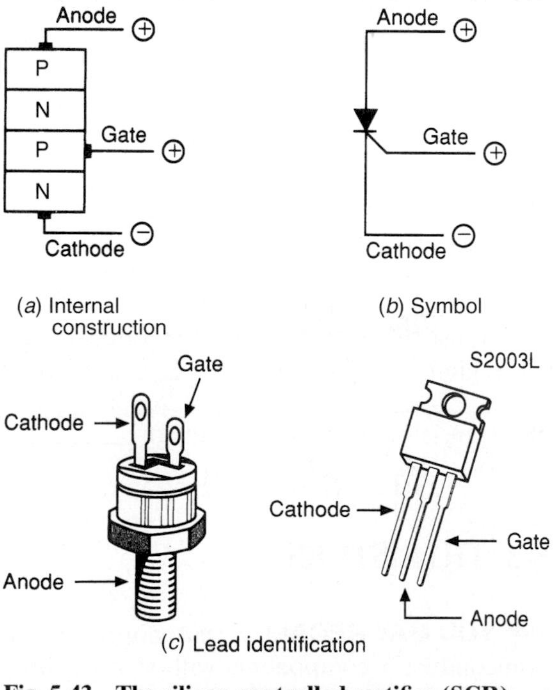

Fig. 5-43 — The silicon-controlled rectifier (SCR)

```
   Anode ⊕
  ┌───┐
  │ P │
  ├───┤
  │ N │
  ├───┤──── Gate ⊕
  │ P │
  ├───┤
  │ N │
  └───┘
  Cathode ⊖
```

(a) Internal construction

```
   Anode ⊕

        ▼
  ──────│──── Gate ⊕

   Cathode ⊖
```

(b) Symbol

S2003L

(c) Lead identification

Cathode, Gate, Anode

Fig. 5-43 **The silicon-controlled rectifier (SCR).**

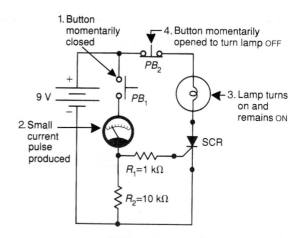

Fig. 5-44 **SCR operated from a dc source.**

The SCR *crowbar circuit* of Fig. 5-45 is designed to deliberately short and blow the input fuse if the supply voltage exceeds a present maximum voltage. This circuitry is used on some special equipment that could be damaged if the input voltage were to go too high.

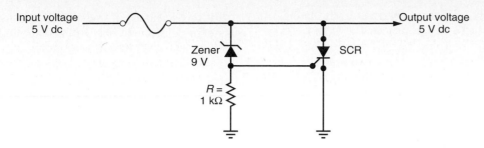

Fig. 5-45 SCR crowbar circuit.

The SCR circuit is normally open at proper input voltages and has no effect on the operation. If the input voltage rises above 9 V, the zener diode will conduct. This will produce a voltage drop across R that is sufficient to gate the SCR into conduction. The fuse will blow, opening the supply line and thereby protecting the circuitry following.

The problem of SCR turnoff does not occur in ac circuits. The SCR is automatically shut OFF during each cycle when the ac voltage across the SCR approaches zero. As zero voltage is approached, anode current falls below the holding current value. The SCR stays OFF throughout the entire negative ac cycle because it is reverse biased.

The SCR can be used for switching current to a load connected to an ac source. Because the SCR is a rectifier, it can conduct only one-half of the ac input wave. The maximum output delivered to the load, therefore, is 50 percent; its shape is that of a half-wave pulsating dc waveform. The schematic of an SCR switch-ing circuit that is operated from an ac source is shown in Fig. 5-46. The anode-cathode circuit can only be switched ON during the half-cycle when the anode is positive (forward biased). With pushbutton PB_1 open, no gate current flows, so the anode-cathode circuit remains OFF. Pressing pushbutton PB_1 continuously closed causes the gate-cathode and anode-cathode circuits to be forward biased at the same time. This produces a half-wave pulsat-ing direct current through the lamp load. When pushbutton PB_1 is released, the anode-cathode current is automatically shut OFF when the ac voltage drops to zero on the sine wave.

When the SCR is connected to an ac voltage source, it can also be used to *vary* the amount of power delivered to a load. Basically, it per-forms the same function as a rheostat, but it is much more *efficient*.

Figure 5-47 on page 148 illustrates the use of an SCR to regulate and rectify the power sup-ply to a dc motor from an ac source. A *trigger circuit* turns on the SCR once at a predeter-

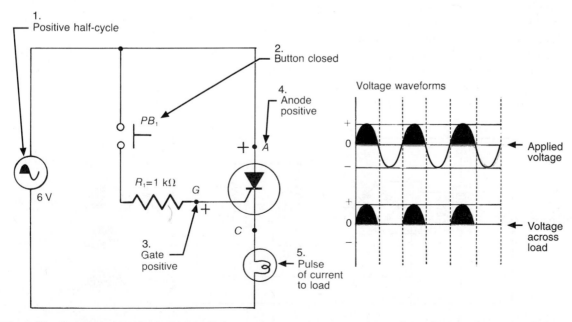

Fig. 5-46 SCR operated from an ac source.

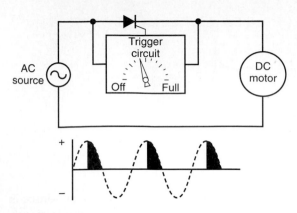

Fig. 5-47 SCR variable output control.

mined point, during each full ac cycle. At the half-power setting of the control knob, the gate trigger pulse occurs at the midpoint of the positive half-cycle. The anode-to-cathode circuit conducts for the remainder of the cycle supplying only half the maximum voltage and current to the load. *The SCR has the ability to be turned* ON *at any time during the positive half of the applied ac cycle.* The amount of time it conducts during each positive half-cycle determines the power that is delivered to the load. The trigger circuit is used to supply trigger pulses of current to the gate. These current pulses are timed by the manual-control knob setting. For full setting of the control knob, a trigger pulse is delivered near the start of the positive half-cycle. The anode-to-cathode circuit then conducts for the full half-cycle supplying maximum voltage and current to the load. The main advantage of this system is that virtually no power is *wasted* by being converted to heat. Heat is generated when current is controlled by resistance, not when it is switched. The SCR is either fully ON or fully OFF. It never throttles current only part way.

Phase shifting is required in order for the SCR to provide a variable output. This means

shifting the phase of the voltage applied to the gate in relation to the voltage applied to the anode. The UJT was developed to do the job of phase-shifting SCR circuits. The schematic of a UJT phase-shifting circuit for an SCR is shown in Fig. 5-48. The transformer and bridge rectifier are used to provide the low-voltage direct current required to operate the UJT. The UJT remains OFF until capacitor C_1 charges to a predetermined voltage level. Once that level is reached, the UJT turns ON and discharges the capacitor through resistor R_2. This discharge produces a pulse of current through R_2, which triggers the gate of the SCR. The charge time of the capacitor and the pulse rate of the UJT are controlled by varying the resistance of R_1. The pulses produced by the UJT will occur early or later in the ac alternation depending on the setting of R_1. As R_1 is decreased, C_1 charges faster, the UJT fires earlier, and the load power is increased.

The SCR circuit of Fig. 5-49 can be used for the "soft start" of a three-phase induction motor. Two SCRs are connected in inverse parallel to obtain full-wave control. In this connection scheme, one SCR controls the voltage when it is positive in the sine wave, and the other controls the voltage when it is negative. Control of current and acceleration is achieved by triggering the firing of the SCR at different times within the half-cycle. If the gate pulse is applied early in the half-cycle, the output is high. If the gate pulse is applied late in the half-cycle, only a small part of the waveform is passed through, and the output is low.

TRIAC

The *triac* is a device very similar in operation to the SCR. When an SCR is connected into an ac circuit, the output voltage is rectified to direct current. The triac, however, is designed

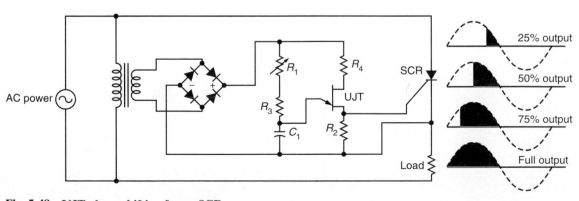

Fig. 5-48 UJT phase shifting for an SCR.

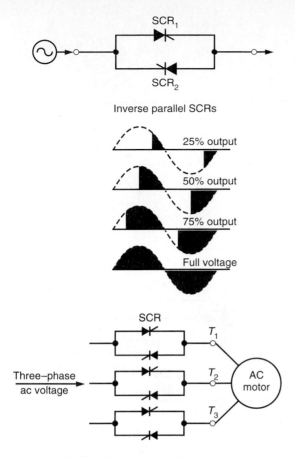

Fig. 5-49 SCR soft start control of an ac motor.

to conduct on both halves of the ac waveform. Therefore, the output of the triac is *alternating current instead of direct current.* Triacs were developed to provide a means for improved control of ac power.

The triac operates as two SCRs in the same case. The equivalent triac circuit shown as two SCRs connected in inverse-parallel is illustrated in Fig. 5-50 on page 150. Note that the SCRs are *inverted.* As such, the triac is capable of conducting with either polarity of terminal voltage. It can also be triggered by *either polarity of gate signal.* The triac has three terminals: main terminal 2 (MT_2), main terminal 1 (MT_1), and the gate G. Terminals MT_2 and MT_1 are designated as such because current flow is bidirectional. Because it interacts with the gate, MT_1 is used as the measurement reference terminal. Current can flow between MT_2 and MT_1 and also between the gate and MT_1.

The triac may be triggered into conduction in either direction by gate current moving in to or out of the gate. Once the direction of main terminal current flow is established, the triac has basically the same internal operating characteristic as the SCR. The triac has four possi-

ble triggering modes. With respect to MT_1, these are:

- MT_2 is positive, and the gate is positive.
- MT_2 is positive, and the gate is negative.
- MT_2 is negative, and the gate is positive.
- MT_2 is negative, and the gate is negative.

Two of these triggering modes are illustrated in Fig. 5-51 on page 150.

Because the triac can conduct on both half-cycles, it is most useful for controlling loads that operate on alternating current. Full efficiency is gained by using both half-waves of the ac input voltage. The schematic of a triac switching circuit operated from an ac source is shown in Fig. 5-52 on page 151. If pushbutton PB_1 is held closed, a continuous trigger current is supplied to the gate. The triac conducts in both directions to switch all of the applied ac voltage to the load. If the pushbutton is opened, the triac turns OFF when the ac source voltage and holding current drop to zero, or reverse polarity. Note that unlike the output of the similar SCR circuit, the output of this circuit is alternating current, not direct current.

One common application of a triac is switching ac current to an ac motor. The triac motor-switching circuit in Fig. 5-53 on page 151 illustrates the ability of a triac to control a large amount of load current with a small amount of gate current. In this application, it operates like a solid-state relay. A 24-V step-down transformer is used to reduce the voltage in the thermostat circuit. The resistor limits the amount of current flow in the gate–MT_1 circuit when the thermostat contacts close to switch the triac and motor ON. The maximum current rating of the thermostat contacts is *much lower* than that of the triac and motor. If the same thermostat were wired in series with the motor to operate the motor directly, the contacts would be destroyed by the heavier current flow.

Triacs can be used to *vary* the average ac current going to an ac load, as illustrated in Fig. 5-54 on page 151. The trigger circuit controls the point of the ac waveform at which the triac is switched ON. The resulting waveform is still alternating current, but the average current is changed. In a lighting circuit, varying the current to an incandescent lamp will vary the amount of light emitted by the lamp. Thus, the triac can be used as a light-dimming control. In some motor circuits, varying the current will change the speed of the motor.

The *diac* is a two-terminal transistorlike device that is used primarily to control the trig-

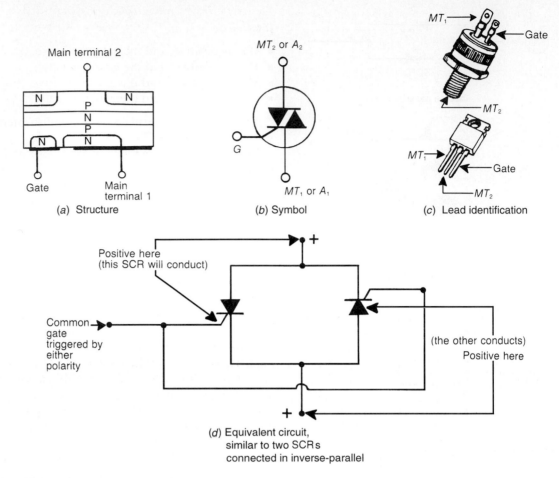

Main terminal 2

(a) Structure

MT_2 or A_2

G

MT_1 or A_1

(b) Symbol

(c) Lead identification

Positive here
(this SCR will conduct)

Common
gate
triggered by
either
polarity

(the other conducts)
Positive here

(d) Equivalent circuit,
similar to two SCRs
connected in inverse-parallel

Fig. 5-50   Triac.

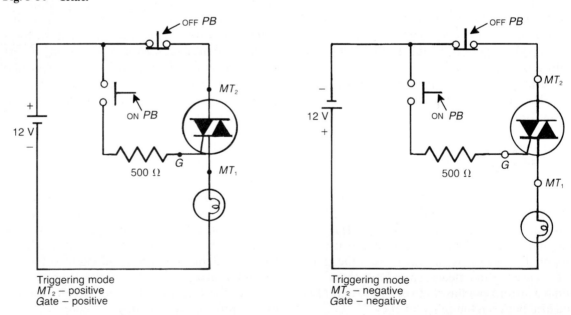

Triggering mode
MT_2 – positive
Gate – positive

Triggering mode
MT_2 – negative
Gate – negative

Fig. 5-51   Triac triggering modes.

gering of SCRs and triacs. Unlike the transistor, the two junctions of a diac are heavily and equally doped. The symbol of the diac shows that it behaves like two diodes pointing in opposite directions. Current flows through the diac (in either direction) when the voltage across it reaches its rated *breakover voltage*. The current pulse produced when the diac

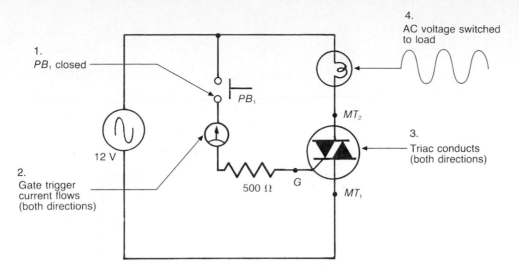

Fig. 5-52 Triac ac switching circuit.

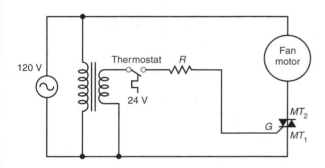

Fig. 5-53 Triac motor-switching circuit.

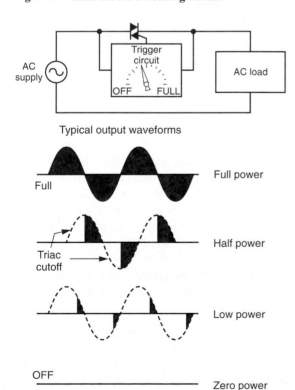

Typical output waveforms

Full power

Half power

Low power

OFF — Zero power

Fig. 5-54 Triac used to vary current.

changes from a nonconducting to a conducting state is used for SCR and triac gate triggering.

An experimental triac/diac lamp dimmer circuit is shown in Fig. 5-55 on page 152. When the variable resistor R_1 is at its lowest value (Bright), capacitor C_1 charges rapidly at the beginning of each half-cycle of the ac voltage. When the voltage across C_1 reaches the breakover voltage of the diac, C_1 is discharged into the gate of the triac. Thus, the triac is ON early in each half-cycle and remains ON to the end of each half-cycle. Therefore, current will flow through the lamp for most of each half-cycle and produce full brightness. As the resistance of R_1 increases, the time required to charge C_1 to the breakover voltage of the diac increases. This causes the triac to fire later in each half-cycle. So, the length of time current flows through the lamp is reduced, and less light is emitted.

Self-Test

Answer the following questions.

39. True or false? Thyristors are primarily designed to amplify signals.

40. The SCR, unlike the diode, requires a momentary _____ voltage applied to the _____ to switch it ON.

41. When operated from an ac source, an SCR can _____ and _____ the power supplied to the load.

42. Phase shifting of an SCR refers to shifting the phase of the voltage applied to the _____ in relation to the voltage applied to the _____.

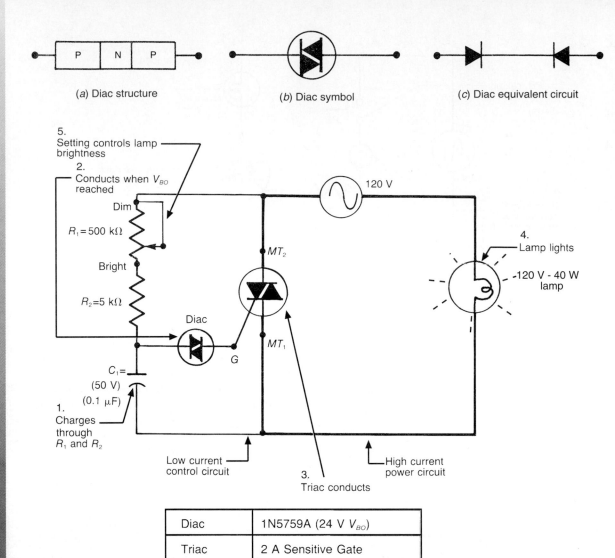

(a) Diac structure

(b) Diac symbol

(c) Diac equivalent circuit

5. Setting controls lamp brightness

2. Conducts when V_{BO} reached

Dim

$R_1 = 500 \text{ k}\Omega$

Bright

$R_2 = 5 \text{ k}\Omega$

Diac

$C_1 =$ (50 V) (0.1 μF)

1. Charges through R_1 and R_2

MT_2

MT_1

G

120 V

4. Lamp lights

120 V - 40 W lamp

Low current control circuit

3. Triac conducts

High current power circuit

Diac	1N5759A (24 V V_{BO})
Triac	2 A Sensitive Gate

(d) Triac/diac lamp dimmer circuit

Fig. 5-55 Triac-diac lamp dimmer.

43. The triac operates as two _____ in the same case.
44. True or false? A triac is capable of conducting with either polarity of terminal or gate voltage.
45. True or false? The output of the triac is alternating current.
46. True or false? The diac is a two-terminal device.
47. True or false? Current flows through a diac in one direction only.

5-6 INTEGRATED CIRCUITS

▶ **YOU MAY RECALL** that an *integrated circuit,* commonly referred to as an *IC,* is a *complete* electronic circuit that is contained within a single chip of silicon. Often no larger than a transistor, an IC can contain as few as several to as many as hundreds of thousands of transistors, diodes, resistors, and capacitors, along with electric conductors processed and contained entirely within a single chip of silicon (Fig. 5-56). Integrated circuits are often called *chips,* which are actually a component part of the IC.

Integrated circuits are made with the same basic materials and techniques used to make transistors. They can be classified according to their application as either *digital ICs* or *analog (linear) ICs.* A digital IC contains ON/OFF *switch-type circuitry.* An analog (linear) IC

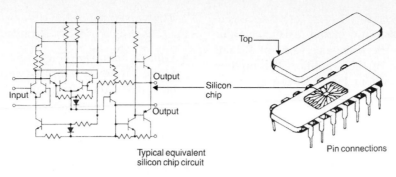

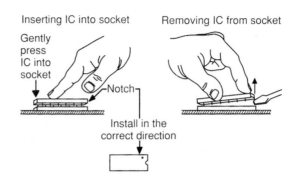

(a) IC chip construction

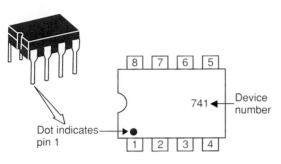

(b) Typical dual-in-line IC package

(c) Removing and inserting IC chips

Fig. 5-56   Integrated circuit (IC).

contains *amplifying-type circuitry*. The analog and digital processes can be seen in a simple comparison between a light dimmer and light switch. A light dimmer involves an analog process, which varies the intensity of light from OFF to fully ON. The operation of a standard light switch, on the other hand, involves a digital process; the switch can be operated only to turn the light OFF or ON.

OPERATIONAL AMPLIFIER ICs

Operational amplifier (op-amp) ICs are among the most widely used *analog* ICs. They take the place of amplifiers that formerly required many components. An op-amp is basically a high-gain amplifier that can be used to amplify weak ac or dc signals.

The schematic symbol for an op-amp is a triangle, shown in Fig. 5-57 on pages 154 and 155. This triangle symbolizes direction and points from input to output. Op-amps have five basic terminals; two for supply voltage, two for input signals, and one for the output signal.

Power supply terminals are labeled $-V$ and $-V$. Operational amplifiers may be operated from a pair of supplies (positive and negative with respect to ground) or from a single supply. The *two* input terminals on the op-amp are

labeled $(-)$ *inverting* and $(-)$ *noninverting input*. The polarity of a voltage applied to the inverting input is reversed at the output. The polarity of a voltage applied to the noninverting input is the same at the output. They are also called *differential input* terminals because the effective input voltage to the op-amp depends on the difference in voltage between them. There is only *one* output terminal in an op-amp. The output is obtained between this output terminal and the common ground. There is a limit to the power that is available from the output. This rating varies with the type of op-amp used.

The op-amp is basically an amplifier with a very high amplification level, or gain (a value of 500,000 or more). Even a very small change in voltage on either input will result in a very large change in output voltage. Because the circuit is so sensitive, the op-amp is quite unstable. As a result, the gain is usually reduced to a more practical level. This reduction is accomplished by *feeding back some of the output signal to the inverting $(-)$ input*. The *negative feedback* decreases the voltage gain. The schematic diagram of a 741 op-amp *inverting voltage amplifier* circuit is shown in Fig. 5-58 on page 155. Two resistors set the voltage gain of the amplifier. Resistor R_{in} is called the *input resistor.* Resistor R_{fb} is called the *feedback resistor.* The ratio of the resistance value of R_{fb} to that of the resistance value of R_{in} sets the voltage gain of the amplifier. The op-amp gain for the circuit in Fig. 5-58 can be determined as:

$$\text{Op-amp gain} = \frac{R_{fb}}{R_{in}}$$

$$\text{Op-amp gain} = \frac{500 \text{ k}\Omega}{50 \text{ k}\Omega} = 10$$

The versatility of the op-amp can be seen in the following variety of applications:

- *Voltage follower* (Fig. 5-59 on page 156). Mainly used as a *buffer amplifier* that isolates one circuit from another. It has a very high input impedance, which ensures that the input circuit is not loaded down. It does not change the input in either phase or amplitude.
- *Current-to-voltage converter* (Fig. 5-60 on page 156).

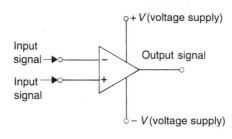

(a) Symbol

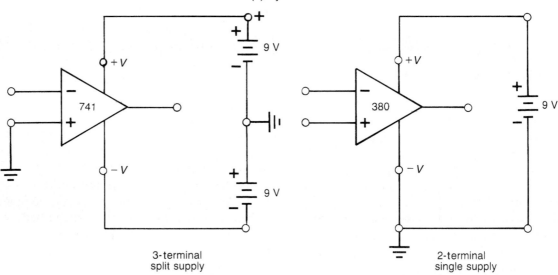

(b) Power supply connections

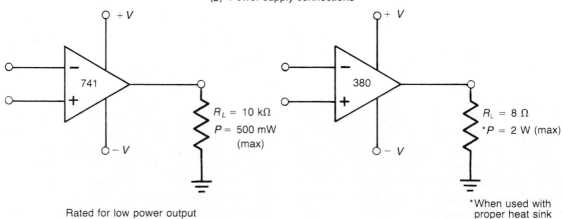

(c) Output connection

Fig. 5-57   Operational amplifier (op-amp).

Noninverting
amplifier

Summing amplifier

Difference
amplifier

Voltage-
comparator circuit

Integration

Active filter

Inverting mode

Noninverting mode

(d) Input connections

**Fig. 5-57 Operational amplifier (op-amp)
(continued).**

- *Noninverting amplifier* (Fig. 5-61 on page
  156). The input signal does not become
  inverted when passing through the amplifier.
- *Summing amplifier* (Fig. 5-62 on page 156).
  Used to *add* ac or dc signals.
- *Difference amplifier* (Fig. 5-63 on page 156).

An op-amp used in a subtracting mode,
which provides an output that is equal to the
difference between two inputs.

- *Voltage-comparator circuit* (Fig. 5-64 on page
  156). Used to compare two different voltage
  levels and to determine which is larger. The
  basic op-amp operated *without* a feedback
  circuit is ideal for this. A difference in one
  input voltage with respect to the other will
  produce a large change in the output voltage. The voltage at input A is compared with
  the voltage at input B. Whenever input A
  voltage is *greater* than input B voltage, the
  LED switches ON to indicate this condition.
- *Integration.* The process of summing the
  input signal during some fixed period of
  time. The integrator in Fig. 5-65 on page 156
  takes a series of pulses coming from the
  pickup coil, and produces an output voltage
  that is directly proportional to the number of
  pulses occurring during that time.
- *Active filter.* Allows current at certain frequencies to pass through the op-amp, while
  preventing the passage of others. Filters that
  use only resistors, capacitors, and inductors

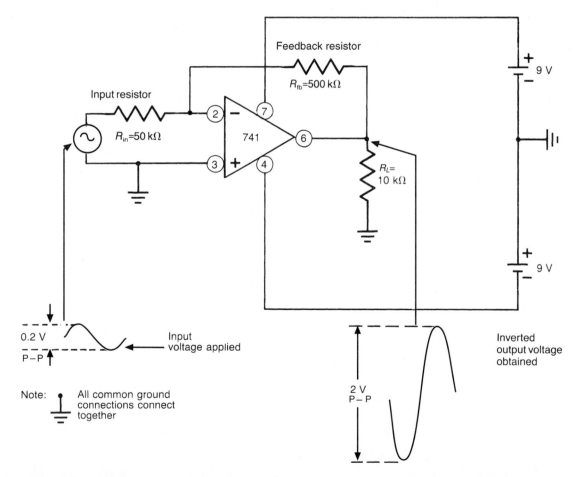

Fig. 5-58   741 op-amp inverting voltage amplifier circuit.

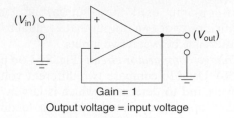

Gain = 1
Output voltage = input voltage

Fig. 5-59   Voltage follower.

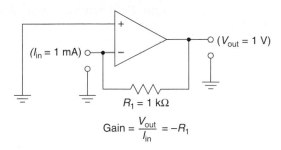

$$\text{Gain} = \frac{V_{out}}{I_{in}} = -R_1$$

Fig. 5-60   Current-to-voltage converter.

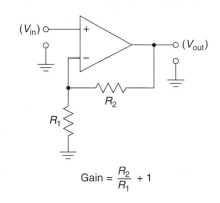

$$\text{Gain} = \frac{R_2}{R_1} + 1$$

Fig. 5-61   Noninverting amplifier.

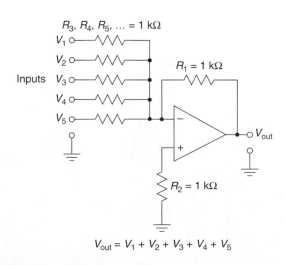

$$V_{out} = V_1 + V_2 + V_3 + V_4 + V_5$$

Fig. 5-62   Summing amplifier.

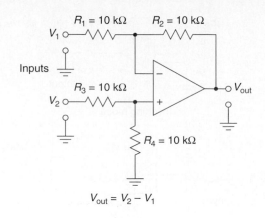

$$V_{out} = V_2 - V_1$$

Fig. 5-63   Difference amplifier.

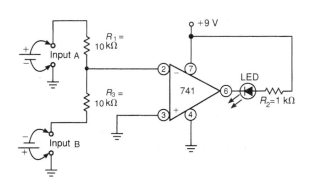

Fig. 5-64   Comparator.

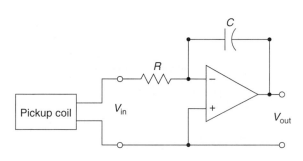

Fig. 5-65   Integration.

are called *passive filters*. With the proper connection of resistors and capacitors, the active filter can block or pass a specific range of frequencies. The circuit shown in Fig. 5-66 is that of a *low-pass filter*.

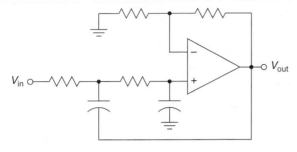

Fig. 5-66   Active filter.

■ *Sine wave oscillator.* An op-amp may be constructed to create an oscillator, as shown in the *Wien-Bridge* (sine wave) oscillator circuit in Fig. 5-67. It operates on the balanced-bridge principle.

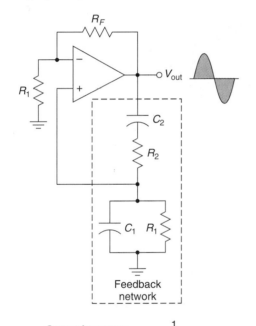

$$\text{Output frequency} = \frac{1}{2\pi\sqrt{R_1 C_1 R_2 C_2}}$$

Fig. 5-67   Sine wave oscillator.

■ *Schmitt trigger.* The op-amp circuit shown in Fig. 5-68 is a waveshape-squaring circuit used to minimize false triggering associated with noisy data signals. The difference between the upper and lower threshold points of the output is called *hysteresis*.

THE 555 TIMER IC

The *555 timer IC* is one of the most popular and versatile ICs ever produced. It is used in

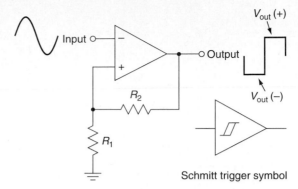

Schmitt trigger symbol

Fig. 5-68   Schmitt trigger.

circuits requiring a time-delay function. It is also used as an oscillator to provide pulses needed to operate digital circuits. Fig. 5-69 on page 158 shows the pin connections and functional block diagram for a 555 timer. The output of a 555 timer IC is a *digital signal*.

Figure 5-70 on page 158 shows the 555 timer connected as a *one-shot timer* (also called a *monostable multivibrator*). This simple circuit consists of only two timing components: R and C. During standby, the trigger input terminal is held higher than one-third V_{CC}, and the output is low. When a trigger pulse appears with a level less than one-third V_{CC}, the timer is triggered, and the timing cycle starts. The output rises to a high level near V_{CC}, and at the same time, C begins to charge toward V_{CC}. When the C voltage reaches two-thirds V_{CC}, the timing period ends with the output falling to zero—and the circuit is ready for another input trigger. The timing period is determined by values of R and C according to the formula:

$$\text{Time (seconds)} = 1.1 \times R \text{ (ohms)}$$
$$\times C \text{ (farads)}$$
$$= 1.1 \, (1 \text{ M}\Omega) \, (10 \, \mu\text{F})$$
$$\text{Time} = 11 \text{ s}$$

The schematic for a typical *555 clock pulse-generator circuit* is shown in Fig. 5-71 on page 159. This circuit operates as an *oscillator* (also called an *astable* or a *free-running multivibrator*). It produces a continuous string of pulses. No input signal is required to operate the circuit. The output signal is a series of rectangular pulses with an approximate frequency range from 1 to 100 Hz. The frequency of the output is changed by varying the setting of R_3. The LED and series resistor R_5 connect across the output to provide a visual indication of the pulse rate.

The output of this clock is not symmetrical; the high (ON) output lasts longer than the low

Passive filters

Low-pass filter

Sine wave oscillator

Schmitt trigger

Hysteresis

555 timer IC

Digital signal

One-shot timer (monostable multivibrator)

Clock pulse-generator circuit

Astable (free-running) multivibrator

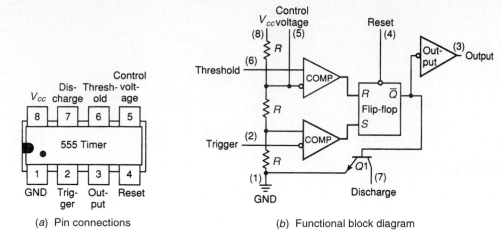

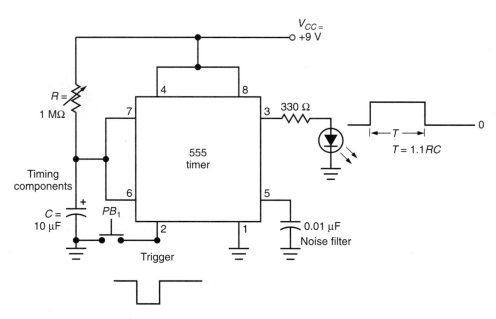

Fig. 5-69   555 timer.

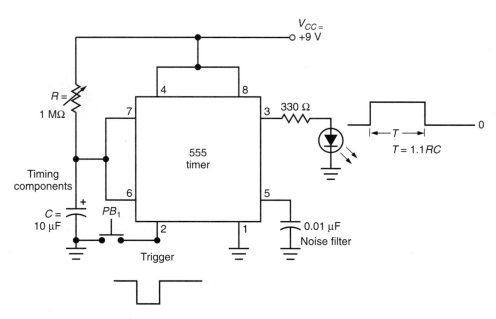

Fig. 5-70   555 one-shot timer circuit.

(OFF) output state. The *duty cycle* of a pulse waveform is the percentage of time the output is high. It can be found by dividing the total period of the waveform into the time the output is high. For Fig. 5-71, the frequency of the output pulses would be determined using the formula:

$$f = \frac{1.44}{[R_1 - 2(R_2 - R_3)]C_1}$$

where f = frequency in hertz
  R = resistance in ohms
  C = capacitance in farads

Depending on the resistances of R_1, R_2, and R_3, the duty cycle is between 50 and 100 percent.

Applying a voltage to the control pin of the 555 timer allows it to be used as a *voltage-controlled oscillator* (*VCO*) or as a *pulse-width modulator* (*PWM*). The external voltage overrides the internal voltage. As a result, you can change the output frequency of the circuit by varying the control voltage. Figure 5-72 shows an experimental circuit for a VCO used to control the speed of a small permanent-magnet dc motor. The circuit, which works by changing the amount of voltage that is applied to the armature of the motor, provides a pulsating dc voltage to the armature. The voltage applied across the armature is the *average* value determined by the length of time that transistor Q_2 is turned ON as compared to the length of time it is turned OFF (duty cycle). Potentiometer R_1

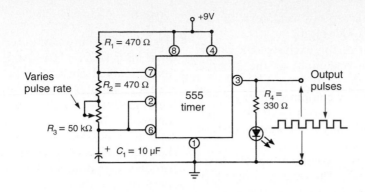

(a) Circuit schematic

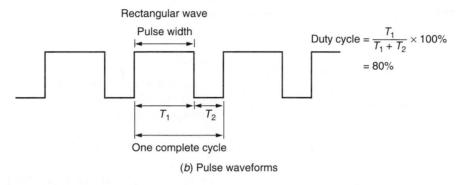

(b) Pulse waveforms

Fig. 5-71   555 clock pulse generator.

controls the length of time the output of the timer will be turned ON, which in turn controls the speed of the motor. If the wiper of R_1 is adjusted close to V_{CC} (higher positive voltage), the output will be turned ON for a longer period of time than it will be turned OFF. This type of control is known as *pulse-width modulation*.

LOGIC GATES

Digital circuits use *logic gates* to perform most of their functions. The various types of logic gates include AND, OR, NOT, XOR, NOR, and NAND. Logic functions include decisions, comparisons, arithmetic, and counting. A logic gate makes a *decision* by producing an output that is HIGH or LOW, depending on its input conditions.

- AND: All inputs for this gate must be in the logic 1, or HIGH, state in order to obtain an output of logic 1, or HIGH.
- OR: The output of this gate is logic 0, or LOW, unless any or all of its inputs are logic 1, or HIGH.

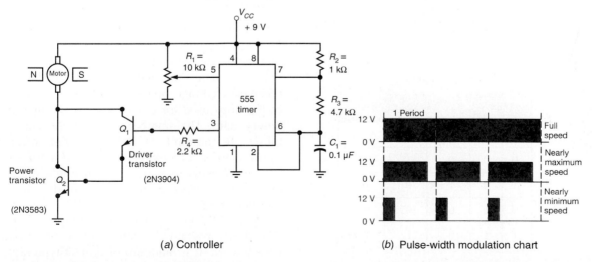

(a) Controller

(b) Pulse-width modulation chart

Fig. 5-72   555 dc motor speed control.

- NOT: This gate inverts or complements the logic state of its *single* input.
- XOR: The output of the exclusive OR gate is HIGH only when one input or the other is HIGH, but *not both*.
- NAND: This gate is a combination of an AND gate and an inverter (NOT) gate. The output is logic 0, or LOW, only if all inputs are 1, or HIGH.
- NOR: This gate is a combination of an OR gate and a NOT gate. The output is logic 1, or HIGH, only if all inputs are 0, or LOW.

COMPARATOR

The basic function of a *comparator* is to compare the relative magnitude of two quantities. Figure 5-73 shows a one-bit comparator that is constructed using a *combinational* network of gates. It has three outputs:

$$A = B \quad (A \text{ equals } B)$$
$$A > B \quad (A \text{ greater than } B)$$
$$A < B \quad (A \text{ less than } B)$$

Only one of the three outputs can be HIGH. The truth table in Fig. 5-73 shows all the combinations of inputs and outputs possible. The HIGH output tells you whether the digital bits are equal or whether one is greater than the other.

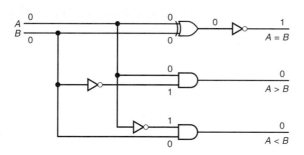

Truth Table

Inputs		Outputs		
A	B	$A > B$	$A = B$	$A < B$
0	0	0	1	0
1	0	1	0	0
0	1	0	0	1
1	1	0	1	0

Fig. 5-73   One-bit comparator.

ADDER

The heart of any computer is the *arithmetic-logic unit (ALU)*. The ALU performs all the mathematical operations that take place within the computer. Mathematical operations include addition, subtraction, multiplication, and division. The only required mathematical operation is addition. All other operations can be accomplished by variations of the addition process. Combination logic can be designed to perform all the arithmetic functions. Figure 5-74 shows the 7483 full four-bit parallel *adder* that will add two 4-bit binary numbers and output the sum.

DECODER

A *decoder* is a combinational logic circuit that recognizes the presence of a specific binary number. In digital electronics, it is often necessary to convert a binary number into some other format. One common decoder application is the conversion of binary numbers into the format required to activate the appropriate segments in a *seven-segment decimal display*. Figure 5-75 on page 162 shows a decoder circuit used to convert a 4-bit binary code into the appropriate decimal digit. Binary-coded numbers are fed into the circuit by pressing the appropriate normally closed (NC) pushbutton. The decoder interprets the incoming binary number and produces an output that drives a seven-segment common-anode LED display. To obtain any digit from zero to nine on the LED display, input the appropriate binary number code by means of the pushbuttons.

REGISTER

Unlike combinational logic circuits, *sequential logic circuits* have *memory*. The main element in a sequential logic circuit is a *flip-flop*. A flip-flop is a circuit that changes state each time it receives a voltage pulse. A *register* is a group of flip-flops used for the temporary storage of binary data. The number of flip-flops determines the amount of data per unit, often referred to as a *computer word, data word,* or just *word*. Registers are used in applications involving the *storage* and *transfer* of data in a digital system.

Figure 5-76 on page 163 shows a typical *storage register,* which will temporarily store a 4-bit binary word. It is made up of four D-type flip-flops. If the word 0111 is presented as input

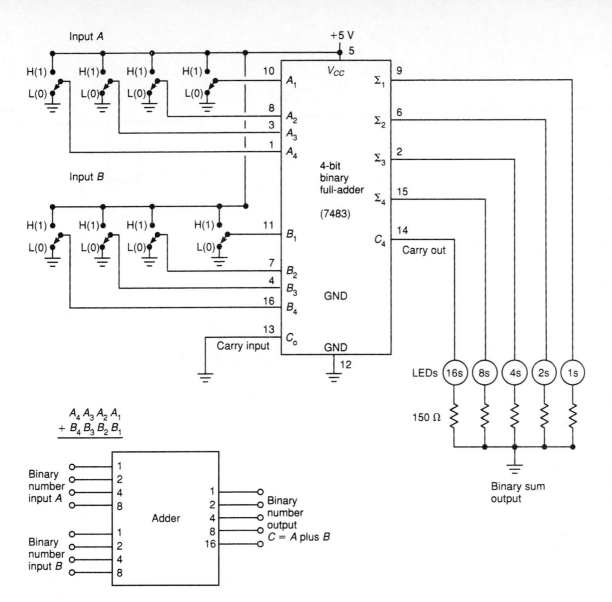

Fig. 5-74 Four-bit adder circuit.

data, it will be transferred through to the Q outputs when the clock pulse appears. If the clock pulse remains at 0 and the binary input data changes, it will have no effect on the output. The output word is latched, or stored in the register, until new data are clocked in.

COUNTER

A *counter* is a sequential logic circuit that is used to count the number of binary pulses applied to it. Binary counters are usually constructed using flip-flops. Figure 5-77 on page 164 shows the wiring diagram for a typical decade counter. The input pulses to be counted are produced by manually operating the count button. Mechanical switches must be debounced prior to connecting them into logic

circuits. If this conditioning of the switch input is not provided, the counter may receive several false make-and-break signals each time the button is closed. A special latch, made up of two inverters from a 7404 IC, supplies the debouncing. Pulses are fed from the debouncing circuit into the 7490 decade counter via pin 14. The pulses cause the four flip-flops inside the 7490 to begin a counting sequence that recycles after ten counts. The binary count in the decade counter is indicated by the LED display. The decimal number value of each lighted LED is indicated in the block diagram.

CONVERTERS

Analog signals must be coded into digital signals before they can be processed by digital

**Analog-to-digital
(A/D) converter**

**Digital-to-analog
(D/A) converter**

Multiplexer (MUX)

**Demultiplexer
(DMUX)**

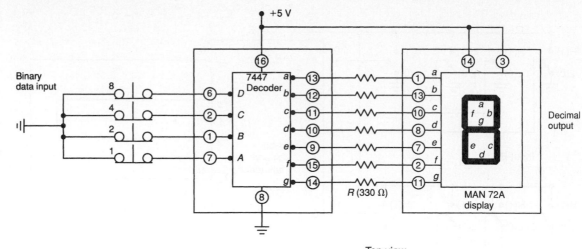

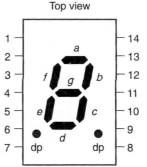

Top view

PIN	FUNCTION
1	CATHODE *a*
2	CATHODE *f*
3	ANODE
4	NO PIN
5	NO PIN
6	CATHODE *dp*
7	CATHODE *e*
8	CATHODE *d*
9	NO CONNECTION
10	CATHODE *c*
11	CATHODE *g*
12	NO PIN
13	CATHODE *b*
14	ANODE

Fig. 5-75   Binary-to-seven segment decoder.

circuits. The circuit that does this is called an *analog-to-digital (A/D) converter.* Its complement, the *digital-to-analog (D/A) converter,* is used to convert the digital code back to an analog signal.

Practically every measurable quantity—including temperature, pressure, speed, and time—is analog in nature. Analog-to-digital conversion is accomplished by assigning binary numerical values to represent the different voltage levels of the analog signal. One method of A/D conversion is shown in Fig. 5-78 on page 164, where (*a*) the analog signal from a sensor is *sampled* (or measured) at regular intervals. Then (*b*) the equivalent binary number for the voltage level is output as a digital signal. The *sampling rate* determines the accuracy with which the sequence of digital codes represents the analog input of the A/D converter.

Digital-to-analog converters are used in digital systems to convert digital signals into proportional voltages or currents to control analog devices. Typical analog output devices include small motors, valves, and analog meters. The input to a D/A converter is usually a parallel binary number. The output is a dc analog voltage that is proportional to the value of the binary input. Figure 5-79 on page 165 is a simplified illustration of a D/A converter. The D/A converter outputs a voltage level corresponding to the input binary code. During the sampling period, the voltage level remains constant; therefore, the output has stepped voltage levels. This causes the analog output voltage to have a staircase appearance.

MULTIPLEXER AND DEMULTIPLEXER

A *multiplexer (MUX)* (also called a *data selector*) is an electronic switch that allows digital information from several sources to be routed onto a single line for transmission over that line to a common destination. Thus, the basic multiplexer has several input lines but only *one* output. It also has select-control inputs that permit digital data on any of the inputs to be switched to the output line.

A *demultiplexer (DEMUX)* (also called *decoder*) is the opposite of a multiplexer in that it has a single input and multiple outputs. Like the multiplexer, the demultiplexer is also a data-routing circuit. Control inputs are required to determine to which output the input will be directed. Demultiplexers are used

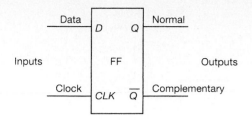

Logic symbol

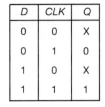

D	CLK	Q
0	0	X
0	1	0
1	0	X
1	1	1

X = previous, or last, state

Truth table

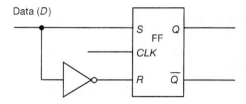

Wiring a D flip-flop using a clocked RS flip-flop and inverter

(a) D-type flip-flop

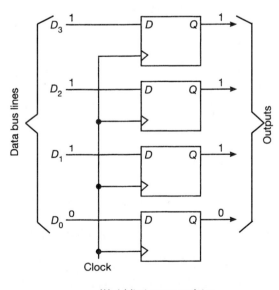

(b) 4-bit storage register

Fig. 5-76 Register.

with multiplexers to convert multiplexed data back to its original form.

Figure 5-80 on page 165 shows an experimental serial data transmission system that uses a 74150 multiplexer and a 74154 demultiplexer, along with a counter. The multiplexer first connects input 0 to the serial data transmission line. The bit is then transmitted to the demultiplexer, which places it at output 0. The multiplexer and demultiplexer proceed to transfer the data at input 1 to output 1, and so on. The bits are transmitted 1 bit at a time. The counter causes the inputs and outputs to be scanned sequentially. If the clock is pulsed very fast, the data can be transmitted quickly.

WORKING WITH DIGITAL ICs

Integrated circuits fall into two broad categories, based on the major device used in the circuit fabrication. The *bipolar* types (also called *TTL* or *transistor-transistor logic*) use the *bipolar-junction transistor* (NPN, PNP) as their principal circuit element. The *metal-oxide semiconductor* (*MOS*) types use *MOS field-effect transistors* (*MOSFETs*) as their principal circuit element. The MOS integrated circuit packages are extremely vulnerable to damage from *electrostatic discharge* (*ESD*).

The *digital logic probe* (Fig. 5-81 on page 166) is an inexpensive tool that can be used in signal-tracing digital signal pulses. After powering the logic probe from the circuit power supply, the needlelike probe is touched to a test point in the circuit. The logic probe will light either the HIGH or the LOW indicator. If neither indicator lights, the voltage is probably somewhere between the HIGH and LOW values.

With digital IC circuit diagrams, no special marking is used when a signal normally goes positive for an *active* or *required* input resulting in a desired output from the IC. If that pin must be negative, however, the input pin is designated with a small circle. In some circuits, letter designations for pins are used. The letters have a bar over them when the pin must be negative (Fig. 5-82 on page 166).

Self-Test

Answer the following questions.

48. Integrated circuits can be classified according to their application as _____ ICs or _____ ICs.

49. An op-amp is basically a high-gain _____.

50. The two input terminals on an op-amp are labeled _____ and _____ input.

Bipolar ICs

Transistor-transistor (TTL) logic

Bipolar-junction transistor

Metal-oxide semiconductor (MOS)

MOS field-effect transistors (MOSFETs)

Electrostatic discharge (ESD)

Digital logic probe

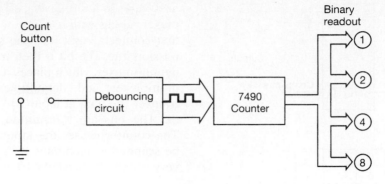

(a) Block diagram

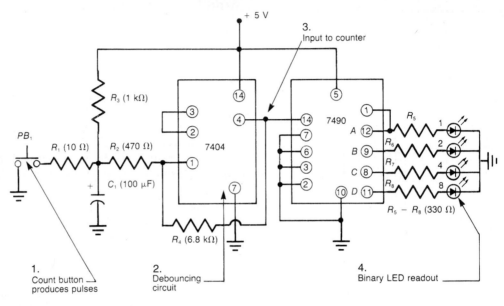

(b) Wiring diagram

Fig. 5-77   Decade counter.

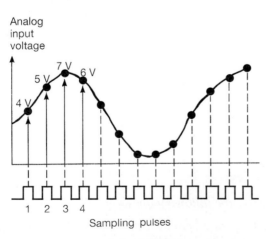

(a) Analog signal is sampled at regular intervals

Sample	Analog input voltage	Binary output
1	4 V	0100
2	5 V	0101
3	7 V	0111
4	6 V	0110

(b) The equivalent binary number is output as a digital signal

Fig. 5-78   Analog-to-digital (A/D) converter.

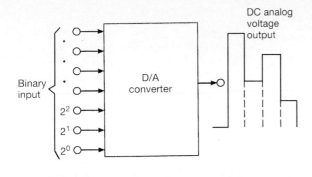

Sample	Binary input	DC analog voltage output
1	0000	0.0 V
2	0001	0.1 V
3	0010	0.2 V
4	0011	0.3 V
5	0100	0.4 V
6	0101	0.5 V

Fig. 5-79   Digital-to-analog (D/A) conversion.

51. Refer to Fig. 5-58. Assume that the value of the feedback resistor is changed to 250 kΩ. What would be the gain of the amplifier?

52. An op-amp voltage-comparator circuit is basically an op-amp operated without a(n) _____ circuit.

53. Integration is the process of _____ the input signal during some fixed period of time.

54. When an op-amp is used as an active filter, it allows current at certain _____ to pass through it.

55. An op-amp Schmitt trigger is used to minimize _____ triggering.

56. True or false? The output of a 555 timer IC is an analog signal.

57. Refer to Fig. 5-70. What resistance value of R would result in a timing period of 2 seconds?

58. The duty cycle of a pulse waveform is the percentage of time the output is _____.

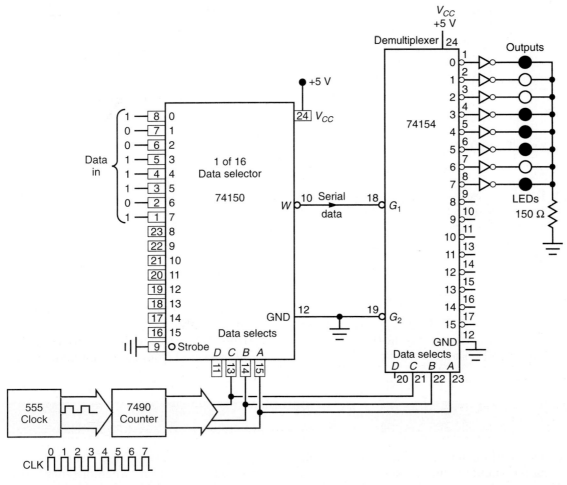

Fig. 5-80   Multiplexer and demultiplexer serial transmission system.

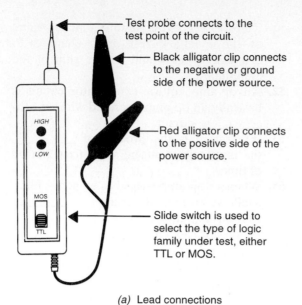

Test probe connects to the test point of the circuit.

Black alligator clip connects to the negative or ground side of the power source.

Red alligator clip connects to the positive side of the power source.

Slide switch is used to select the type of logic family under test, either TTL or MOS.

HIGH

LOW

MOS

TTL

(a) Lead connections

Logic probe

OH
●L

1
0 ————▷o— 0

Logic probe

●H
OL

0
0 ————▷o— 1

(b) Using a logic probe to test a gate

Fig. 5-81 Digital logic probe.

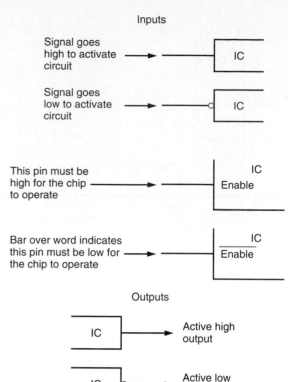

Inputs

Signal goes high to activate circuit → — IC

Signal goes low to activate circuit → —o IC

This pin must be high for the chip to operate → — IC Enable

Bar over word indicates this pin must be low for the chip to operate → — IC $\overline{\text{Enable}}$

Outputs

IC — → Active high output

IC o— → Active low output

Fig. 5-82 Digital signal coding.

59. True or false? The output of a logic OR gate is HIGH or 1 if any or all of its inputs are logic 1, or HIGH.

60. True or false? An inverter gate has only one input.

61. Refer to Fig. 5-73. List the number and type of gates used to construct this comparator circuit.

62. Refer to Fig. 5-74. Assume all input A and B switches are set to HIGH or 1. What binary LEDs should be ON?

63. Unlike combinational logic circuits, sequential logic circuits have _____.

64. A(n) _____ is a group of flip-flops used for the temporary storage of binary data.

65. Binary counters are usually constructed using _____.

66. Analog-to-digital conversion is accomplished by assigning _____ numerical values to represent the different voltage levels of the analog signal.

67. The analog voltage output of a D/A converter has a(n) _____ appearance.

68. True or false? The basic multiplexer has a single input and multiple outputs.

69. Transistor-transistor logic (TTL) ICs use _____ transistors as their principal circuit element.

70. True or false? Digital logic probes are usually powered from the circuit under test.

SUMMARY

1. The multimeter is a combination voltmeter, ammeter, and ohmmeter in a single instrument.

2. Multimeter front-panel controls may include a function switch, a range switch, and jacks.

3. Important multimeter specifications and features may include input impedance, accuracy and resolution, battery life, protection, combination digital and analog display, auto ranging, auto polarity, hold feature, response time, diode test, averaging or true rms, and multifunction meters.

4. The oscilloscope is used to measure and display voltages, as well as to provide information about the shape, time period, and frequency of a voltage.

5. Oscilloscope front-panel controls may include the following: INTENSITY, FOCUS, Y SHIFT, X SHIFT, VOLTS/DIV, TIME/DIV, AC/DC, and TRIGGER CONTROLS.

6. A dual-trace oscilloscope can display two different traces at the same time.

7. Digital-storage oscilloscopes have additional measurement features, which may include SAVE/RECALL, AUTOSET, STEP SEQUENCE, CURSORS, COUNTER/TIMER, and DMM.

8. A bridge circuit consists of four sections connected to form two parallel circuits.

9. The balanced-bridge or null detector method of making measurements is extremely accurate.

10. Rectifier diodes are used in the process of changing alternating current to direct current.

11. Full-wave rectifiers produce less pulsating and more powerful direct current than do half-wave rectifiers.

12. Three-phase rectifiers are used for heavier load industrial applications.

13. The ripple produced by changing ac voltage into dc voltage can be made much smoother by the use of filters.

14. Large power semiconductors must be heat sinked in order to operate at their listed power ratings.

15. Clamping or despiking diodes are used in inductive circuits to limit the voltage created by the collapsing magnetic field.

16. Zener diodes are used in voltage-regulator circuits.

17. The zener diode is designed to conduct in the reverse-bias direction once its rated voltage value is reached.

18. Light-emitting diodes (LEDs) are used as signaling lamps and in displays.

19. The LED emits light when current flows through it.

20. Laser diodes are specially made LEDs that are capable of operating as lasers.

21. Photodiodes are used for light detection.

22. The photodiode operates in the reverse-bias mode. The leakage current increases as the light level increases.

23. In a bipolar-junction transistor (BJT), a small base current is used to control a large collector current.

24. In a field-effect transistor (FET), no input current is used. Instead, output current is controlled by a varying electric field.

25. Transistors can function as amplifiers of electric signals, variable resistors, or as switches.

26. The FET transistor can be divided into two main groups: the junction field-effect transistor (JFET) and the metal-oxide semiconductor field-effect transistor (MOSFET).

27. Unlike the gate of a JFET, the gate of a MOSFET has no electric contact with the source and drain.

28. All JFETs are normally ON devices, whereas MOSFETs can be normally ON (depletion type) or normally OFF (enhancement type).

29. Unijunction transistors (UJTs) cannot amplify; they are used strictly as switches that trigger circuit conduction.

30. A Darlington transistor is two bipolar transistors connected so that a smaller transistor provides the base current for a larger transistor.

31. An insulated-gate bipolar transistor (IGBT) is a MOSFET and BJT connected so that the MOSFET drives the BJT.

32. A phototransistor is used for highly sensitive light detection and is basically a bipolar transistor with a light-sensitive collector-base junction.

33. Transistor amplifiers operate in the linear region between cutoff and saturation.

34. Transistor oscillators are used to convert pure direct current to alternating current or to pulsating (varying) direct current.

35. Power inverters are oscillators that are used to convert dc power to ac power.

36. The operation of a thyristor is similar to that of a switch having only an ON or OFF state. It cannot amplify.

37. A silicon-controlled rectifier (SCR) electrically resembles a diode except that it requires a momentary positive voltage applied to the gate in order to switch it ON.

38. Once the SCR is switched ON, it stays ON, even after the gate voltage is removed.

39. Phase shifting is required in order for the SCR to provide a variable output.

40. A triac operates as two SCRs in the same case.
41. A triac, unlike an SCR, can conduct on both half-cycles, making it useful for controlling loads that operate on alternating current.
42. The diac is a two-terminal transistorlike device used primarily to control the triggering of SCRs and triacs.
43. Current flows through a diac (in either direction) when the voltage across it reaches its rated breakover value.
44. An integrated circuit (IC) is a complete electronic circuit that is contained within a single chip of silicon.
45. A digital IC contains ON/OFF switch-type circuitry.
46. An analog (linear) IC contains amplifying-type circuitry.
47. An op-amp is basically a high-gain amplifier that amplifies weak ac or dc signals.
48. Op-amp applications include inverting and noninverting amplifier, voltage follower, current-to-voltage converter, summing amplifier, difference amplifier, comparator, integrator, active filter, oscillator, and Schmitt trigger.
49. The 555 timer is a digital IC commonly used in circuits requiring a time-delay function. It is also used as an oscillator.
50. A logic gate makes a decision by producing an output that is high or low, depending on its input conditions.
51. Types of logic gates include: AND, OR, NOT, XOR, NOR, and NAND.
52. The basic function of a comparator is to compare the relative magnitude of two quantities.
53. In actuality, the only true mathematical operation that is performed by a computer is addition.
54. A decoder is a combinational logic circuit that recognizes the presence of a specific binary number.
55. Unlike combinational logic circuits, sequential logic circuits have memory.
56. The main element in a sequential logic circuit is a flip-flop.
57. A register is a group of flip-flops used for temporary storage of binary data.
58. A counter is a sequential logic circuit that is used to count the number of binary pulses applied to it.
59. Converters are used to change analog signals to digital codes or vice versa.
60. A multiplexer is an electronic switch that allows digital information from several sources to be routed onto a single line.

CHAPTER REVIEW QUESTIONS

Answer the following questions.

5-1. State the purpose of each of the following parts of a multimeter front-panel switching arrangement:
  a. Function switch **b.** Range switch **c.** Jacks
5-2. Give two reasons why digital meters are preferred over analog types.
5-3. Give a brief explanation of each of the following multimeter specifications or functions:
  a. Input impedance **e.** Hold
  b. Resolution **f.** Response time
  c. Auto ranging **g.** Diode test
  d. Auto polarity **h.** Averaging or true rms
5-4. State the general function of each of the following oscilloscope front-panel controls:
  a. INTENSITY **e.** VOLTS/DIV
  b. FOCUS **f.** TIME/DIV
  c. Y SHIFT **g.** AC/DC
  d. X SHIFT **h.** TRIGGER CONTROLS

5-5. What precaution must be observed when connecting the test probe of an oscilloscope to test a circuit that has one side grounded?

5-6. Describe the main feature of a digital storage oscilloscope.

5-7. Explain each of the following oscilloscope specifications:
    a. Bandwidth
    b. Rise time
    c. Sensitivity

5-8. Answer each of the following questions with reference to a Wheatstone bridge circuit:
    a. How many separate resistance sections are required for this circuit?
    b. When a balanced-bridge null condition is indicated by the galvanometer, what are the resistance and voltage relationships of the circuit?

5-9. **a.** What determines whether a diode is forward or reverse biased?
    b. Under what condition is a diode considered to be connected in forward bias?

5-10. What is the purpose of a rectifier diode?

5-11. Compare half-wave and full-wave rectifier circuits with regard to the value and amount of pulsation of the dc output voltage.

5-12. **a.** What are two advantages of a three-phase, half-wave rectifier over a single-phase, full-wave rectifier?
    b. How many diodes are required for a three-phase, full-wave bridge rectifier?
    c. From what type(s) of transformer system(s) can a three-phase bridge rectifier be operated?

5-13. What is the purpose of a clamping or despiking diode?

5-14. In what way is the operation of a zener diode different from that of a conventional diode?

5-15. State two ways in which high current power semiconductor packages differ from the low current types.

5-16. How is an infrared-emitting diode (IRED) different from a conventional light-emitting diode (LED)?

5-17. What type of light beam is emitted from a laser diode?

5-18. Under what conditions is a photodiode designed to conduct current?

5-19. What are the two main families of transistors?

5-20. Compare the control of output current of a bipolar-junction transistor (BJT) with that of a field-effect transistor (FET).

5-21. Why are FETs said to be unipolar?

5-22. What are the two main types of FETs?

5-23. Why are junction field-effect transistors (JFETs) classified as normally ON devices?

5-24. What is the basic difference in construction between a JFET and a metal-oxide semiconductor field-effect transistor (MOSFET)?

5-25. **a.** What is the basic function of a unijunction transistor (UJT)?
    b. To what junction of the UJT is the input signal applied?

5-26. What is the main advantage of using Darlington and IGBT transistors?

5-27. **a.** Briefly explain how a phototransistor operates to control current.
    b. Why is the phototransistor more sensitive than the photodiode to changes in light intensity?

5-28. Compare the operation of transistor switching and amplifying circuits.

5-29. When transistors are used like variable resistors, what advantages do they have over the rheostats that they replace?

5-30. For what are ac transistor amplifiers used?

5-31. **a.** What type of circuit is a power inverter?
    b. For what is it used?

5-32. State three common industrial applications for SCR and triac circuits.

5-33. Compare the operation of an SCR with that of a conventional diode.

5-34. Why does the problem of SCR turnoff not occur in ac circuits?

5-35. An SCR circuit, operated from an ac source, uses the pulses produced by a UJT oscillator to trigger the gate. Why are the pulses produced by the UJT considered to be "phase shifted"?

5-36. **a.** Why is the triac more efficient than an SCR when operated from an ac source?

b. The SCR has only one triggering mode. What is it?

c. State the four possible triggering modes of a triac.

5-37. Under what conditions will a diac conduct a current?

5-38. List four common electronic component parts that can be formed on an integrated circuit (IC) chip.

5-39. List the five basic terminals of an op-amp.

5-40. For what is an op-amp voltage follower circuit used?

5-41. For what is an op-amp summing amplifier used?

5-42. Compare the output pulses produced by a 555 timer IC when connected as a monostable and as an astable multivibrator.

5-43. Why is an AND gate called an *all-or-nothing gate?*

5-44. Why is an OR gate called an *any-or-all gate?*

5-45. Why is the NOT gate called an *inverter?*

5-46. What information about the logic gate is provided by its truth table?

5-47. Explain the function of the decoder in a binary-to-decimal decoder circuit.

5-48. In what types of applications are registers used?

5-49. What is a binary counter?

5-50. Why are A/D and D/A converters of great importance in digital systems?

5-51. Explain the basic function of a demultiplexer (DEMUX) circuit.

5-52. **a.** Explain how the two leads and probe of a typical digital logic probe are connected to the circuit to be tested.

b. What is the most important application of a digital logic probe?

CRITICAL THINKING QUESTIONS

5-1. Refer to Fig. 5-12. Assume that the bridge is balanced with light shining on the photoconductive cell. When light to the cell is blocked, what change in voltage (increase, decrease, or no change) appears across:

a. R_1 **b.** R_2 **c.** R_3 **d.** R_4

5-2. Refer to Fig. 5-16(a). Assume that diode D_4 has been incorrectly connected backward in the bridge configuration. Explain what effect this would have on the resultant dc output voltage and the current flow through the diodes.

5-3. Refer to Fig. 5-17. You suspect that one of the diodes has burned open. What tests can be made to verify this?

5-4. Refer to Fig. 5-27(a). It is necessary to increase the timing period to twice that of the original circuit.

a. What one component should be changed to accomplish this?

b. What should be the approximate value of the replacement component be?

5-5. Refer to Fig. 5-33(a). Assume that resistor R_1 has burned open. Explain how this would affect the operation of the circuit.

5-6. Refer to Fig. 5-39(b). Assume that the gate lead of the MOSFET has become grounded. Explain how this would affect the operation of the lamp.

5-7. Refer to Fig. 5-40. Assume that resistor R_2 has become short-circuited. How would this affect:

a. The dc collector current?   **b.** The ac output signal obtained?

5-8. Refer to Fig. 5-41. A second LED is to be wired into the circuit that will be ON when the other LED is OFF and vice versa. Explain how this second LED would be connected into the existing circuit.

5-9. Refer to Fig. 5-48. Which resistance setting of R_1 (minimum or maximum) will produce full output? Why?

5-10. Refer to Fig. 5-53. Assume that the fan motor will not operate with the thermostat switched ON. Assume also that one of the components is faulty. Outline the test you would make of each component to determine if it is at fault.

5-11. Refer to Fig. 5-61. If R_1 is 5 kΩ, what value of R_2 is required to produce a gain of 26?

5-12. Refer to Fig. 5-71(a). Calculate the frequency of the output pulses when R_3 is set to 10 kΩ.

5-13. Refer to Fig. 5-72(a). Assume that the motor operates at full speed regardless of the setting of R_1. Assume also that the 555 timer chip, the driver transistor, or the power transistor is at fault. Outline the test you would make of each of these components to determine which is at fault.

5-14. Develop a combination logic circuit using AND, OR, and NOT gates that will produce an output *only* when the following set of conditions has been met: input *A or B* is high *and* input *D or not* input *C* is high.

5-15. Refer to Fig. 5-75. Assume a logic probe is used to test the state of each pin connection on the 7447 decoder IC chip. With none of the pushbuttons pressed, indicate the state (HIGH or LOW) of each pin connection shown.

Answers to Self-Tests

1. voltmeter; ammeter; ohmmeter
2. false
3. auto ranging
4. true
5. voltages
6. true
7. TIME/DIV
8. comparing
9. true
10. bandwidth
11. four
12. 375 Ω
13. 1500 Ω, or 1.5 kΩ
14. The two voltages would be equal in value.
15. alternating current; direct current
16. false
17. false
18. true
19. false
20. filter
21. heat sinked
22. reverse
23. false
24. light

25. false
26. true
27. amplifiers; variable resistors; switches
28. true
29. base; gate
30. true
31. true
32. true
33. true
34. amplify
35. base current
36. MOSFET
37. cutoff; saturation
38. direct current; alternating current
39. false
40. positive; gate
41. regulate; rectify
42. gate; anode
43. SCRs
44. true
45. true
46. true
47. false
48. digital; analog

49. amplifier
50. inverting; noninverting
51. 5
52. feedback
53. summing
54. frequencies
55. false
56. false
57. 182 kΩ
58. HIGH
59. true
60. true
61. XOR gate (1); AND gate (2); inverter (3)
62. 2s, 4s, 8s, and 16s
63. memory
64. register
65. flip-flops
66. binary
67. staircase
68. false
69. bipolar
70. true

CHAPTER 6

Industrial Motors and Generators

∎

CHAPTER OBJECTIVES

This chapter will help you to:

1. *Describe* the construction, connection, and basic operating characteristics of ac generators.
2. *Describe* the construction, connection, and basic operating characteristics of dc generators.
3. *Discuss* the different types of dc motors and describe their operating characteristics.
4. *Explain* the principle of operation of various types of three-phase motors.
5. *Discuss* the specifications and the installation and maintenance procedures for basic motors.

In these days of rapid electronic advancement, we sometimes overlook the importance of the electric motor and generator as key system elements. Technicians who wish to be completely functional in the industrial environment must be familiar with the various motors and generators used in industry. This chapter presents a condensed version of the topic, with emphasis on the important operating principles of motors and generators.

∎

6-1 ALTERNATING CURRENT GENERATORS

Alternating current (ac) generators (also called *synchronous generators* or *alternators*) are the primary source of all the electric energy we consume. These machines are the largest energy converters in the world.

⫸ **YOU MAY RECALL** that a *generator* is a machine that uses magnetism to convert mechanical energy into electric energy. The *generator principle,* simply stated, is that a voltage is *induced* in a conductor whenever the conductor is moved through a magnetic field so as to cut lines of force. A generator is driven by some type of mechanical machine (steam or water turbine, gasoline engine, or electric motor). It requires mechanical energy for its operation.

The *left-hand rule for a generator* (Fig. 6-1) shows the relationship between the direction

the conductor is moving, the direction of the magnetic field, and the resultant direction of the induced current flow. When the thumb is pointed in the direction of the conductor's motion, and the index finger is pointed in the direction of the flux, the middle finger will point in the direction of the induced electron flow. The rule is also applicable when the magnet, instead of the conductor, is moved. In this case, however, the thumb must point in the direction of *relative* conductor motion.

The amount of voltage induced in a conductor as it moves through a magnetic field depends upon:

- The strength of the magnetic field. The stronger the field the more voltage induced.
- The speed at which the conductor cuts through the flux. Increasing the conductor speed increases the amount of voltage induced.
- The angle at which the conductor cuts the flux. Maximum voltage is induced when the

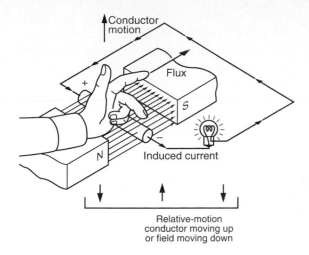

Fig. 6-1   Left-hand generator rule.

conductor cuts the flux at 90°, and less voltage is induced when the angle is less than 90°.

- The length of the conductor in the magnetic field. If the conductor is wound into a coil of several turns, its effective length increases, and so the induced voltage will increase.

Alternating current generators are built with either a stationary or a rotating dc magnetic field. The *stationary-field* type is usually of relatively small kilovolt-ampere capacity and low-voltage rating. It resembles a dc generator in appearance, except that it has slip rings instead of a commutator (Fig. 6-2). The salient poles create the dc field, which is cut by a revolving armature. The armature possesses windings whose terminals are connected to slip rings mounted on the shaft. A set of brushes, sliding on the slip rings, enables us to connect the armature to an external load. An ac generator cannot supply its own field current. The field excitation must be direct current and, therefore, must be supplied from an external source. The armature is driven by a source of mechanical power such as a diesel engine.

The *revolving-field* type of ac generator simplifies the problems of insulating generated voltages, which are commonly as high as 18,000 to 24,000 V. A revolving-field ac generator has a stationary armature called a *stator*. The three-phase stator winding is directly connected to the load without going through slip rings and brushes. This makes it easier to insulate the windings because they are not subjected to centrifugal forces. Different methods of field excitation have been developed and used. The revolving-field ac generator shown in Fig. 6-3 uses a *brushless exciter system* in which a small ac generator mounted on the same shaft as the main generator is used as an *exciter*. The ac exciter has a rotating armature. The output of the armature is rectified by solid-state diodes, which are also mounted on the main shaft. The rectified output of the ac exciter is fed directly by means of insulated connections along the shaft to the rotating synchronous generator field. The field of the ac exciter is stationary and is supplied from a separate dc source. The output of the ac exciter and, consequently, the generated voltage of

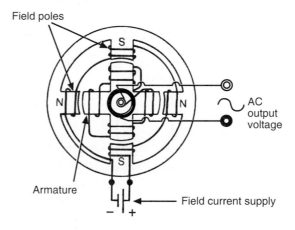

Fig. 6-2   Stationary-field single-phase ac generator.

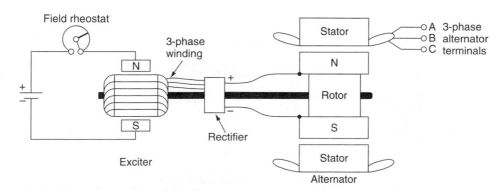

Fig. 6-3   Revolving-field three-phase ac generator.

From page 172:

Synchronous generators (alternators)

Generator principle

Left-hand generator rule

On this page:

Stationary-field generator

Revolving-field generator

Stator

Brushless exciter system

the synchronous generator is controlled by varying the field strength of the ac exciter through adjustment of the field rheostat.

As the armature of a simple, two-pole alternator is rotated through one complete revolution, a *sine wave voltage* is produced at its output terminals. This generator sine wave voltage varies in both voltage value and polarity (Fig. 6-4). The sine wave is the most basic and widely used ac waveform. The *frequency* of an ac sine wave (in hertz) is the number of cycles produced per second. The standard frequency of alternating current in the United States is 60 Hz. One *cycle* is one complete wave of alternating voltage or current. The *peak value* of a sine wave refers to the maxi-

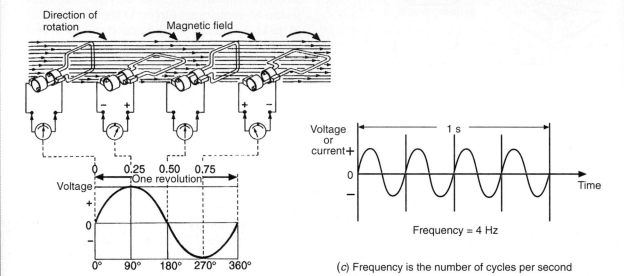

(a) Generation of a sine wave

(c) Frequency is the number of cycles per second

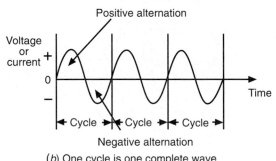

(b) One cycle is one complete wave

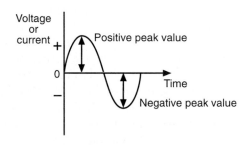

(d) The peak value is the maximum voltage or current value

$$V_{peak} = V_{rms} \times 1.414$$
$$V_{peak} = (120 \text{ V})(1.414)$$
$$V_{peak} = 170 \text{ V}$$

$$I_{rms} = I_{peak} \times 0.707$$
$$I_{rms} = (10 \text{ A})(0.707)$$
$$I_{rms} = 7.07 \text{ A}$$

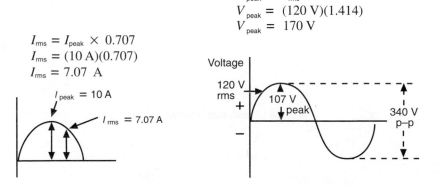

(e) The effective or rms value of a sine wave

Fig. 6-4 AC sine wave.

mum voltage or current value. The *effective* or *rms value* of a sine wave is the one most extensively used when referring to an ac voltage or current. It is common practice to assume that *all* ac voltage and current readings are effective values unless otherwise stated.

When a coil makes one revolution in a generator with two poles, one cycle of voltage is generated. When a coil makes one revolution in a generator with four poles, however, two cycles of voltage are generated. Therefore, a distinction is made between mechanical and electrical degrees (Fig. 6-5):

- *Mechanical degrees.* When a coil or armature conductor makes one complete revolution, it has passed through 360 mechanical degrees. In Fig. 6-5(*a*), the mechanical degrees equal the electrical degrees.
- *Electrical degrees.* When either an emf or an alternating current passes through one cycle, it has passed through 360 electrical time degrees. In Fig. 6-5(*b*), the number of electrical degrees in one complete revolution would be 720.

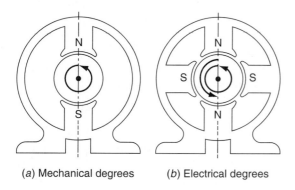

(*a*) Mechanical degrees   (*b*) Electrical degrees

Fig. 6-5 Comparison of mechanical and electrical degrees.

The speed and number of poles of an ac generator determine the frequency of the generated voltage. If a generator has two poles (north and south) and the coils rotate at a speed of one revolution per second, the frequency will be one cycle per second. A formula to determine the frequency of an ac generator is:

$$f = \frac{pn}{120}$$

where f = frequency of the induced voltage (Hz)
p = number of poles on the rotor
n = speed of the rotor (r/min)

The amount of any generated voltage depends on the rate at which magnetic flux lines are cut, or, in the case of the ac generator, the amount of voltage depends on the field strength and the speed of the rotor. Because most generators are operated at a constant speed, the amount of electromotive force (emf) generated becomes dependent on the field excitation.

As load is added to an ac generator operating at a constant speed and with a constant field excitation, the terminal voltage changes. The amount of change depends on the machine design and on the power factor of the load. The formula to determine the percentage voltage regulation is:

% regulation (at a stated power factor)

$$= \frac{\text{No-load voltage} - \text{Full-load voltage}}{\text{Full-load voltage}} \times 100$$

The terminal voltage of an ac generator varies with changes in load; therefore, there is usually some means for maintaining the constant voltage required for the operation of most electric equipment. A common way of doing this is to use a *voltage regulator* to control the amount of dc field excitation supplied to the generator. When the generator terminal voltage drops because of changes in load, the voltage regulator automatically increases the field excitation, which restores normal-rated voltage. Similarly, when the terminal voltage increases because of load changes, the regulator restores normal-rated voltage by decreasing the field excitation.

Figure 6-6 on page 176 shows the basic arrangement used to generate single-phase (1ϕ) and three-phase (3ϕ) ac voltages. Single-phase systems are used for small power demands. Almost all generation and distribution systems used by power utilities are three-phase.

The three sets of stator coils of the three-phase alternator may be connected in *wye* (also known as *star*) or *delta* form. Figure 6-7 on page 176 shows the connection of a wye alternator. The *three-phase, 4-wire wye system* is very common and is the standard system supplied by many power utilities to commercial and industrial customers. It is very versatile because power utilities can deliver both single- and three-phase power within the 4-wire system.

In a three-phase wye-connected alternator, the phase-to-neutral voltage is equal to the voltage generated in each coil. The phase-to-

Effective (rms) value of sine wave

Mechanical degrees

Electrical degrees

Voltage regulator

Wye (star) system

Delta system

Three-phase, 4-wire wye system

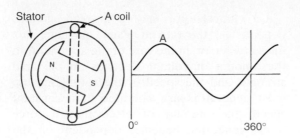

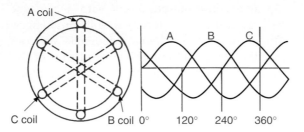

(a) Single-phase

(b) Three-phase

Fig. 6-6   Generation of single-phase and three-phase ac voltages.

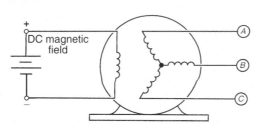

(a) Three-wire wye-connected alternator

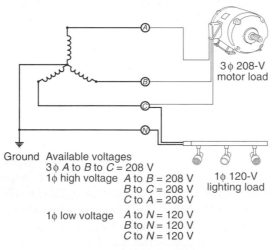

Ground   Available voltages
3φ *A* to *B* to *C* = 208 V
1φ high voltage   *A* to *B* = 208 V
                  B to *C* = 208 V
                  C to *A* = 208 V
1φ low voltage   *A* to *N* = 120 V
                 B to *N* = 120 V
                 C to *N* = 120 V

3φ 208-V motor load

1φ 120-V lighting load

(b) Four-wire wye-connected alternator

Fig. 6-7   Wye-alternator connection.

phase voltage is obtained by multiplying the phase-to-neutral voltage by 1.73 because the coils are set 120 electrical degrees apart. With a three-phase load connected to the alternator, the current in the line is the same as the current in the coil (phase) winding.

Figure 6-8 illustrates the connection of a delta alternator. In a three-phase delta system, the voltage measured across any two lines is equal to the voltage that is generated in the coil winding.

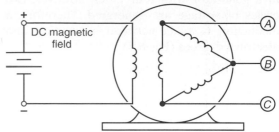

Fig. 6-8   Delta-alternator connection.

$$V_{\text{line-to-line}} = V_{\text{coil}} \text{ (phase)}$$

This is because the voltage is being measured directly across the coil winding. The coils, however, are 120 electrical degrees apart, as in the wye system; therefore, the line current will be the vector sum of the two coil currents. With a three-phase load connected to the alternator, the line current is then equal to 1.73 times the current in one of the coils.

$$I_{\text{line}} = 1.73 \times I_{\text{coil}} \text{ (phase)}$$

Mechanical energy must be applied to the generator armature shaft in order to spin it around and produce electricity. The source of the mechanical energy is called the *prime mover.* Engine-driven alternators are commonly used to provide emergency power in the event of a power failure. Prime movers for these on-site alternators may be powered by gasoline, diesel, or gaseous fuel engines.

It may be necessary at times to add another generator in parallel to increase the total power available. Before two three-phase ac generators may be paralleled, the following conditions must be fulfilled:

- Their phase sequence must be the same.
- Their terminal voltages must be equal.
- Their voltages must be in phase.
- Their frequencies must be equal.

When two generators are operating so that these requirements are satisfied, they are said to be in *synchronism.* The operation of getting the machines into synchronism is called *syn-*

chronizing. This is normally accomplished with an array of control and monitoring equipment.

Cogeneration (Fig. 6-9) is the simultaneous production of electric and thermal energy from a single fuel. The fact that both the electricity and heat produced by the system are used brings overall efficiency up to 80 percent or higher—this in contrast to the 30 percent or so efficiency of coal-fired generating stations, where much of the thermal energy goes up the stack. In applications where there is a constant need for electricity and hot water, the cost of the power produced can be much lower than that sold by the utility.

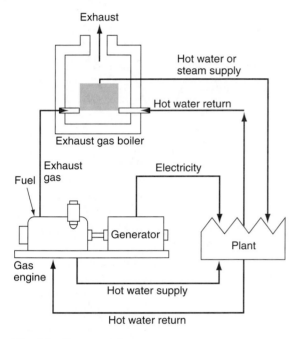

Fig. 6-9 Cogeneration.

Self-Test

Answer the following questions.

1. A voltage is _____ in a conductor whenever it cuts lines of force.
2. The _____ -field type of ac generator is normally used for generation of relatively high voltages.
3. True or false? The field excitation for an ac generator must be direct current.
4. True or false? The sine wave is the most basic and widely used ac waveform.
5. How many electrical degrees must a coil pass through to produce one cycle?
6. A four-pole ac generator is required to produce 60 Hz. At what speed should it be rotated?

7. When full load is applied to an alternator operating at constant speed and field excitation, the voltage drops from 264 V to 240 V. What is the percentage voltage regulation?
8. A voltage regulator keeps the output voltage of a generator constant by varying the amount of _____ current for different load conditions.
9. The stator coils of a three-phase alternator are connected in _____ or _____.
10. True or false? In a three-phase wye system, the voltage across any two lines is equal to the voltage generated in the coil winding.
11. True or false? The mechanical energy required by a generator to produce electricity is supplied by the prime mover.
12. True or false? Alternating current generators may be operated in parallel by connecting the stator windings of the two machines in parallel.
13. Cogeneration is the simultaneous production of electric and _____ energy from a single fuel.

6-2 DIRECT CURRENT GENERATORS

Generating direct current is basically the same as generating alternating current.

IIII➡ YOU MAY RECALL that the voltage produced in any generator is inherently alternating. It only becomes direct current after it has been rectified.

Direct current generators are not as common as they used to be because direct current, when required, is mainly produced by solid state rectifier diodes. In the past industrial plants sometimes used *motor-generator sets* for converting alternating current to direct current. In this application an ac motor is used to drive a dc generator. Alternating current is applied to the motor, and dc voltage is obtained from the generator (Fig. 6-10 on page 178).

Figure 6-11 on page 178 shows a simplified dc generator. The shape of the voltage generated in the loop is still that of an ac sine wave. Notice, however, that the two slip rings of the ac generator have been replaced by a single segmented rotating contact called a *commutator*. This commutator acts like a mechanical switch or rectifier to automatically convert the

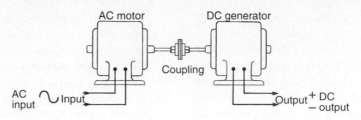

Fig. 6-10 Motor-generator set.

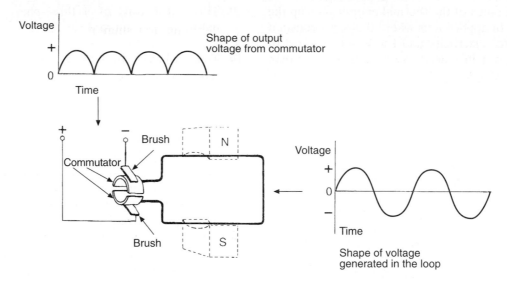

Fig. 6-11 Simplified dc generator.

generated ac voltage into a dc voltage. As the armature starts to develop a negative alternation, the commutator switches the polarity of the output terminals via the brushes. This keeps all positive alternations on one terminal and all negative alternations on the other. Then the only essential difference between an ac and a dc generator is the use of slip rings on the one and a commutator on the other.

A single-loop generator produces a dc output that is pulsating. Most applications require a dc-generated voltage that has a steady value with minimum ripple or variation. This is accomplished by adding more coils and commutator segments in the armature (Fig. 6-12). All the coils are connected in series between the two brushes. The voltage appearing at the brushes is the sum of the voltages in the separate coils. The result is a dc output that is almost steady.

Most dc generators use electromagnetic field coils rather than permanent magnets. The direct current used to energize the field windings is called the *exciting current*. Direct current generators are classified according to the method by which current is supplied to these field coils. The two major classifications are

separately excited and *self-excited generator* types.

A dc generator that has its field current supplied by an outside source is called a *separately excited generator* (Fig. 6-13). The outside source may be a battery or any other type of dc supply. With the speed held constant, the output of this generator may be varied by controlling the current through the field coils. This is accomplished by inserting a rheostat in series with the dc source and field windings. The output voltage of the generator will then vary in direct proportion to the field current flow.

The inconvenience of a separate dc source for field excitation led to the development of self-excited generators. *Self-excited generators* use part of the generated current to excite the field. They are classified according to the method by which the field coils are connected. Self-excited generators may be series-connected, shunt-connected, or compound-connected.

In a *shunt generator,* the shunt field windings are connected in parallel with the armature (Fig. 6-14 on page 180). The shunt field windings consist of many turns of relatively small wire and actually use only a small part of the

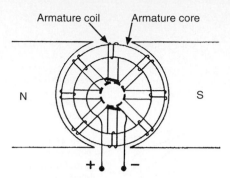

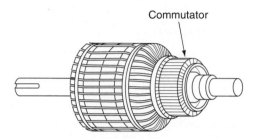

(a) Armature coils connected in series between the two brushes

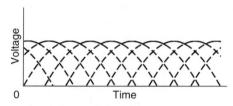

(b) Armature winding with attached commutator

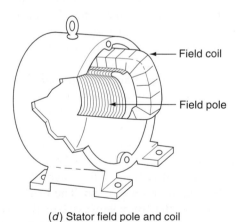

(c) Output voltage appearing at the brushes

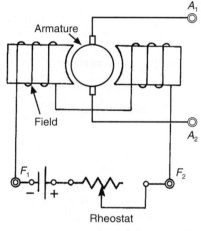

(d) Stator field pole and coil

Fig. 6-12   Practical dc generator.

generated current. The initial voltage the generator is required to build up is produced by the residual magnetism in the iron of the field poles. Residual magnetism is the magnetism that remains in the poles when no current flows through the field coils. As the generated voltage increases, the current through the field

coils also increases. This strengthens the magnetic field and allows the generator to build up to its rated output voltage. A rheostat connected in series with the field coils is used to vary the field current, which in turn controls the generator output voltage. As load is added to the generator, the output voltage will drop unless some provision is made to keep it constant.

A *compound generator* (Fig. 6-15 on page 180) is similar to a shunt generator, except that it has additional field coils connected in series with the armature. These *series field coils*, placed on the same poles as the shunt field winding, are composed of a few turns of heavy wire, big enough to carry the armature current. The compound generator was developed to prevent the terminal voltage of a dc generator from decreasing with increasing load. When the generator runs at no-load, the current in

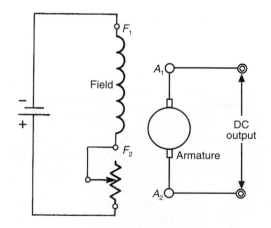

(a) Wiring diagram

(b) Schematic diagram

Fig. 6-13   Separately excited generator.

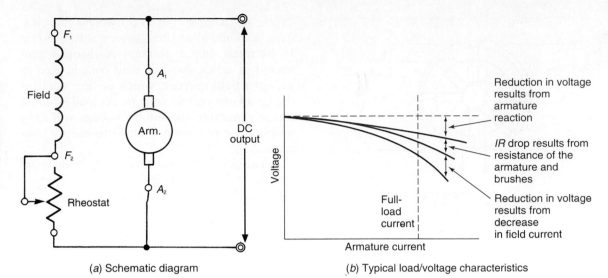

(a) Schematic diagram

(b) Typical load/voltage characteristics

Fig. 6-14   Shunt generator.

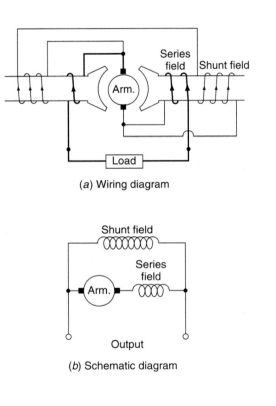

(a) Wiring diagram

(b) Schematic diagram

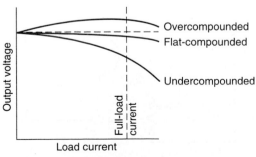

(c) Typical load/voltage characteristics

Fig. 6-15   Compound generator.

the series coils is zero. As the generator is loaded, the terminal voltage tends to drop, but load current now flows through the series field coils. The magnetism then developed by these series coils acts to strengthen the magnetic field and reduce voltage drop. The voltage of an overcompound generator increases when full load is applied, whereas the voltage of a flat-compound generator remains constant and that of an undercompound generator drops slightly, so that it resembles a shunt generator.

Neutral planes are those places on the surface of the armature where the flux density is zero. When the generator operates at no-load, the neutral planes are located exactly between the poles, as shown in Fig. 6-16(*a*). No voltage is induced in a coil that cuts through the neutral plane. Normally the brushes are set so that they are in contact with coils that are momentarily in a neutral plane.

Commutation is the process of reversing the direction of the current in an armature coil as the commutator segments to which the coil is connected pass under a brush. The brushes are positioned so that they short-circuit the armature coil when it is not cutting through the magnetic field. At this instant no current flows, and there is no sparking at the brushes. Arcing at the brushes, if allowed to occur, causes excessive brush and commutator wear.

Current flowing in the armature coils creates a powerful magnetomotive force that distorts and weakens the flux coming from the poles. This distortion and field weakening effect is called *armature reaction*. Figure 6-16(*b*) shows how the armature field distorts the main field,

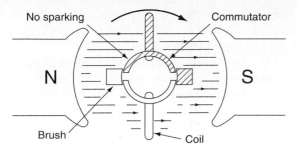

No sparking   Commutator

N   S

Brush   Coil

(*a*) The neutral plane is located exactly between the poles

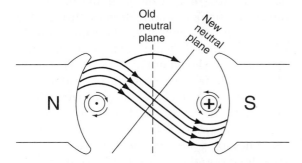

(*b*) The neutral plane shifts in the direction of rotation

Fig. 6-16   **Armature reaction.**

which results in the neutral plane's being shifted in the *direction of rotation*. If the brushes are allowed to remain in the old neutral plane, they will be short-circuiting coils that have voltage induced in them. To prevent this, the brushes must be shifted to the new neutral plane.

When the load on the generator fluctuates, the neutral plane shifts back and forth between the no-load and full-load positions. For small dc generators, the brushes are set in an intermediate position to produce acceptable commutation at all loads. In larger generators, *interpoles* (also called *commutating poles*) are placed between the main field poles to minimize the effects of armature reaction (Fig. 6-17). These narrow poles have a few turns of large wire connected in series with the armature. The magnetic field generated by the interpoles is designed to be equal to and opposite that produced by the armature reaction for all values of load current and improves commutation.

A typical dc generator nameplate contains important manufacturer's specifications, such as:

Power – 120 kW
Voltage – 240 V
Exciting current – 20 A
Temperature rise – 50°C

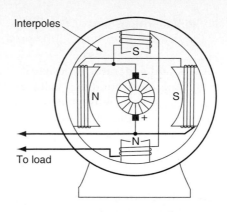

Fig. 6-17   **Correcting armature reaction.**

Speed – 1200 rpm
Type – Compound
Class – B

These specifications tell us that the machine can deliver, continuously, a power of 120 kW at a voltage of 240 V, without exceeding a temperature rise of 50°C. It can therefore supply a load current of 500 A (120,000/240). It possesses a series winding, and the current in the shunt field is 20 A. In practice, the terminal voltage is adjusted to a value close to its rating of 240 V. We may draw any amount of power from the generator, as long as it does not exceed 120 kW. The class B designation refers to the class of insulation used in the machine.

Self-Test

Answer the following questions.

14. True or false? The voltage generated in a dc generator is at first alternating current.
15. The _____ of a dc generator works like a mechanical rectifier.
16. True or false? The output ripple voltage of a dc generator is reduced by adding more coils to the armature winding.
17. The current used to energize the field windings of a dc generator is known as the _____ current.
18. A dc generator that has its field supplied by an outside source is referred to as a _____ generator.
19. True or false? The output voltage of a dc generator is inversely proportional to the field current flow.
20. Self-excited dc generators may be _____-connected, _____-connected, or _____-connected.
21. True or false? The shunt field winding of a dc generator has a much higher resistance than the series field winding.

22. The initial voltage required for a self-excited generator to build up is produced by _____ magnetism.
23. A compound generator is designed to prevent generated output voltage from _____ with increasing load.
24. No voltage is induced in an armature coil that cuts through the _____ plane.
25. The process of reversing the direction of the current in an armature coil is called _____.
26. True or false? In a dc generator, sparking at the brushes resulting from armature reaction can be reduced by shifting the brushes in the direction of rotation of the armature.
27. True or false? Interpoles are used to minimize the drop in voltage when full load is applied to a dc generator.
28. The amount of load current that a generator rated at 250 kW and 125 V can supply is _____.

6-3 DIRECT CURRENT MOTORS

Over 50 percent of the electricity produced in the United States is used to power motors. Motors are used to turn the wheels of industry.

➠ **YOU MAY RECALL** that the electric motor utilizes electric energy and magnetic energy to produce mechanical energy. The operation of a motor depends on the interaction of two magnetic fields. Simply stated, an electric motor operates on the principle that *two magnetic fields can be made to interact to produce motion.* The purpose of a motor is to produce a rotating force (torque).

Direct current motors are seldom used in normal industrial applications because all electric utility systems furnish alternating current. For special applications, however, it is advantageous to transform the alternating current into direct current in order to use dc motors. Direct current motors are used where a wide range of torque and speed control is required to match the needs of the application. The brush-commutator arrangement, however, presents the problems of brush maintenance and electrical arcing.

A current-carrying conductor, placed in and at right angles to a magnetic field, tends to move at right angles to the field. The amount of force exerted to move it varies directly with

the strength of the magnetic field, the amount of current flowing through the conductor, and the length of the conductor.

To determine the direction of movement of a conductor carrying current in a magnetic field, the *right-hand motor rule* is used [Fig. 6-18(a)]. The thumb and first two fingers of the right hand are arranged so that they are at right angles to each other, with the forefinger pointing in the direction of the magnetic lines of force of the field and the middle finger pointing in the direction of current flow (− to +) in the conductor. The thumb will be pointing in the direction of movement of the conductor, as shown in Fig. 6-18(b). This figure illustrates how motor torque is produced by a current-carrying coil or loop of wire placed in a magnetic field. The interaction between the two magnetic fields causes a bending of the lines of force. When the lines tend to straighten out, they cause the loop to undergo a rotating motion. The left conductor is forced downward, and the right conductor is forced upward, causing a counterclockwise rotation of the armature.

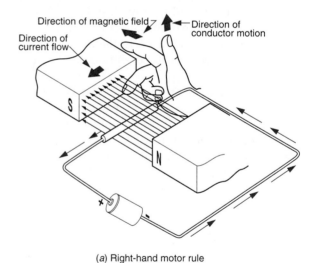

(a) Right-hand motor rule

(b) Developing motor torque

Fig. 6-18 Motor principle.

Direct current motors and generators are built the same way; consequently, a dc machine can operate either as a motor or as a generator. A *permanent-magnet dc motor* is a motor in which the main field flux is produced by permanent magnets. An electromagnet is used for the secondary field or armature flux. Figure 6-19 illustrates the operation of the permanent-magnet motor. Current flow through the armature coil from the dc voltage source causes the armature to act as a magnet. The armature poles are attracted to field poles of opposite polarity, causing the armature to rotate. In Fig. 6-19(*a*), the armature rotates in a clockwise direction. When the armature poles are in line with the field poles, the brushes are at a gap in the commutator and no current flows in the armature. Thus forces of magnetic attraction and repulsion stop, as illustrated in Fig. 6-19(*b*). Then inertia carries the armature past this neutral point. The *commutator reverses the armature current* when unlike poles of the armature and field are facing each other, thus reversing the polarity of the armature field. Like poles of the armature and field then repel each other, causing continuous armature rotation, as shown in Fig. 6-19(*c*).

The direction of rotation of a permanent-magnet dc motor is determined by the direction of the current flow through the armature coil. Reversing the leads to the armature will reverse the direction of rotation. One of the features of a dc motor is that its speed can be easily controlled. The speed of a permanent-magnet motor is directly proportional to the value of the voltage applied to the armature. The greater the armature voltage, the faster the speed of the motor.

Permanent-magnet servomotors (Fig. 6-20) are used on machines requiring exact positioning of an object or component, where high starting and operating torque are required, and where a constant torque is required. Other applications include the operating of a valve under pressure, precise positioning of dampers, and other specific operations in various control systems.

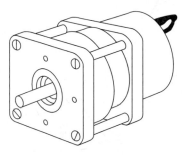

Fig. 6-20   DC permanent-magnet servometer.

The *series-type dc motor* (Fig. 6-21 on page 184) consists of a series field (identified by S_1 and S_2) made of a few turns of heavy wire connected in series with the armature (identified by A_1 and A_2). This type of dc motor has a high starting torque and a variable speed characteristic. This means that the motor can start very heavy loads but the speed will increase as the load is decreased. The series dc motor can develop this high starting torque because the same current that passes through the armature also passes through the field. Thus, if the armature calls for more current (developing more torque), this current also passes through the field, increasing the field strength. Therefore, a series motor runs fast with a light load and runs slower as the load is increased.

The most important feature of a series dc motor is its ability to start very heavy loads. For this reason, these motors are often used in cranes, hoists, and elevators. To reverse the direction of a dc series motor, reverse the current flow through either the series field or the

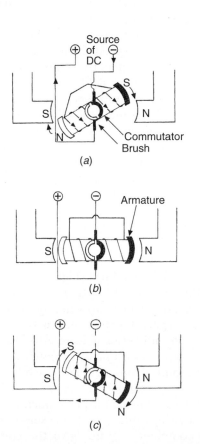

Fig. 6-19   Permanent-magnet dc motor operation.

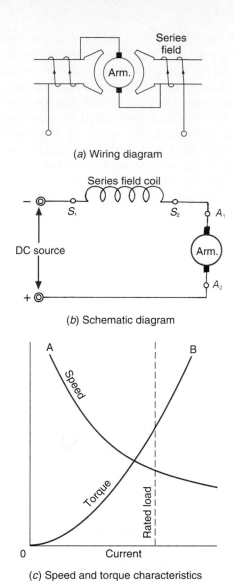

(a) Wiring diagram

(b) Schematic diagram

(c) Speed and torque characteristics

Fig. 6-21   Series dc motor.

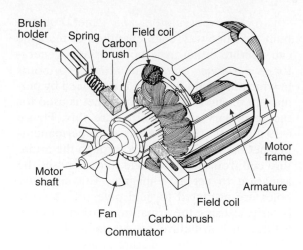

Fig. 6-22   Universal motor.

armature. This type of dc motor is also known as the *universal motor* because it can be operated on either direct current or alternating current (Fig. 6-22). The reason for this is that a dc motor will continue to turn in the same direction if the current through the armature and the current through the field are reversed at the same time.

Belt drives are never used with series motors. These motors are always coupled directly to the load. The reason is that series motors should never be operated without a load. The only field flux present at no-load is that caused by residual magnetism; therefore, the field is very weak. Operating the motor without a load will allow the rotor to reach such high speeds that the centrifugal force will cause the windings to tear free. In small dc series motors brush friction, bearing friction, and winding loss may provide sufficient load to hold the speed down to a safe level. When used as a portable power tool, the gear train in the unit or the attached tool will provide sufficient load.

In the *shunt-type dc motor* (Fig. 6-23) the shunt field coils (identified by F_1 and F_2) are made of many turns of small wire and, therefore, have *high resistance*. The shunt motors have their armatures and field circuits wired in parallel, giving essentially constant field strength and motor speed. The shunt motor is used where good speed regulation is desirable in the driven machine. By adding a rheostat in series with the shunt field circuit, the speed of the motor can be controlled above base speed. The speed of the motor will be inversely proportional to the field current. This means that a shunt motor runs fast with a low field current and runs slower as the current is increased. *Shunt motors can race to dangerously high speeds if current to the field coil is lost.* In order to reverse a dc shunt motor, reverse the current flow through either the shunt field or the armature.

A *compound-type dc motor* uses both series and shunt field windings (Fig. 6-24), which are normally connected so that their fields *add cumulatively.* This two-winding connection produces characteristics intermediate to the shunt field and series field motors. The speed of these motors varies a little more than that of shunt motors, but not as much as series motors. Also, compound-type dc motors have a fairly large starting torque—much more than shunt motors, but less than series motors. These combined features give the compound motor a wide variety of uses.

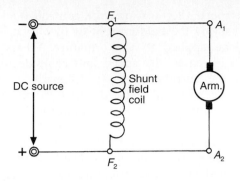

(a) Shunt motor connection

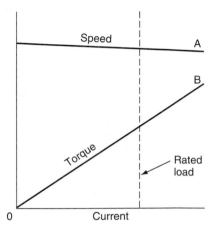

(b) Speed and torque characteristics

Fig. 6-23   Shunt dc motor.

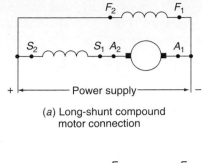

(a) Long-shunt compound motor connection

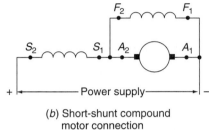

(b) Short-shunt compound motor connection

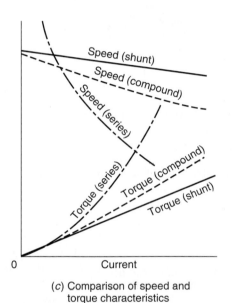

(c) Comparison of speed and torque characteristics

Fig. 6-24   Compound dc motor.

Armature reaction

Counter electromotive force (cemf)

Ordinarily, a motor is set up to do a particular job that requires a fixed direction of rotation, but there are times when you may find it necessary to change the direction of rotation. The *direction of rotation* of a dc motor depends on the direction of the field and the direction of current flow in the armature. If either the direction of the field or the direction of current flow through the armature is reversed, the rotation of the motor will reverse. If both of these two factors are reversed at the same time, however, the motor will continue rotating in the same direction. To reverse direction of rotation in a compound motor, reverse leads to the armature *or* to *both* the series and shunt fields. The industrial standard is to reverse the current through the armature.

The dc motor, like the dc generator, has *armature reaction*. Because the motor armature has current flowing through it, a magnetic field will be generated around the armature coils as a result of this current. The armature field causes distortion of the main field in a motor, causing the neutral plane to be shifted.

The direction of distortion in a motor, however, is just the opposite of that in a generator. In a motor, armature reaction shifts the neutral commutating plane *against* the direction of rotation. Interpoles are used in some dc motors to prevent sparking at the brushes.

As the armature rotates in a dc motor, the armature coils cut the magnetic field and induce a voltage or emf in these coils. This is sometimes referred to as the generator action of a motor. Because this induced voltage opposes the applied terminal voltage, it is called *counter electromotive force (cemf)*. The effective voltage acting in the armature circuit of a motor is the applied or terminal voltage minus the cemf (Fig. 6-25). The armature current by Ohm's law is:

$$I_A = \frac{V_{MT} - E}{R_A}$$

where I_A = armature current
V_{MT} = motor terminal voltage
E = cemf
R_A = armature-circuit resistance

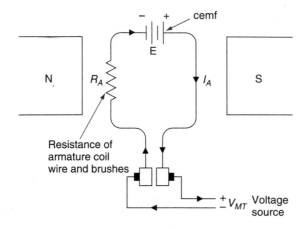

Fig. 6-25 **Motor counter electromotive force (cemf).**

The strength of cemf in a motor is determined by the strength of the field, the number of armature conductors in series between the brushes, and the speed of the armature. At the moment a motor starts, the armature is not rotating, so there is no cemf. Consequently, full line voltage is applied across the motor, and it draws current through the armature circuit in accordance with Ohm's law. The only factor limiting current is the resistance of the motor windings. As the armature increases speed or accelerates, it generates cemf, and this limits the flow of current into the motor. As the *speed increases, cemf increases,* and the current drawn by the motor *decreases.*

When a motor reaches its full no-load speed, it is designed to be generating a cemf *nearly equal to line voltage.* Only enough current can flow to maintain this speed. If a load is applied to the motor, its speed will be decreased, which will reduce the cemf, and more current will be drawn to drive the load. Thus, the load of a motor regulates the speed by affecting the cemf and current flow.

The *speed of a dc motor* depends on the strength of the magnetic field and the voltage applied to the armature, as well as on the load.

The speed, therefore, may be controlled either by varying the field current or by varying the voltage applied to the armature. When the load increases, the speed and cemf decrease and the current increases. Likewise, as the load decreases, the speed and cemf increase and the current decreases. A motor is designed to produce its rated horsepower at full-load speed. Its normal (full-load) speed is known as the *base speed* of the motor. The base speed is obtained when rated armature voltage and full field current is applied (Fig. 6-26).

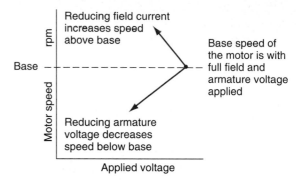

Fig. 6-26 **Control of dc motor speed.**

The speed of a dc motor is proportional to the cemf. A weakening of the main field flux reduces the cemf. The lower cemf allows more current to flow in the armature circuit. This increase in the armature current provides a stronger magnetic field in the armature, which causes an increase in the armature speed. The speed increases until the cemf can limit the armature current to a new value. This value is determined by the main field strength. At this point, the motor drives the load at a constant speed. In shunt motors the control of speed may be provided by connecting a rheostat in series with the shunt field winding. An increase in the resistance in series with the field reduces the field current, and so weakens the magnetic field. A *decreased* field strength means the motor must turn *faster* (Fig. 6-27). This method of speed control is frequently used because it is simple and inexpensive. Speeds greater than base speed are available only at reduced torque. Field weakening reduces a motor's torque capability.

If the main field flux is kept constant, *at full field strength,* only the armature voltage will affect the speed. By raising or lowering the armature voltage, the motor speed will rise or fall proportionally (Fig. 6-28). In Fig. 6-28(*a*), armature speed is controlled using a rheostat.

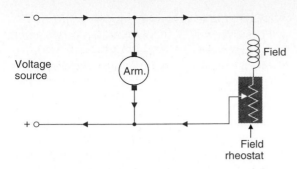

Fig. 6-27 Field speed control.

The rheostat *reduces* the speed below its nominal speed. The use of a rheostat is recommended only for small motors. It is not usually an efficient method of speed control because the armature is a high-current circuit. Thus, a lot of power and heat is wasted in the rheostat, and the speed regulation is poor.

The *Ward-Leonard system of armature speed control* is shown in Fig. 6-28(*b*). The motor armature (M) is connected to a separately excited dc generator (G). The generator out-

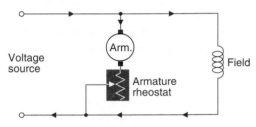

(*a*) Armature speed control using a rheostat

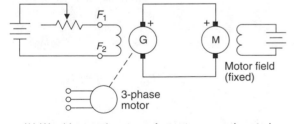

(*b*) Ward-Leonard system of armature speed control

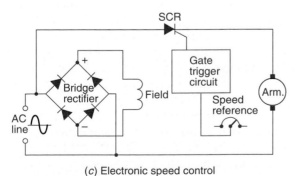

(*c*) Electronic speed control

Fig. 6-28 Armature speed control.

put voltage can be varied from zero to maximum. Consequently, the motor speed can be varied from zero to maximum.

With electronic speed control, illustrated in Fig. 6-28(*c*), the conduction of the SCR is controlled by the setting of the speed-reference potentiometer, which varies the ON time of the SCR per each positive half-cycle and so varies the voltage applied to the armature.

Self-Test

Answer the following questions.

29. An electric motor uses _____ energy and _____ energy to produce mechanical energy.

30. True or false? Direct current motors can operate directly from the power supplied by the electric utility system.

31. True or false? A current-carrying conductor, placed in and at right angles to a magnetic field, tends to move parallel to the field.

32. Torque in a motor is produced by interaction between two _____.

33. True or false? The commutator in a dc motor reverses the armature coil to produce continuous armature rotation.

34. True or false? One of the features of a dc motor is that its speed can be easily controlled.

35. True or false? Direct current permanent-magnet servomotors are used in cranes to start heavy loads.

36. True or false? A dc series motor runs fast with a light load.

37. True or false? The speed of a dc shunt motor is inversely proportional to the field current.

38. A compound-type dc motor uses both _____ and _____ field windings.

39. True or false? The direction of rotation of a dc motor can be changed by reversing the direction of current in both the field coils and the armature.

40. True or false? Unlike a dc generator, a dc motor armature does not produce armature reaction.

41. The effective voltage acting on the armature circuit of a dc motor is equal to the applied voltage minus the _____.

42. At the moment a dc motor starts, the only factor limiting current is the _____ of the motor windings.

43. The _____ speed of a dc motor is obtained when rated armature voltage and full field current are applied.

44. True or false? Increasing the voltage applied to the armature of a dc motor will cause its speed to decrease.

6-4 ALTERNATING CURRENT MOTORS

Over 90 percent of all motors run on alternating current. Both ac and dc motors have inherent characteristics that govern their uses.

Characteristics of ac motors

- Lower cost for use.
- Less maintenance required.
- Various enclosures readily available for different operating environments.
- Ability to withstand harsh operating environments.
- Physically smaller than dc motors of the same horsepower.
- Less costly repairs.
- Ability to run at speeds above the nameplate rating.

Characteristics of dc motors

- High torque at low speed.
- Good speed control over full range (no low-end cogging).
- Better overload capability.

- More expensive than ac motors.
- Physically larger than ac motors of the same horsepower.
- More routine maintenance and repairs required.

A common feature of all ac motors is a *rotating magnetic field* that is set up by the stator windings. This concept can be illustrated for three-phase motors by considering three coils placed 120 electrical degrees apart. Each coil is connected to one phase of a three-phase power supply (Fig. 6-29). When a three-phase current passes through these windings, a rotating magnetic field effect that travels around the inside of the stator core is set up. The speed of this rotating magnetic field depends on the number of stator poles and the frequency of the power source. The speed, called the *synchronous speed,* is determined by the formula:

$$S = \frac{120f}{P}$$

where S = synchronous speed in rpm
f = frequency in Hz of the power supply
P = number of poles wound in each of the single-phase windings

Two poles wound in
each single-phase winding

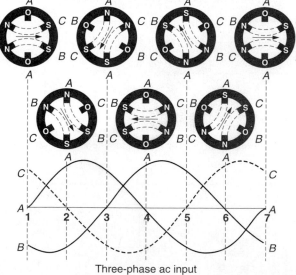

Resultant magnetic field

Three-phase ac input

Direction of rotation of the magnetic field is reversed by interchanging any two of the three-phase input leads to the stator

(*a*) Placement of electromagnetic stator winding

(*b*) Resultant rotating magnetic field pattern

Fig. 6-29 Development of a rotating magnetic field.

In Fig. 6-29(*b*), the synchronous speed could be calculated as:

$$S = \frac{120f}{P}$$

$$= 120 \times \frac{60}{2}$$

$$= 3600 \text{ rms}$$

Alternating current motors are classified by operating principle as either induction or synchronous motors. The *ac induction motor* is by far the most commonly used motor because it is relatively simple and can be built at less cost than other types. Induction motors are made in both three-phase and single-phase types. The induction motor is so named because *no external voltage is applied* to its rotor. Instead, the ac current in the stator *induces* a voltage across an air gap and into the rotor winding to produce rotor current and magnetic field. The stator and rotor magnetic fields then interact and cause the rotor to turn (Fig. 6-30).

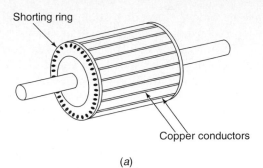

Fig. 6-31 is referenced (a)

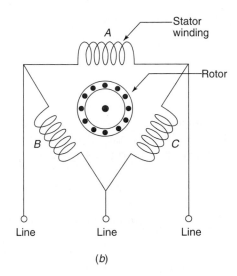

(*b*)

Fig. 6-31   Three-phase squirrel-cage induction motor.

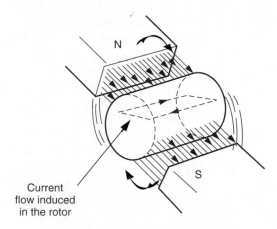

Fig. 6-30   Induced rotor current.

The majority of industrial applications usually involve the use of a *three-phase, squirrel-cage induction motor* (Fig. 6-31). Characteristics of the squirrel-cage motor include the following:

- The rotor consists of copper conductors embedded in a solid core with the ends shorted to resemble a squirrel cage.
- Speed is essentially constant.
- Large starting currents required by this motor can cause voltage fluctuations.
- Direction of rotation can be reversed by interchanging any two of the three main power lines to the motor.
- Power factor tends to be poor for reduced loads.
- When voltage is applied to the stator winding, a rotating magnetic field is produced, which induces a voltage in the rotor. This voltage, in turn, creates a large current flow in the rotor. The high current in the rotor creates a magnetic field of its own. The rotor field and the stator field tend to attract each other. This situation creates a torque, which spins the rotor in the same direction as the rotation of the magnetic field produced by the stator.
- Once started, the motor will continue to run with a phase loss as a single-phase motor. The current drawn from the remaining two lines will almost double, and the motor will overheat.

Squirrel-cage motors are usually chosen over other types of motors because of their simplicity, ruggedness, and reliability. Because of these unique features, squirrel-cage motors have become the accepted standard for ac all-purpose, constant-speed motor applications.

The rotor of an induction motor does not rotate at synchronous speed, but *lags* it slightly. For example, an induction motor having a synchronous speed of 1800 rpm will often have a rated speed of 1750 rpm at rated horsepower. The lag is usually expressed as a percentage of the synchronous speed, called the *slip*.

$$\% \text{ slip} = \frac{\left(\begin{array}{c}\text{Synchronous}\\\text{speed}\end{array}\right) - \left(\begin{array}{c}\text{Running}\\\text{speed}\end{array}\right)}{\text{Synchronous speed}} \times 100$$

$$= \frac{1800 - 1750}{1800} \times 100$$

$$= 2.78\%$$

The speed of the rotor of an induction motor depends upon the synchronous speed and the load that it must drive. The rotor does not revolve at synchronous speed, but tends to slip behind. If the rotor turned at the same speed at which the field rotates, there would be no relative motion between the rotor and the field and no voltage induced. Because the rotor *slips* with respect to the rotating magnetic field of the stator, voltage and current are induced in the rotor. Thus, a normal motor with, say, 2.8 percent slip and 1800 rpm synchronous speed would have a slip of 50 rpm and a full load speed of 1750 rpm (1800 − 50 = 1750 rpm). It is this full-load speed that will be found on the motor's nameplate.

The induction motor is basically a transformer in which the stator is the primary and the rotor is a short-circuited secondary. The *no-load current* is similar to the exciting current in a transformer. Thus, it is composed of a magnetizing component that creates the revolving flux and a small active component that supplies the windage and friction losses in the rotor plus the iron losses in the stator.

When the induction motor is *under-load,* the rotor current develops a flux that opposes and, therefore, weakens the stator flux. This allows more current to flow in the stator windings, just as an increase in the current in the secondary of a transformer results in a corresponding increase in the primary current. The exciting current and reactive power under-load remain about the same as at no-load. The active power (kW) absorbed by the motor, however, increases in proportion to the mechanical load. It follows that the power factor of the motor improves dramatically as the mechanical load increases. At full-load, it ranges from 0.70 for small machines to 0.90 for large machines. The efficiency at full-load is particularly high; it can attain 98 percent for very large machines.

The *locked-rotor current* to *starting current* of the induction motor is five to six times the full-load current. As soon as the rotor is released, it rapidly accelerates in the direction of the rotating field. As it picks up speed, the relative velocity of the field with respect to the rotor diminishes progressively. This causes both the value and the frequency of the induced voltage to decrease because the rotor bars are cut more slowly. The rotor current, very large at first, decreases rapidly as the motor picks up speed. The rotor must therefore never remain locked for more than a few moments.

Whereas a three-phase induction motor sets up a rotating field that can start the motor, a *single-phase motor needs an auxiliary means of starting.* Once a single-phase induction motor is running, it develops a rotating magnetic field. Single-phase induction motors are larger in size for the same horsepower than a three-phase motor; they find limited applications where three-phase power is not available. When running, the torque produced by a single-phase motor is pulsating and irregular, contributing to a much lower power factor and efficiency than for a polyphase motor.

The direction of the rotation of the stator field of a three-phase induction motor depends on the phase sequence. The rotor field is attracted by the field of the stator and therefore revolves in the same direction as the stator field. Interchanging any two of the three-phase leads supplying current to the stator will reverse the phase sequence and cause the rotor to reverse direction.

It is important to remember how the speed of an induction motor is determined—that is, by the number of poles and the frequency of the power supply (*not the supply voltage*). The speed of the standard squirrel-cage induction motor is inherently constant. Special multi-speed squirrel-cage motors, however, are manufactured with stator windings in which the number of poles may be changed by changing the external connections. These multispeed motors are available in two or more fixed speeds, which are determined by the connections made to the motor. Two-speed motors usually have one winding that may be connected to provide two speeds, one of which is half the other (Fig. 6-32).

A *wound-rotor induction motor* is an induction motor with a *wire-round* rotor that is used

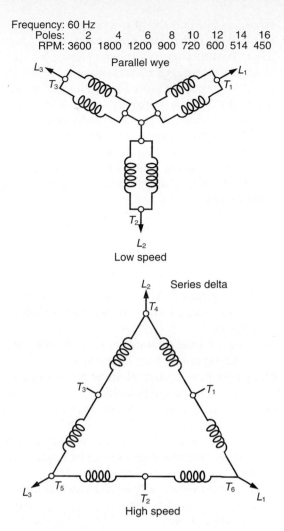

Frequency: 60 Hz
Poles: 2 4 6 8 10 12 14 16
RPM: 3600 1800 1200 900 720 600 514 450

Parallel wye

L_3 T_3 T_1 L_1

T_2

L_2

Low speed

Series delta

L_2

T_4

T_3 T_1

L_3 T_5 T_2 T_6 L_1

High speed

Fig. 6-32 Multispeed squirrel-cage motor-winding connection.

for variable-speed applications (Fig. 6-33). The stator consists of three single-phase windings placed 120 electrical degrees apart and connected to a three-phase power source. The three-phase rotor has leads that come out to

the slip rings. The speed of the wound rotor can be varied by placing various amounts of resistance in the rotor circuit through the slip rings. The more resistance placed in the rotor circuit, the slower the motor will go; when all the resistance is removed from the rotor circuit, the motor will run at full speed. Placing resistance in the rotor circuit reduces the starting current and provides high starting torque. The power factor of this type of motor is low at no-load and high at full-load. To reverse this type of motor, interchange any two voltage supply leads.

Some of the advantages and disadvantages of wound-rotor induction motors include:

Advantages

- High starting torque with low starting current.
- Smooth acceleration under heavy loads.
- No abnormal heating during the starting period.
- Good speed adjustment when operating under a constant load.

Disadvantages

- Greater initial cost and maintenance costs than those of the squirrel-cage motor.
- Poor speed regulation when operating with resistance in the rotor circuit.

The *synchronous motor,* as its name suggests, is a motor that runs at a constant speed from no-load to full-load. Its speed is the *same* as that of the rotating magnetic field. A synchronous motor uses a single- or three-phase stator to generate a rotating magnetic field and an electromagnetic rotor that is supplied with direct current. The rotor acts like a magnet and is attracted by the rotating stator field. This attraction will exert a torque on the rotor

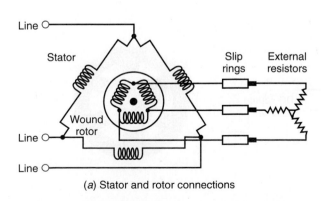

Line

Stator

Wound rotor

Line

Line

Slip rings

External resistors

(a) Stator and rotor connections

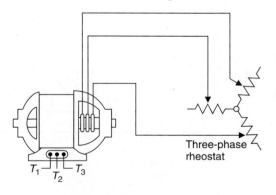

Three-phase rheostat

T_1 T_2 T_3

(b) Rotor connected to a three-phase rheostat for variable speed control

Fig. 6-33 Three-phase wound-rotor induction motor.

and cause it to rotate with the field. Synchronous motors are not self-starting and must be brought up to near synchronous speed before they can continue rotating by themselves.

In a *three-phase synchronous motor* (Fig. 6-34), the rotor generally has two windings: an ac winding, which may be of the squirrel-cage or the wound-rotor type, and a dc winding. The ac rotor winding brings the motor up to near synchronous speed, at which point the dc rotor winding is energized and the motor locks in step with the rotating field. The stator windings are similar to those of the polyphase, squirrel-cage, and wound-rotor motors.

A synchronous motor *cannot* be started with the dc field energized. Under this condition, an alternating torque is produced in the rotor. As the stator field sweeps across the rotor, it tends to cause the rotor to try to turn—first in the direction opposite to that of the rotating field, and then in the same direction. This action takes place so rapidly that the rotor remains stationary.

To start the synchronous motor, the rotor is left deenergized. The motor is started in the same manner as the squirrel-cage or wound-rotor motor, depending on the rotor construction. When the rotor reaches approximately 95 percent of synchronous speed, direct current is applied to the exciter winding. The direct current produces definite north and south poles in the rotor, which lock on to the rotating magnetic field of the stator and turn the rotor at synchronous speed.

Three-phase synchronous motors can be used for power factor corrections. Motors operated in this manner are referred to as *synchronous capacitors*. The squirrel-cage motor and the wound-rotor motor are induction-type motors that cause a lagging power factor. The lagging power factor can be corrected by

overexciting the rotor of a synchronous motor. This will produce a leading power factor, canceling out the lagging power factor of the induction motor. An underexcited dc field will produce a lagging power factor (seldom used). When the field is normally excited, the synchronous motor will run at a unity power factor. Synchronous motors are generally used for driving loads requiring constant speeds and infrequent starting and stopping. Common types of loads are dc generators, blowers, and compressors.

Self-Test

Answer the following questions.

45. True or false? Alternating current motors require more routine maintenance than dc motors.

46. True or false? Direct current motors are more expensive than ac motors.

47. A common feature of all ac motors is a(n) _____ magnetic field.

48. What is the synchronous speed of a 12-pole, 60-Hz ac motor?

49. The ac induction motor is so named because no external voltage is applied to its _____.

50. The most commonly used industrial motor is the _____-phase, _____-cage induction motor.

51. Calculate the percentage of slip of a 60-Hz, 8-pole induction motor operating at 840 rpm.

52. True or false? Unlike three-phase motors, single-phase motors need an auxiliary means of starting.

53. The direction of rotation of a three-phase induction motor is dependent on the _____ sequence.

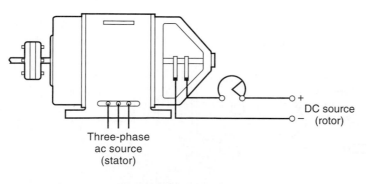

(a) Stator and rotor connections

(b) Rotor windings

Fig. 6-34   Three-phase synchronous motor.

54. True or false? The speed of a squirrel-cage induction motor is changed by varying the supply voltage.

55. True or false? The speed of a synchronous motor is the same as that of the rotating magnetic field.

56. True or false? Three-phase synchronous motors are not self-starting.

57. A synchronous motor with a(n) _____ dc field will produce a leading power factor.

6-5 MOTOR SELECTION, INSTALLATION, AND MAINTENANCE

The mechanical power rating of motors is expressed in either horsepower (hp) or watts (W): 1 hp = 746 W. Two important factors that determine mechanical power output are torque and speed. *Torque* is the amount of twist or turning power; it is often stated in *pound-feet* (lb/ft). *Motor speed* is commonly stated in *revolutions per minute* (rpm).

$$\text{Horsepower} = \frac{\underset{\text{(in rpm)}}{\text{Speed}} \times \underset{\text{(in lb/ft)}}{\text{Torque}}}{5252}$$

Thus, for any given motor, the horsepower depends upon the speed. The slower the motor operates, the more torque it must produce to deliver the same amount of power. To withstand the greater torque, slow motors need stronger components than those of higher speed motors of the same power rating. Slower motors are generally larger, heavier, and more expensive than faster motors of equivalent power rating. The amount of torque produced by a motor generally varies with speed and depends on the type and design of the motor. Figure 6-35 shows a typical motor torque-speed graph. Some important factors indicated by the graph include:

- *Starting torque.* This is the torque produced at zero speed.
- *Pull-up torque.* This is the minimum torque produced during acceleration from standstill to operating speed.
- *Breakdown torque.* This is the maximum torque that the motor can produce before stalling.

The *power efficiency* [Fig. 6-36(*a*)] of an electric motor is defined as follows:

$$\text{Efficiency (\%)} = \frac{\text{Output}}{\text{Input}}$$

$$= \frac{\text{Power output}}{\text{Power output} + \text{losses}}$$

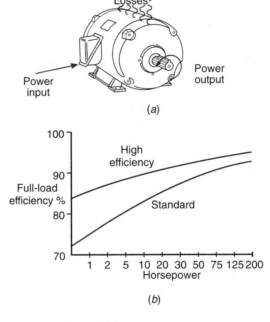

(*a*)

(*b*)

Fig. 6-36   Motor efficiency.

Because of losses, the useful mechanical output of a motor is less than the electrical input.

Heat is the final determining factor in setting a motor's horsepower rating. The power input to the motor is either transferred to the shaft as power output or lost as heat through the body of the motor. The efficiency of electric motors ranges between 75 and 98 percent. *Energy-efficient motors* [Fig. 6-36(*b*)] cost less to operate and have reduced heat losses, so they require less electric power input to provide the same mechanical power output. This

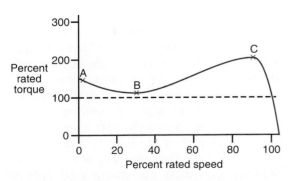

Fig. 6-35   Typical motor torque-speed graph.

Torque

Motor speed

Starting torque

Pull-up torque

Breakdown torque

Power efficiency

Energy-efficient motors

improvement in efficiency is basically accomplished by using more material in the motor, using better material, and implementing design changes. Specific losses associated with the operation of a motor include:

- *Core loss.* Core loss represents the energy required to magnetize the core material (hysterisis) and losses owing to the creation of small electric currents that flow in the core (eddy currents).
- *Stator loss.* The I^2R heating loss in the stator winding as current I flows through the winding conductors of resistance R.
- *Rotor loss.* The I^2R loss in the rotor winding. (In squirrel-cage induction motors, the "winding" is actually conductive bars running axially along the rotor and joined at the ends.)
- *Stray load loss.* This is the result of leakage flux induced by load current; it varies as the square of load current.
- *Windage and friction loss.* This loss represents air and bearing friction against the rotating rotor.

The kilowatts required are approximately the same regardless of the size of motor used. The kiloVAR required, however, climb rapidly as a larger motor than necessary is used. As a result, the kilovolt-amperes required, which determine the size of switchgear and electrical cables used, will also increase. A reasonable oversizing margin is to operate the motor between 75 and 100 percent load. This is usually accomplished by recognizing that induction motors are only available in so many sizes. Remember that when a motor has a service factor (SF) greater than 1.0, it has been designed to operate satisfactorily at the service factor load (e.g., a service factor of 1.15 can operate at 115 percent load continuously), although at a slightly reduced efficiency. Short-term overloads can often be accommodated by this service factor capability rather than by using a larger horsepower motor.

Motor enclosures are designed to provide adequate protection, depending on the environment in which the motor has to operate. The most common enclosures are:

- *ODP (open drip-proof).* ODP enclosures are used for clean environments; they will tolerate dripping liquids no greater than 15° from vertical. Ambient air is drawn through the motor for cooling.
- *TEFC (totally enclosed fan cooled).* TEFC enclosures are used for dusty and corrosive environments; air is cooled by an external integral fan.
- *Explosion proof.* This is a TEFC motor used in flammable environments. It is able to withstand internal gas explosion without igniting external gas (will not allow internal sparking or fire to escape).

Induction motors have been standardized according to their torque characteristics as *NEMA* (*National Electrical Manufacturers Association*) design A, B, C, D, or F. The design you select must have adequate torque to start the load and to accelerate it to full speed. Table 6-1 lists the torque characteristics for different NEMA designs.

The *squirrel-cage induction motor* is the simplest, most reliable motor because of its rugged rotor "cage" winding and the absence of brushes. Large starting currents required by this motor can cause voltage fluctuations. The general-purpose, squirrel-cage induction motor (*NEMA design B*) is the most commonly used induction motor. NEMA design B motors are used to drive fans, centrifugal pumps, and so forth.

High starting-torque motors (*NEMA design C*) are used when starting conditions are difficult. Elevators and hoists that have to start under-load are two typical applications. In general, these motors have a *double-cage rotor.*

High-slip motors (*NEMA design D*) are designed to provide high starting torque and low starting current. They have a high rotor resistance and operate at between 85 and 95 percent of synchronous speed. These motors drive high inertia loads (such as centrifugal dryers), which take a relatively long time to reach full speed. The high-resistance squirrel cage is made of brass, and the motors are usually designed for intermittent operation to prevent overheating.

Motor insulation is classified by letter according to the temperatures each is capable of withstanding without a serious deterioration of its insulating properties. Insulation temperature ratings are based on 40°C ambient temperature. Table 6-2 shows temperature *rises above ambient* for classes of insulation.

Consider that water boils at 100°C, and the temperature under which motors operate can be appreciated. Although most insulation will not burn or melt if the limit is exceeded, the useful life of the insulation is seriously reduced. The most commonly used type of motor insulation is class B.

Table 6-1 Torque Characteristics

Induction Motor Design	Starting Torque	Starting Current	Full-Load Slip	Breakdown Torque
A	Normal	Normal	Low	Higher
B	Normal	Normal	Low	Normal
C	High	Normal	Low	Normal
D	High	Low	High	High
F	Low	Low	Low	Low

Table 6-2 Motor Insulation Classification

	Insulation		
	Class B	Class F	Class H
Motors without SF. Temperature rise at rated load.	80°C	105°C	125°C
Motors with 1.15 SF. Temperature rise at 115% load.	90°C	115°C	135°C

There are two types of bearings used in electric motors: sleeve bearings and ball bearings. *Sleeve bearings* consist of a bronze or brass cylinder, a wick, and a reservoir. The shaft of the motor rotates in the bronze or brass sleeve and is lubricated with oil from the reservoir by the wick, which transfers oil from the reservoir to the sleeve. Sleeve bearings are used on smaller light-duty motors and should be lubricated at least every 6 months or to manufacturer's specifications.

The *ball bearing* type consists of ball bearings in an inner and outer race with an inner and outer seal. Ball bearings are used for heavy loads and come in three different styles: permanently lubricated, hand-packed, and bearings that require lubrication through fittings. Not lubricating the bearings can damage a motor, for obvious reasons; overlubricating, however, can also damage a motor. If you overlubricate a ball bearing (especially with grease fittings), you will blow the inner seal and fill the motor with grease, causing the motor to overheat.

In addition to lubrication, another important factor in bearing wear is the correct *alignment of motors and loads* (Fig. 6-37). For direct-coupled motors, the load shaft must line up perfectly with the drive shaft, as shown in Fig. 6-37(*a*). For belt-drive systems there must be proper alignment and proper belt tension: approximately 1 in. of play in the belt for most single belt applications. The belt should be

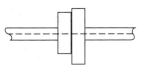

(*a*) The load shaft must line up perfectly with the drive shaft

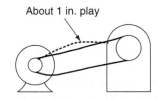

(*b*) The motor shaft pulley must be accurately aligned with the pulley on the device to be driven

Fig. 6-37 Motor and load alignment.

adjusted only enough to prevent slippage. Overtightening can reduce bearing life. When changing belts on a multibelt system it is important to change all the belts even if only one breaks because they share the load equally. Once you have correctly aligned the system and set the belt tension, always check the operating current of the motor to see that it does not exceed its full-load current rating. If a bearing is bad and is left in the motor, it could wear enough to cause the rotor to grind into the stator and destroy the motor.

Belt-drive systems are very useful because a standard speed motor can be used to drive a load at a given speed. This is done by using the pulley/speed ratio. The formula is used to calculate speed and pulley size.

$$\frac{\text{Motor rpm}}{\text{Equipment rpm}} = \frac{\text{Equipment pulley diameter}}{\text{Motor pulley diameter}}$$

EXAMPLE 6-1

You have a motor to drive a load. The motor operates at 1725 rpm and has a 2-in. drive pulley; the load must rotate at 1150 rpm. What size of pulley is needed for the load?

$$\frac{1725}{1150} = \frac{\text{Equipment pulley diameter}}{2}$$

$$\frac{1725 \times 2}{1150} = \text{3-in. pulley}$$

The motor must be connected to a power source corresponding to the voltage and frequency rating shown on the motor nameplate. Wiring should be done in accordance with the electrical code that governs local wiring. Undersized wire between the motor and power source will limit starting abilities and cause overheating of the motor. Both your motor and the equipment or apparatus to which it is connected should be grounded as a precaution against the possibility of electric shock.

Motors subjected to overload, locked rotor, current surge, or inadequate ventilation conditions may experience rapid heat buildup, presenting risks of motor damage or fire. To minimize such risks, the use of motors with proper overload protectors is advisable for most applications.

- *Automatic overload protector.* A built-in automatic reset overload device safeguards the motor by disconnecting the motor from the line before the temperature becomes dangerously high. The contacts of the device remain open long enough for the motor to cool, then automatically close and reconnect the motor to the line. Normal operation is resumed if the overload is only temporary.
- *Manual overload protector.* The built-in manual reset overload device is *recommended for use where automatic restarting would create a hazard for the operator or result in damage to the equipment.* This type of protector operates similarly to the auto-

matic reset device, except that resetting is accomplished by pressing the reset button on the motor after suitable time has been allowed for the motor to cool.

Electric motors are the workhorse of industry and, when properly applied, will provide years of service with a minimum of maintenance. Here are a few suggestions that should be part of a motor maintenance program:

- *Schedule periodic inspections.* At least once per year each electric motor should be inspected for things such as shaft alignment, motor base tightness, and vee belt condition and tension. If it is a dc motor, remove the covers and perform checks on brush wear, spring tension, commutator wear or scoring, and, if possible, the condition of windings.
- *Keep your motors clean.* If screens or filters become clogged, air flow will be reduced and the motor could overheat. This will eventually lead to insulation failure.
- *Keep your motors dry.* Motors that are used continuously are not prone to moisture problems. It is the intermittent use or standby motor that may have difficulties. Try to run the motor for at least a few hours each week to drive off moisture. Be careful that steam and water are not directed into open drip-proof motors.
- *Check lubrication.* Make sure that the right amount of lubricant is used. Too much or too little lubricant can cause bearing damage. Be sure not to use contaminated or improper lubricants.
- *Utilize vibration analysis.* A noticeable increase in motor vibration is an indication of trouble.
- *Excessive starting is a prime case of motor failures.* The high current flow during start-up contributes a great amount of heat to the motor. For motors 200 hp and below, the maximum acceleration time a motor connected to a high inertia load can tolerate is about 20 s. The motor should not exceed more than about 150 "start-seconds" per day.

Self-Test

Answer the following questions.

58. The input to a motor is 20 kW. What is the horsepower input?
59. True or false? Slower motors are generally smaller than faster motors of equivalent power rating.

60. The power input to a motor is either transferred to the shaft or lost as _____.

61. True or false? Core losses in a motor result from the creation of eddy currents in the core material.

62. What type of motor enclosure would you recommend for a saw in a lumber mill?

63. What is the most commonly used NEMA design for induction motors?

64. True or false? A motor with a class B insulation temperature rating is designed to safely withstand higher operating temperatures than one with a class H rating.

65. The two types of bearings used in electric motors are _____ and _____.

66. True or false? Overtightening of motor belts can reduce bearing life.

67. A motor operates at a speed of 3250 rpm. The machine it is driving requires a speed of 650 rpm. If the machine pulley is 20 in. in diameter, what diameter of pulley must be installed on the motor?

68. True or false? All motors should normally be grounded.

69. True or false? Resetting of an automatic built-in motor overload device is accomplished by pressing the reset button.

70. True or false? If a motor is not kept clean it could overheat.

SUMMARY

1. A generator is a machine that uses magnetism to convert mechanical energy into electric energy.

2. The amount of voltage induced in a generator armature conductor depends on the strength of the magnetic field, the speed and length of the conductor, and the angle at which the conductor cuts the flux.

3. The ac generator resembles the dc generator except that it has slip rings instead of a commutator.

4. Stationary-field ac generators are not as common as revolving-field generators and are limited to small kilovolt-ampere and low-voltage ratings.

5. The revolving-field ac generator simplifies the problems of insulating high-generated voltages because the windings are not subjected to centrifugal forces.

6. Alternating current generators produce sine wave-shaped voltages.

7. A formula to determine the frequency of an ac generator is:

$$f = \frac{pn}{120}$$

8. The formula to determine the percentage voltage regulation of a generator is:

% regulation =

$$\frac{\text{No-load voltage} - \text{Full-load voltage}}{\text{Full-load voltage}} \times 100$$

9. A voltage regulator is used to maintain a constant generator terminal voltage by varying the field excitation in accordance with different load conditions.

10. The stator coils of three-phase alternators may be connected in wye or delta.

11. In a three-phase wye-connected alternator, the phase-to-neutral voltage is equal to the voltage generated in each coil, and the phase-to-phase voltage is equal to the phase-to-neutral voltage times 1.73.

12. In a three-phase delta-connected alternator, the voltage measured across any two lines is equal to the voltage generated in the coil winding.

13. The source of the mechanical energy supplied to a generator is called the prime mover.

14. Before two three-phase ac generators may be paralleled, their phase sequence must be the same, their voltages must be in phase, and their frequencies and terminal voltages must be equal.

15. Cogeneration is the simultaneous production of electric and thermal energy from a single fuel.

16. The voltage generated in any generator is inherently alternating; it only becomes direct current after it has been rectified.

17. The commutator of a dc generator acts like a mechanical rectifier to convert the generated ac voltage into dc voltage.

18. The direct current used to energize field windings of a generator is called the exciting current.

19. A separately excited dc generator has its field current supplied by an outside source.

20. Self-excited generators may be series-, shunt-, or compound-connected.

21. The compound generator prevents the terminal voltage of a dc generator from decreasing with increasing load.

22. No voltage is induced in a coil that cuts through a neutral plane.

23. Brushes are set so that they are in contact with coils that are momentarily in a neutral plane.

24. Commutation is the process of reversing the direction of the current in an armature coil.

25. In a dc generator, armature reaction creates a magnetomotive force that distorts the main field flux and shifts the neutral plan in the direction of the rotation of the armature.

26. Interpoles are placed between the main field poles to minimize the effects of armature reaction.

27. An electric motor operates on the principle that two magnetic fields can be made to interact to produce motion.

28. Direct current motors are used where a wide range of torque and speed control is required.

29. A dc machine can operate either as a motor or as a generator.

30. Direct current permanent-magnet servomotors are used on machines requiring exact positioning of an object.

31. The series-type dc motor can start very heavy loads, but the speed will increase as the load is decreased.

32. Direct current series motors should never be operated without a load.

33. The shunt-type dc motor is used where constant speed is desirable in the driven machine.

34. Direct current shunt motors can race to dangerously high speeds if current to the field coil is lost.

35. The compound-type dc motor uses both series and shunt field windings and produces characteristics intermediate to the series and shunt motors.

36. In a dc motor, armature reaction shifts the neutral commutating plane against the direction of rotation.

37. Generator action in a motor produces a cemf that opposes the applied terminal voltage.

38. As the speed of a dc motor increases, the cemf increases and the current drawn by the motor decreases.

39. The speed of a dc motor is controlled either by varying the field current or by varying the voltage applied to the armature.

40. Decreasing the field strength of a dc motor will cause it to turn faster.

41. Raising or lowering the armature voltage of a dc motor will cause the motor to rise and fall in proportion.

42. Compared to dc motors, ac motors cost less, require less maintenance, are more rugged, and are physically smaller than dc motors of the same horsepower.

43. A common feature of all ac motors is a rotating magnetic field.

44. The formula to determine the synchronous speed of an ac motor is:

$$S = \frac{120f}{P}$$

45. In an induction motor, the alternating current in the stator induces a voltage into the rotor winding to produce rotor current and magnetic field.

46. The formula to determine the percentage slip of an induction motor is:

$$\% \text{ slip} = \frac{\left(\begin{array}{c}\text{Synchronous}\\\text{speed}\end{array}\right) - \left(\begin{array}{c}\text{Running}\\\text{speed}\end{array}\right)}{\text{Synchronous speed}} \times 100$$

47. Compared to polyphase motors, single-phase motors require an auxiliary means of starting, are larger in size for the same horsepower, deliver irregular torque, and operate at a lower power factor and efficiency.

48. The direction of rotation of a three-phase induction motor may be reversed by interchanging any two of the three-phase leads that supply current to the stator.

49. The speed of an induction motor is determined by the number of poles and frequency of the power supply (not the supply voltage).

50. A wound-rotor induction motor is an induction motor with a wound rotor that is used for variable speed applications.

51. The more resistance that is placed in the rotor circuit of a wound-rotor induction motor, the slower the motor will turn.

52. An ac synchronous motor runs at constant speed from no-load to full-load.

53. A three-phase synchronous motor requires some auxiliary means for starting.

54. Three-phase synchronous motors produce a leading power factor when operated with an overexcited rotor.

55. The mechanical power rating of motors is expressed in either horsepower or watts.
56. Two factors that determine mechanical power output are torque and speed.
57. Specific losses associated with the operation of a motor include core loss, stator loss, rotor loss, stray load loss, and windage and friction loss.
58. Motor enclosures are designed to provide protection, depending on the environment in which the motor has to operate.
59. The NEMA design classification indicates the torque characteristics of different induction motors.
60. Motor insulation is classified by letter according to the temperatures they are capable of withstanding.
61. Both sleeve and ball bearings are used in electric motors.
62. Excessive bearing wear can be caused by improper lubrication, incorrect alignment of motors and loads, and excessive belt tension.
63. Belt-drive systems are very useful because a standard speed motor can be used to drive a load at a given speed.
64. Undersized wire between the motor and power source will limit starting abilities and cause overheating of the motor.
65. All motors should be grounded as a precaution against electric shock.
66. Use of motors with proper overload protectors is advisable.
67. A motor maintenance program should include scheduled periodic inspections, keeping motor clean and dry, and vibration analysis.

CHAPTER REVIEW QUESTIONS

Answer the following questions.

6-1. List four factors that determine the amount of voltage induced in a conductor as it moves through a magnetic field.

6-2. What is the advantage of a revolving-field ac generator over that of a stationary-field type?

6-3. Explain the operation of a brushless exciter system.

6-4. Define each of the following as they apply to a sine wave:
   a. Cycle
   b. Frequency
   c. Peak value
   d. RMS value

6-5. In the case of the ac generator, what two factors determine the amount of voltage generated?

6-6. Explain how a voltage regulator maintains the terminal voltage of an ac generator constant under varying load conditions.

6-7. List the different 1ϕ and 3ϕ voltages available from a standard 4-wire wye-connected ac system.

6-8. List four conditions that must be fulfilled before two three-phase ac generators may be paralleled.

6-9. What type of applications are best suited for cogeneration systems?

6-10. Explain in what way the armature circuit of a stationary-field ac generator is different from that of a dc generator.

6-11. In what way is the self-excited dc generator different from the separately excited type?

6-12. List three reasons why the terminal voltage of a shunt generator drops when full-load is applied.

6-13. Why was the compound generator developed?

6-14. In what position are the brushes of a dc generator set? Why?

6-15. Define *commutation*.

6-16. a. What is the cause and effect of armature reaction in a dc generator?
   b. State two ways in which the effect of armature reaction in a generator can be minimized.

6-17. Why are dc motors seldom used in normal industrial applications?

6-18. What special application may warrant the use of a dc motor?

6-19. When a current-carrying conductor is placed in a magnetic field, what three factors determine the amount of force exerted on it to move?

6-20. **a.** How is the direction of rotation of a permanent-magnet dc motor changed?

b. How is the speed of a permanent-magnet motor controlled?

6-21. Explain the function of the commutator in the operation of a dc motor.

6-22. Outline the unique operating characteristics of a dc series motor.

6-23. Why should large dc series motors not be operated without a load?

6-24. How is the speed of a dc shunt motor controlled above base speed?

6-25. How are the series and shunt field windings of a dc compound motor normally connected?

6-26. What two factors determine the direction of rotation of a dc motor?

6-27. **a.** Where and how is cemf produced in a dc motor?

b. What happens to the cemf and the current drawn by the motor as the motor accelerates?

6-28. Under what operating conditions is the base speed of a dc motor obtained?

6-29. List four advantages of ac motors over dc types.

6-30. How is the rotating magnetic field produced in a three-phase motor?

6-31. Calculate the synchronous speed of the rotating magnetic field of a four-pole, three-phase ac motor feed from a standard voltage source.

6-32. Why is the induction motor so named?

6-33. Outline the principle of operation of a three-phase, squirrel-cage induction motor.

6-34. **a.** How is the speed of a three-phase wound-rotor induction varied?

b. How is its direction of rotation reversed?

6-35. What is the major difference between the starting requirement for a three-phase and a single-phase induction motor?

6-36. **a.** Outline the starting sequence for a three-phase synchronous motor.

b. State two advantages of using synchronous motor drives in an industrial plant.

6-37. What two factors determine the mechanical power output of a motor?

6-38. List five specific heat losses associated with the operation of a motor.

6-39. What determines the type of motor enclosure selected for a particular application?

6-40. What NEMA design type motor would be selected for driving a pump that requires a high starting torque at normal starting current?

6-41. How can a motor be damaged by overlubrication of a ball bearing?

6-42. Outline the general procedure to follow when installing a belt on a belt-drive system.

6-43. How can undersized wire between the motor and power source affect the operation of a motor?

6-44. In what type of application is it advisable to use a built-in manual reset motor overload device?

CRITICAL THINKING QUESTIONS

6-1. With reference to the generation of a sine wave voltage in a coil:

a. Why is the voltage zero two times during each cycle of rotation?

b. Why does the induced voltage reverse in polarity during each cycle of rotation?

6-2. **a.** An ac sine wave voltage is measured using an ac voltmeter and found to be 220 V. What is the peak value of this voltage?

b. What is the effective, or rms, value of this voltage?

6-3. A prime mover turning at 200 rpm is connected to an ac generator. If the induced voltage has a frequency of 60 Hz, how many poles does the revolving-field rotor have?

6-4. If the phase voltage in a three-phase, 4-wire, wye-connected alternator is 80 kV, what is the line voltage?

6-5. Why are all dc generators constructed with stationary fields?

6-6. A self-excited dc generator (known to be in working order) is driven by a prime mover, but fails to build up to full rated voltage. Suggest four possible causes of this trouble.

6-7. The output of a dc self-excited shunt generator is accidentally short-circuited. What will happen to the output voltage and current of the generator? Why?

6-8. Why can the universal motor operate from either an ac or a dc source?

6-9. Assume the terminal identification numbers on a compound motor are wrong or missing.
  a. How can the armature, shunt field leads, and series field leads be identified using an ohmmeter?
  b. What operating test could be made to ensure cumulative connection of the shunt and series fields?

6-10. A six-pole induction motor is operated from a three-phase, 60-Hz source. If the full-load speed is 1140 rpm, calculate the slip.

6-11. Give reasons why it would not be recommended to use a 40-hp induction motor to drive a 10-hp load.

6-12. Assume one of the lines to a three-phase squirrel-cage motor becomes open.
  a. Will the motor continue to rotate?
  b. In what way could the motor be damaged?
  c. Will the motor be able to start on such a line?

6-13. A motor appears hot to the touch. Does this always indicate it is operating at too high a temperature? Explain.

6-14. If a drive motor ran at 1200 rpm with a 4-in. pulley, and a 10-in. pulley was connected to the driven machine, what is the speed of the driven machine?

6-15. A noticeable increase in motor vibration is noted during a routine inspection. What possible faults does this indicate?

Answers to Self-Tests

1. induced
2. revolving
3. true
4. true
5. 360 degrees
6. 1800 rpm
7. 10 percent
8. field
9. wye; delta
10. false
11. true
12. false
13. thermal
14. true
15. commutator
16. true
17. exciting
18. separately excited

19. false
20. series; shunt; compound
21. true
22. residual
23. decreasing
24. neutral
25. commutation
26. true
27. false
28. 2000 A
29. electric; magnetic
30. false
31. false
32. magnetic fields
33. true
34. true
35. false

36. true
37. true
38. series; shunt
39. false
40. false
41. cemf
42. resistance
43. base
44. false
45. false
46. true
47. rotating
48. 600 rpm
49. rotor
50. three; squirrel
51. 6.7 percent
52. true
53. phase

54. false
55. true
56. true
57. overexcited
58. 26.8 hp
59. false
60. heat
61. true
62. TEFC
63. design B
64. false
65. sleeve; ball
66. true
67. 4 in.
68. true
69. false
70. true

CHAPTER 7

Relays

∎

CHAPTER OBJECTIVES

This chapter will help you to:

1. *Compare* the electromagnetic and solid-state relay in terms of construction and operation.
2. *Identify* relay symbols used on schematic diagrams.
3. *Illustrate* uses of relays for industrial applications.
4. *Explain* how relays are rated.
5. *Describe* the operation of ON-delay and OFF-delay timer relays.
6. *Discuss* the use of relays in logic circuits.

Most applications in industry and in process control require relays as critical control elements. Relays are used primarily as switching devices in a circuit. This chapter explains the operation of different types of relays along with the advantages and limitations of each type. Relay specifications are also presented to show how to determine the correct relay type for different applications.

∎

7-1 ELECTROMECHANICAL CONTROL RELAYS

An *electromechanical relay* (*EMR*) is a magnetic switch. It turns a load circuit ON or OFF by energizing an electromagnet, which opens or closes contacts in the circuit. The EMR has a large variety of applications in both electric and electronic circuits. For example, EMRs may be used in the control of fluid power valves and in many machine sequence controls such as drilling, boring, milling, and grinding operations.

A relay will usually have only one coil, but it may have any number of different contacts. A typical EMR is shown in Fig. 7-1. Electromechanical relays contain both stationary and moving contacts. The moving contacts are attached to the plunger. Contacts are referred to as normally open (NO) and normally closed (NC). When the coil is energized, it produces an electromagnetic field. Action of this field, in turn, causes the plunger to move through the coil, closing the NO contacts and opening the NC contacts. The distance that the plunger moves is generally short—about ¼ in. or less.

Normally open contacts are *open* when no current flows through the coil but *closed* as soon as the coil conducts a current or is energized. *Normally closed contacts* are *closed* when the coil is deenergized and *open* when the coil is energized. Each contact is usually drawn as it would appear with the coil *deenergized.* Most machine control relays have some provision for changing contacts normally open to normally closed, or vice versa. It ranges from a simple flip-over contact to removing the contacts and relocating with spring location changes.

Many EMRs contain several sets of contacts operated by a single coil. Such relays are used to *control several switching operations by a single, separate current.* A typical control relay used to control two pilot lights is shown in Fig. 7-2. With the switch *open,* coil ICR is deenergized. The circuit to the green pilot light is completed through NC contact ICR 2, so this light will be ON. At the same time, the circuit to the red pilot light is opened through NO contact ICR 1, so this light will be OFF. With the switch closed, the coil is energized. The NO contact ICR 1 closes to switch the red pilot light ON. At

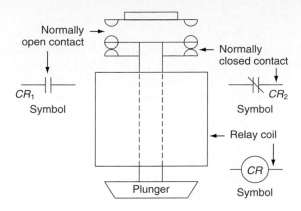

Fig. 7-1   Electromechanical relay (EMR).

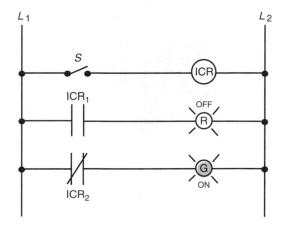

(a) Switch open—coil de-energized

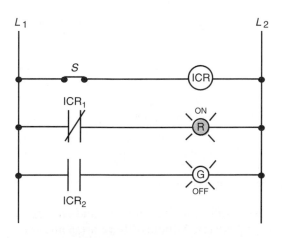

(b) Switch closed—coil energized

Fig. 7-2   Relays used to control several switching operations by a single, separate current.

the same time, the NC contact ICR 2 opens to switch the green pilot light OFF.

In general, control relays are used as auxiliary devices to switch control circuits and loads such as small motors, solenoids, and pilot

lights. The EMR can be used to *control a high-voltage load circuit with a low-voltage control circuit.* This is possible because the coil and contacts of the relay are electrically insulated from each other. From a safety point of view, this circuit provides extra protection for the operator. For example, assume that you want to use a relay to control a 120-V lamp circuit with a 12-V control circuit. The lamp would be wired in series with the relay contact to the 120-V source (Fig. 7-3). The switch would be wired in series with the relay coil to the 12-V source. Operating the switch would energize or deenergize the coil. This, in turn, would close or open the contacts to switch the lamp ON or OFF.

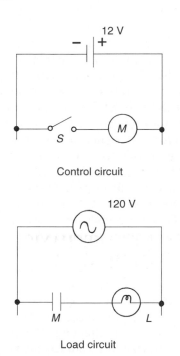

Control circuit

Load circuit

Fig. 7-3   Using a relay to control a high-voltage load circuit with a low-voltage control circuit.

Another basic application for a relay is to *control a high-current load circuit with a low-current control circuit.* This is possible because the current that can be handled by the contacts can be much greater than what is required to operate the coil. Relay coils are capable of being controlled by low-current signals from integrated circuits and transistors, as shown in Fig. 7-4 on page 204. In this circuit, the electronic control signal switches the transistor ON or OFF, which in turn causes the relay coil to energize or deenergize. The current in the control circuit, which consists of the transistor and relay coil, is quite small. The current in the

From page 202:

Electromechanical relay (EMR)

Normally open contacts

Normally closed contacts

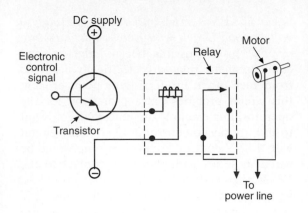

Fig. 7-4 Using a relay to control a high-current load circuit with a low-current control circuit.

power circuit, which consists of the contacts and small motor, is much larger in comparison.

The level of voltage at which the relay coil is energized, resulting in the contacts switching, is called the *pick-up voltage*. After the relay is energized, the level of voltage on the relay coil at which the contacts return to their unoperated condition is called the *drop-out voltage*. Relay coils are designed to not drop out until the voltage drops to a minimum of approximately 85 percent of the rated voltage. The relay coils also will not pick up (energize) until the voltage rises to 85 percent of the rated voltage. Generally, coils will operate continuously at 110 percent of the rated voltage without damage to the coil. Relay coils are now being made of a molded construction. This aids in reducing moisture absorption and increases mechanical strength.

There is also a difference in the current in the relay coil from the time the coil is first energized to when the contacts are completely operated. When the coil is energized, the plunger is in an out position. Because of the open gap in the magnetic path (circuit), the initial current in the coil is high. The current level at this time is known as *in-rush current*. As the plunger moves into the coil, closing the gap, the current level drops to a lower value. This lower value is called *sealed current*. The in-rush current approximates six to eight times the sealed current.

Electromechanical relays are made in a variety of types for different applications. Relay coils and contacts have *separate* ratings. Relay coils are usually rated for type of operating current (dc or ac), normal operating voltage or current, resistance, and power. Very sensitive relay coils, rated in the low milliampere range, are often operated from transistor or inte-

grated circuits. Figure 7-5(*a*), an open-type relay, is not enclosed; the contacts, coil, and all moving parts are exposed and thus readily visible. With the enclosed-type relay a plastic cover keeps the contacts from being exposed to corrosive environments. The plug-in type shown in Fig. 7-5(*b*) can be changed without disturbing the circuit wiring. When a relay is used for a particular application, one of the first steps should be to determine the control (coil) voltage at which the relay will operate. Coils are available to cover most standard voltages.

(*Courtesy of Allen-Bradley Company, Inc.*)

(*a*) Typical industrial control relay

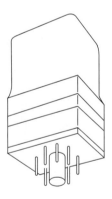

(*b*) Enclosed-type relay

Fig. 7-5 Typical control relays.

Relays differ in the number and arrangement of *contacts*. Although there are some *single-break contacts* used in industrial relays, most of the relays used in machine tool control have *double-break contacts* (Fig. 7-6). All contacts bounce on closing, and in rapid-operating relays this can be a source of trouble. The use of double-break contacts reduces this problem.

The most important relay contact specification is its *current rating*. This indicates the maximum amount of current the contacts are capable of handling. The three current ratings generally specified are:

- In-rush or "make contact" capacity.
- Normal or continuous carrying capacity.
- The opening or break capacity.

(a) Double-break contact is open
at two points

(b) Single-break contact is open
at one point

Fig. 7-6   Relay contact arrangements.

Contacts are also rated for the maximum ac- or dc-voltage level at which they can operate. Most relays are used in control circuits; therefore, their lower contact ratings (0 to 15 A maximum to 600 V) show the reduced current levels at which they operate. Although control relays from various manufacturers differ in appearance and construction, they are interchangeable in control wiring systems if their specifications are matched to the requirements of the system.

Most contacts today are made of a silver alloy rather than copper. This material is used because of the excellent conductivity of silver. Silver oxide, which forms on the contacts, is also a good conductor. Even when the contacts appear dull or tarnished, they are still capable of operating normally.

Self-Test

Answer the following questions.

1. The two basic parts of an electromechanical relay are the _____ and the _____.
2. Normally open contacts are defined as those contacts that are open when the coil is _____.
3. True or false? When no current flows through the coil of a relay, it is said to be energized.
4. True or false? The coil and contacts of a relay are not normally electrically insulated from each other.
5. True or false? The relay can be used to control a high-current load circuit with a low-current control circuit.

6. True or false? The pick-up and drop-out voltage of an EMR is normally the same value.
7. True or false? The in-rush current of an EMR is always greater than its sealed current.
8. True or false? It is possible for a relay coil to be rated for 6 Vdc and its contacts rated for 240 Vac.
9. True or false? Contact bounce on closing is normal in an EMR.
10. Control relays are interchangeable provided their _____ are matched.
11. Most relay contacts are made of a _____ alloy.

7-2   SOLID-STATE RELAYS

After performing switching tasks for several decades, the EMR is being replaced in some applications by a new type of relay, the *solid-state relay* (*SSR*) (Fig. 7-7 on page 206). Although EMRs and solid-state relays are designed to perform similar functions, each accomplishes the final results in different ways. Unlike EMRs, SSRs do not have actual coils and contacts. Instead, they use semiconductor switching devices such as bipolar transistors, MOSFETs, silicon-controlled rectifiers (SCRs), or triacs. The solid-state relay has no moving parts, it is resistant to shock and vibration, and it is sealed against dirt and moisture.

Like EMRs, SSRs find application in isolating a low-voltage control circuit from a high-power load circuit. A block diagram of an *optically coupled solid-state relay* is shown in Fig. 7-8(*a*) on page 206. A light-emitting diode (LED) incorporated in the input circuit glows when the conditions in that circuit are correct to actuate the relay. The LED shines on a phototransistor, which then conducts, causing the trigger current to be applied to the triac. Thus, the output is isolated from the input by the simple LED and phototransistor arrangement, just as the electromagnet isolated the input from the switching contacts in the conventional EMR. Because a light beam is used as the control medium, no voltage spikes or electrical noise produced on the load side of the relay can be transmitted to the control side of the relay. Most often, the *black-box approach* is used to symbolize an SSR. That is, a square or rectangle will be used on the schematic to represent the relay. The internal circuitry will not be shown, and only the input and output connections to the box will be given.

(a) Printed circuit board mounting

(b) Bulkhead mounting

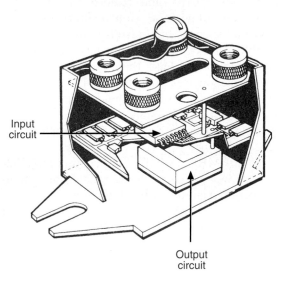

(c) Internal construction

(*Courtesy of Grayhill, Inc.*)

Fig. 7-7 Typical solid-state relays (SSRs).

Solid-state relays can be used to control ac or dc loads (Fig. 7-9). If the relay is designed to control an ac load, a *triac* is used to connect the load to the line. Solid-state relays intended for use as dc controllers have a power *transistor*, rather than a triac, connected to the load circuit. When the input voltage turns on the LED, a photodetector connected to the base of the transistor turns the transistor ON and connects the load to the line.

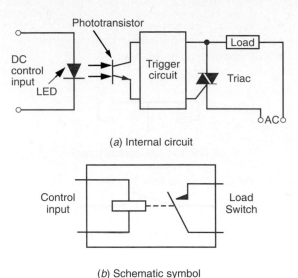

(a) Internal circuit

(b) Schematic symbol

Fig. 7-8 Optically coupled SSR.

The control voltage for SSRs can be direct current or alternating current and usually ranges from 3 to 32 V for the dc versions and 80 to 280 V for the ac versions. Maximum load circuit amps of up to 50 A are possible at input line voltage ratings of 120, 240, and 480 Vac. In most applications, SSRs are used to interface between a low-voltage control circuit and a higher ac line voltage.

Many SSRs used to control ac loads have a feature known as *zero switching* (Fig. 7-10). Zero switching ensures that the relay is turned ON or OFF at the beginning of the ac voltage wave at the zero crossover point. Zero voltage switching is often needed to reduce in-rush current and radio frequency interference (RFI).

Also available are hybrid SSRs that incorporate a small reed relay to serve as the actuating device (Fig. 7-11). A small set of reed contacts are connected to the gate of the triac. The control circuit is connected to the coil of the reed relay. When the coil is energized by the control current, a magnetic field is produced around the coil of the relay. This magnetic field closes the reed contacts, causing the triac to turn ON. In this type of SSR, a magnetic field, rather than a light beam, is used to isolate the control circuit from the load circuit.

The SSR has several advantages over the EMR. The SSR is more reliable and has a longer life because it has no moving parts. It is compatible with transistor and IC circuitry and does not generate as much electromagnetic interference. The SSR is more resistant to shock and vibration, has a much faster response time, and does not exhibit contact bounce.

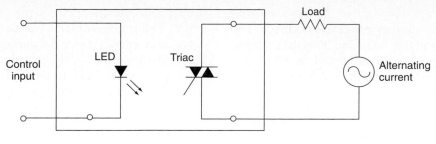

(a) Triac used to control ac load

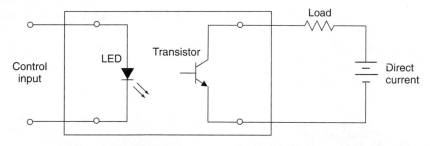

(b) Power transistor used to control dc load

Fig. 7-9   Controlling ac and dc loads.

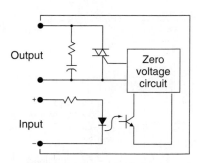

AC SSR utilizing zero voltage turn-on
and zero current turn-off of the load

Fig. 7-10   Zero switching.

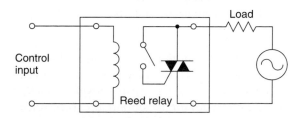

**Fig. 7-11   Hybrid solid-state relay incorporating a
small reed relay.**

As with every device, SSRs do have some disadvantages. The SSR contains semiconductors that are susceptible to damage from voltage and current spikes. Unlike the EMR contacts, the SSR switching semiconductor has a significant ON-state resistance and OFF-state leakage current.

Self-Test

Answer the following questions.

12. True or false? Solid-state and electromechanical relays perform similar functions.
13. An SSR has no _____ parts.
14. In an optically coupled SSR, the output is isolated from the input by a(n) _____.
15. True or false? An SSR designed to control an ac load may use a transistor to connect the load to the line.
16. True or false? The control voltage for SSRs can be direct current or alternating current.
17. _____ switching refers to a feature of SSRs that ensures the relay is turned ON or OFF at the beginning of the ac voltage wave.
18. Hybrid SSRs may incorporate a small _____ relay to serve as the actuating device.
19. True or false? Solid-state relays do not exhibit contact bounce.
20. True or false? Solid-state switching semiconductors have a significant ON-state resistance and OFF-state leakage current.

7-3   TIMING RELAYS

There are very few industrial control systems that do not need at least one or two timed functions. Typical applications include machines in which the start of an event must be delayed until another event has occurred. For example, a mixing machine may be delayed until the liquid has been heated, or a fan may remain OFF until a heating coil has heated the surrounding air.

Timing relays are conventional relays that are equipped with an additional hardware mechanism or circuitry to delay the opening or closing of load contacts. Timing relays are similar to other control relays in that they use a coil to control the operation of some number of contacts. The difference between a control relay and a timing relay is that the contacts of the timing relay delay changing their position when the coil is energized or deenergized.

A *pneumatic (air) timing relay* uses mechanical linkage and an air-bellows system to achieve its timing cycle (Fig. 7-12). The bellows design allows air to enter through a needle valve at a predetermined rate to provide the different time-delay increments and to switch a contact output. Pneumatic timing relays are popular throughout industry because they are rugged and dependable. They are adjustable over a wide range of time periods, they are relatively unaffected by temperature or voltage variations, and they have good repeat accuracy.

(Courtesy of Allen-Bradley Canada)

Fig. 7-12   Pneumatic (air) timing relay.

Some circuits require both timing contacts as well as instantaneous contacts operated by the same energized relay coil. The *instantaneous contacts* operate when the coil is energized or deenergized, independent of the timing mechanism. The timing contacts can be arranged to delay after energizing or deenergizing the coil. Figure 7-13 shows the construction of an ON-delay pneumatic (air) timer with two timed and two instantaneous contacts. When the coil is energized, the timed contacts are prevented from opening or closing. When the coil is deenergized, however, the timed contacts return immediately to their normal state. The instantaneous contacts change their positions immediately when the coil is energized and change back to their normal positions immediately when the coil is deenergized.

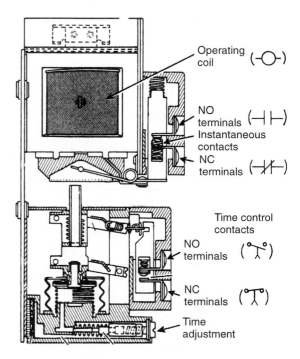

Fig. 7-13   Pneumatic ON-delay timer with both timed and instantaneous contacts.

A *solid-state timing relay* (Fig. 7-14) uses electronic circuitry to achieve its timing cycle. Some of these timers use a *resistor/capacitor (RC) time constant* to obtain the time base, and others use quartz clocks as the time base. An *RC* oscillator network generates a highly stable and accurate pulse that is used to provide the different time-delay increments and switch a contact output. The length of the time delay can be set by adjusting a control knob or potentiometer located on the front of the timer. Timing indication is provided by an LED that flashes during timing, glows steadily

(Courtesy of Allen-Bradley Canada)

(a) Timing relay

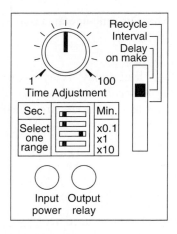

Type: digital CMOS circuitry.
Time ranges: 0.1 s to 1000 min in six ranges; selectable four-position dipswitch.
Repeat accuracy: ±0.1 percent, or ±16 ms, whichever is greater.
Modes of operation: Six-position switch selects Delay On Make, Interval, Single Shot, Recycling, Delay On Break, and Retriggerable Single Shot.
Operating voltage: 19 to 264 V alternating current 50–60 Hz and 19–30 volts direct current.
Status indication: Two LEDs indicate input power and output status.
Output: 10 amp resistive at 240 Vac, SPDT.
Operating temperature: –20° to +65°C.

(b) Typical time control adjustments and specifications

Fig. 7-14 Solid-state timing relay.

(a) On-delay NO and NC contacts

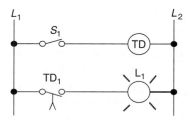

Sequence of operation

S_1 opens, TD deenergizes, TD_1 opens, L_1 is OFF.
S_1 closes, TD energizes, timing period starts. TD_1 is still open, and L_1 is still OFF.
After 10 s, TD_1 closes, and L_1 is switched ON.
S_1 opens, TD deenergizes, TD_1 opens instantly, and L_1 is switched OFF.

(b) ON-delay timer circuit with NO contact

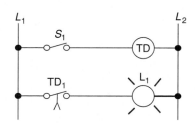

Sequence of operation

S_1 opens, TD deenergizes, TD_1 closes, L_1 is ON.
S_1 closes, TD energizes, timing period starts, TD_1 is still closed, and L_1 is still ON.
After 10 s, TD_1 opens and L_1 is switched OFF.
S_1 opens, TD deenergizes, TD_1 closes instantly, and L_1 is switched ON.

(c) ON-delay timer circuit with NC contact

Fig. 7-15 ON-delay relay.

after timing, and is OFF when the timer is deenergized.

Time-delay relays can be classified into two basic groups: ON-delay and OFF-delay. The ON-delay relay (Fig. 7-15) is often referred to as DOE, which stands for "delay on energize." When power is connected to the coil of an ON-

delay timer, the contacts delay changing position for some period of time. For this example a time delay of 10 s is assumed. When voltage is removed and the coil is deenergized, the contacts will immediately change back to their normal positions. The contact symbols shown are standard NEMA symbols. Time-delay relays can have NO, NC, or a combination of NO and NC contacts.

The OFF-delay relay (Fig. 7-16) is often referred to as DODE, which stands for "delay

(a) OFF-delay NO and NC contacts

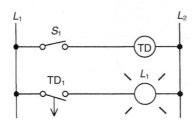

Sequence of operation

S_1 opens, TD deenergizes, TD_1 opens, and L_1 is OFF.
S_1 closes, TD energizes, TD_1 closes instantly, and L_1 is switched ON.
S_1 opens, TD deenergizes, timing period starts, TD_1 is still closed, and L_1 is still ON.
After 10 s, TD_1 opens and L_1 is switched OFF.

(b) OFF-delay timer circuit with NO contact

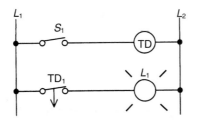

Sequence of operation

S_1 opens, TD deenergizes, TD_1 closes, and L_1 is ON.
S_1 closes, TD energizes, TD_1 opens instantly, and L_1 is switched OFF.
S_1 opens, TD deenergizes, timing period starts, TD_1 is still open, and L_1 is still OFF.
After 10 s, TD_1 closes and L_1 is switched ON.

(c) OFF-delay timer circuit with NC contact

Fig. 7-16   OFF-delay relay.

on deenergize." The operation of the OFF-delay timer is the opposite of the operation of the ON-delay timer. When voltage is applied to the coil of the OFF-delay timer, the contacts will change position immediately. When the coil is deenergized, however, there is a time delay before the contacts change to their normal positions.

The abbreviations *TO* and *TC* are used with standard contact symbols on some control schematics to indicate a time-operated contact

(Fig. 7-17). The abbreviation *TO* stands for "time opening," and *TC* stands for "time closing."

(a) Time-closing contact (must be connected to ON-delay relay if it is to be time-delayed when closing)

(b) Time-opening contact (must be connected to OFF-delay relay if it is to be time-delayed when opening)

Fig. 7-17   Alternate time-operated contact symbols.

Solid-state time-delay relays can also be designed for *interval-ON* and *recycle* operation. Different timing functions are illustrated in Fig. 7-18.

Self-Test

Answer the following questions.

21. Timing relays are used to _____ the opening or closing of contacts for circuit control.
22. A(n) _____ timing relay uses mechanical linkage and an air-bellows system to achieve its timing cycle.
23. The _____ contacts of a timing relay operate independently of the timing mechanism.
24. A solid-state timing relay uses _____ circuitry to achieve its timing cycle.
25. The ON-delay timer is sometimes referred to as DOE, which represents _____.
26. An OFF-delay timing relay provides time delay when its coil is _____.
27. The abbreviation *TC* used with standard contact symbols represents _____.

7-4   LATCHING RELAYS

Electromechanical latching relays are designed to hold the relay closed after power has been removed from the coil. Latching relays are used where it is necessary for contacts to stay open and/or closed even though the coil is

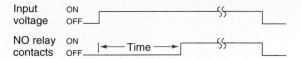

The delay period begins when input voltage is applied. The relay remains deenergized until the end of the delay period, at which time it energizes and operates the output contacts. Reset is accomplished by removal of the input voltage.

(a) Delay on operate

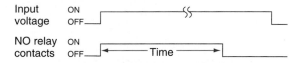

The relay operates immediately upon application of input voltage. At the end of the adjustable, predetermined interval, the relay drops out, even though input voltage is still applied. The relay remains dropped out until removal and reapplication of input voltage.

(b) Interval on

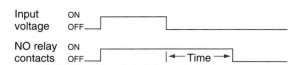

The relay operates immediately upon application of input voltage. Timing begins when input voltage is removed. When timing is complete, the relay deenergizes. Reset occurs when input voltage is reapplied.

(c) Delay on dropout

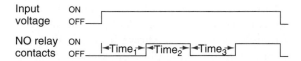

The first delay period begins when input voltage is applied. At the end of the first delay, or OFF period, the internal relay pulls in, and the second delay, or ON period, begins. When the second delay period ends, the relay drops out. This recycling sequence continues until the input voltage is removed. At this time, relay drops out.

(d) Recycle timing

Fig. 7-18 Timing functions.

energized only momentarily. Figure 7-19(*a*) shows a mechanically held latching relay that uses two coils. The *latch coil* is momentarily energized to set the latch and hold the relay in the latched position. The *unlatch* or *release coil* is momentarily energized to disengage the

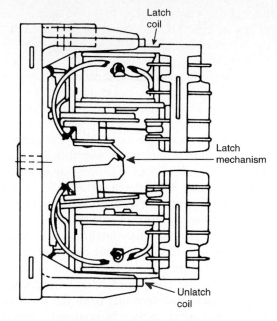

(a) Mechanically held latching relay

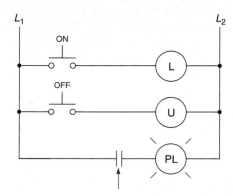

(b) Latching-type relay circuit (contact with relay in unlatched position)

Fig. 7-19 Latching relay.

mechanical latch and return the relay to the unlatched position.

Figure 7-19(*b*) shows the schematic diagram for an electromagnetic latching-type relay circuit. The contact is shown with the relay in the *unlatched* position. In this state, the circuit to the pilot light is open, so the light is OFF. When the ON button is *momentarily* actuated, the latch coil is energized to set the relay to its latched position. The contacts close, completing the circuit to the pilot light, so the light is switched ON. Note that the relay coil *does not* have to be continuously energized to hold the contacts closed and keep the light ON. The only way to switch the lamp OFF is to actuate the OFF button, which will energize the unlatch coil and return the contacts to their open, unlatched

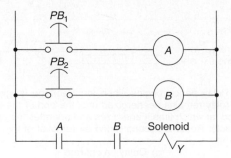

(a) Safety circuit for punch press

(b) Computer logic symbol for an AND gate

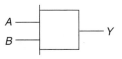

(c) NEMA logic symbol for an AND gate

Fig. 7-20   Relay logic AND gate.

state. In cases of power loss, the relay will remain in its original latched or unlatched state when power is restored. This arrangement is often referred to as a *memory relay*.

The latching relay has several advantages in electrical circuit design. For example, it is common in a control circuit to have to remember when a particular event takes place and not permit certain functions once this event occurs. Running out of a part on an assembly line may signal the shutdown of the process by momentarily energizing the unlatch coil. The latch coil would then have to be momentarily energized before further operations could occur.

Another use for the latching-type relay involves power failure. Circuit continuity during power failures is often important in automatic processing equipment, where a sequence of operations must continue from the point of interruption after power is resumed—rather than return to the beginning of the sequence. In addition, latching relays are often used in machine tool control circuits. These relays can be latched and unlatched through the operation of pilot devices such as limit switches and pushbuttons.

Self-Test

Answer the following questions.

28. True or false? Latching relays are designed to hold the relay closed after power has been removed from the coil.
29. In a latching relay that uses two coils, the coils are identified as the _____ coil and _____ coil.
30. True or false? Latching relays can provide circuit continuity during power failures.

7-5   RELAY LOGIC

A relay can be considered digital in nature because it is basically an ON/OFF, two-state device. The coil is the input and the contacts are the output. Whereas magnetic relays are single-input, multi-output devices, solid-state logic gate circuits are multi-input, single-output devices.

Control circuitry that requires two or more functions to be completed as a precondition for another event to take place describes the AND circuit. Figure 7-20 illustrates the relay equivalent circuit of a logic AND gate. This is an example of a safety interlock found on

many punch presses. Both pushbuttons PB_1 and PB_2 must be pushed at the same time if the solenoid is to energize and operate the punch. The controls are placed on opposite sides of the punch press in such a way that the operator must use both hands to operate the machine. This placement of the switches eliminates any possibility of injury to the operator's hands.

Control circuitry in which one condition or another separate condition can cause an event to take place describes the OR circuit. The example in Fig. 7-21 shows a circuit that can turn a light ON if the photosensor senses darkness *or* someone turns the switch to the ON position. Note that either the sensor or the switch can allow the light to be illuminated and that they can occur independent of each other. These are the basic criteria for the OR circuit.

The requirement of a NOT or inverting gate is that it will produce an output when an input is *not* present. There are occasions when an event is taking place and some indication is desired to specify the negative or opposite indication. The example in Fig. 7-22 shows a circuit that indicates the open (OFF light) and closed (ON light) state of the pressure switch.

Memory is used in logic circuits to recall or store past events. Figure 7-23 illustrates an example of OFF-return memory, which remembers the state of its output until the power is

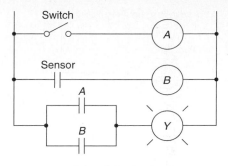

(a) OR light circuit

(b) Computer logic symbol for an OR gate

(c) NEMA logic symbol for an OR gate

Fig. 7-21   Relay logic OR circuit.

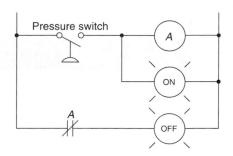

(a) NOT circuit used to indicate state of pressure switch

(b) Computer logic symbol for a NOT gate

(c) NEMA logic symbol for a NOT gate

Fig. 7-22   Relay logic NOT circuit.

turned OFF and then always goes to the OFF condition. The operation of the circuit is similar to that of a magnetic starter with a 3-wire control circuit. Momentarily actuating PB_2 turns the motor ON. It will remain ON until PB_1 is momentarily actuated. This memory ele-

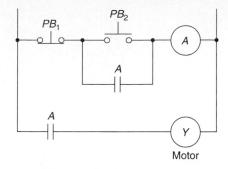

(a) OFF-return memory circuit

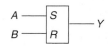

(b) Computer logic symbol for memory

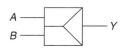

(c) NEMA logic symbol for memory

Fig. 7-23   Memory with relays.

ment remembers the conditions of its output as long as the power remains ON.

Memory that is lost with a power failure is called *volatile* memory. Memory that is retained with a power loss is called *nonvolatile*. Nonvolatile relay memory is implemented using a latching relay that mechanically remembers its last position.

Logic control can be implemented by three different technologies. Relay control was first, and it is still being used. The second technology was solid-state logic modules [Fig. 7-24(a) on page 214], a minor player because it led to the development of inexpensive electronic computers. The third, and latest, technology is the microcomputer, which has many important advantages over the other two. Microcomputer technology has been responsible for the programmable logic controller (PLC), which is the device of choice for many control systems [Fig. 7-24(b)].

Self-Test

Answer the following questions.

31. A relay can be considered digital in nature because it is basically a(n) _____ state device.

32. Control circuitry that requires two or more functions to be completed for another

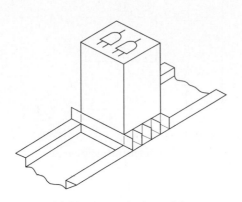

(a) Plug-in type logic module

Output from programmable logic controller

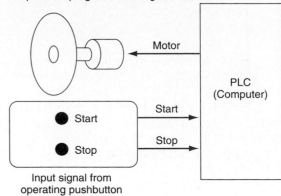

(b) PLC (programmable logic controller)

Fig. 7-24 Solid-state logic control.

event to take place describes the logic _____ circuit.

33. Control circuitry in which one condition or another separate condition can cause an event to take place describes the logic _____ circuit.

34. The requirement of a logic NOT circuit is that it will produce an output when a(n) _____ is not present.

35. Nonvolatile relay memory is implemented using a(n) _____ relay.

36. True or false? Relay logic control is the most popular choice for complex control systems.

SUMMARY

1. An electromechanical relay (EMR) operates by energizing an electromagnet, which in turn opens or closes contacts.

2. Normally open (NO) relay contacts are open when the coil is deenergized but closed as soon as the coil is energized.

3. Normally closed (NC) relay contacts are closed when the coil is deenergized but open as soon as the coil is energized.

4. Relays with multiple contacts are used to control several switching operations by a single separate current.

5. Because the coil and contacts of a relay are electrically insulated from each other, the relay can control a high-voltage load circuit with a low-voltage control circuit.

6. Because the current that can be handled by the contacts can be much greater than what is required to operate the coil, the relay can be used to control a high-current control circuit.

7. The level of voltage at which a relay coil is energized is called the pick-up voltage; it is always greater than the drop-out voltage.

8. The in-rush current of a relay coil is always greater than the sealed current.

9. Relay coils are rated for voltage, current, resistance, and power.

10. Relay contacts may be of the single-break or double-break type.

11. In addition to maximum operating voltage, relay contacts are rated for in-rush current, continuous current, and opening current.

12. Most relay contacts are made of a silver alloy.

13. Electromechanical and solid-state relays are designed to perform similar functions.

14. Solid-state relays (SSRs) use bipolar transistors, MOSFETs, SCRs, or triacs for switching load circuits.

15. In an optically coupled SSR, the output is isolated from the input by an LED.

16. Solid-state relays use triacs to control ac loads and transistors to control dc loads.

17. Zero switching is a feature of SSRs that ensures the relay is turned ON or OFF at the beginning of the ac voltage wave.

18. Hybrid SSRs incorporate a small reed relay to serve as the actuating device.

19. Advantages of SSRs over EMRs include (a) higher reliability, (b) longer life, (c) more compatibility with solid-state cir-

cuits, (d) the generation of less electro-magnetic interference, (e) more resistance to vibration, (f) faster response time, and (g) no contact bounce.

20. Unlike contacts, SSR switching semiconductors have significant ON-state resistance and OFF-state leakage current.

21. Timing relay contacts delay changing position when a coil is energized or deenergized.

22. A pneumatic timing relay uses mechanical linkage and an air-bellows system to achieve its timing cycle.

23. A timing relay may contain both timing contacts as well as instantaneous contacts operated by the same energized relay coil.

24. Solid-state timing relays use an *RC* time constant or quartz clock to achieve their timing cycle.

25. The ON-delay timer is often called DOE, which means "delay on energize."

26. The OFF-delay timer is often called DODE, which means "delay on deenergize."

27. Latching relays are used where it is necessary for contacts to stay open and/or closed even though the coil is energized only momentarily.

28. A relay is digital in nature because it is basically an ON/OFF, two-state device.

29. AND relay logic requires two or more functions to be completed as a precondition for another event to take place.

30. With OR relay logic, one condition or another separate condition can cause an event to take place.

31. A NOT relay logic circuit will produce an output when an input is not present.

32. Memory is used in logic circuits to recall or store past events.

33. Nonvolatile relay memory is implemented using a latching relay.

CHAPTER REVIEW QUESTIONS

Answer the following questions.

7-1. Explain the basic operating principle of an EMR.

7-2. Define the terms *normally open* and *normally closed contact* as they apply to a relay.

7-3. List three basic ways in which control relays are put to use in electric and electronic circuits.

7-4. What are the three current ratings generally specified for control relay contacts?

7-5. What electronic component is used to control the output of an SSR that controls a dc voltage?

7-6. What electronic component is used to control the output of an SSR that controls an ac voltage?

7-7. Explain opto-isolation as it applies to an SSR.

7-8. Explain zero switching as it applies to an SSR.

7-9. Explain how isolation between the control and load circuit is obtained in SSRs that incorporate a reed relay.

7-10. List the advantages of SSRs over EMRs.

7-11. List three limitations of SSRs.

7-12. In what way is a timing relay different from a standard control relay?

7-13. What are instantaneous contacts?

7-14. **a.** What are the two basic classifications of timers?
   b. Explain the operations of each.

7-15. Name two methods used by electronic timers to obtain their time base.

7-16. Describe the operation of a latching relay that uses two coils.

7-17. Why can a relay be considered digital in nature?

7-18. What type of relay logic circuit would be used to:
   a. Produce an output, when an input is not present.
   b. Produce an output when each of several input conditions is met.
   c. Produce an output when any one of several input conditions is met.
   d. Upon failure of power, have output remain in its original state when power is restored.

7-19. List the three different technologies that can be used to implement logic control.

7-1. Why is the pick-up voltage of an EMR normally higher than its drop-out voltage?

7-2. A typical industrial relay has the following contact ratings: a 10-A noninductive continuous load (ac), and a 6-A inductive load at 120 Vac.

   a. What type of load would be classified as noninductive?

   b. What type of load would be classified as inductive?

7-3. Explain why contact bounce can be a source of trouble.

7-4. Explain what each of the following catalog specifications indicate for a typical SSR:

   a. *Duty:* continuous

   b. *Operating temperature:* −30°C to +85°C

   c. *Isolation:* 4000 V rms

   d. OFF-*state leakage current:* 10 mA max

   e. ON-*state voltage:* 1.6 V max

   f. *Turn-on time:* 8.3 ms max

   g. *Turn-off time:* 8.3 ms max

7-5. Explain the operation of the digital logic circuit in Fig. 7-25.

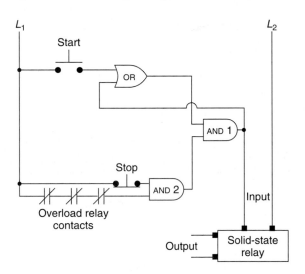

Fig. 7-25   Digital logic circuit.

Answers to Self-Tests

1. coil; contacts	10. specifications	19. true	28. true
2. deenergized	11. silver	20. true	29. latch; unlatch
3. false	12. true	21. delay	30. true
4. false	13. moving	22. pneumatic	31. two
5. true	14. LED	23. instantaneous	32. AND
6. false	15. false	24. electronic	33. OR
7. true	16. true	25. delay on energize	34. input
8. true	17. zero	26. deenergized	35. latching
9. true	18. reed	27. time closing	36. false

CHAPTER 8

Contactors and Motor Starters

∎

CHAPTER OBJECTIVES

This chapter will help you to:

1. *List* the basic uses for the contactor.
2. *Explain* the reason for using arc suppression on ac and dc contactors.
3. *Discuss* the major factors in selecting the size of a contactor and type of enclosure.
4. *Explain* the difference between a contactor and a motor starter.
5. *Explain* the function and operation of motor overload relays.
6. *Compare* NEMA- and IEC-rated contactors and motor starters.
7. *Describe* the operation of a solid-state contactor.

Often there are industrial applications dealing with large current and high voltages. In those cases, the technician is required to have a good knowledge and understanding of contactors and motor starters. This chapter explains how electromagnets applied to contactors can be used to make magnetic contactors and magnetic motor starters. Both NEMA and IEC standards are discussed. Operation of solid-state contactors is also included.

∎

8-1 MAGNETIC CONTACTOR

The magnetic contactor is similar in operation to the electromechanical relay (EMR). Both have one important feature in common: *contacts operate when the coil is energized.*

▷ YOU MAY RECALL   that relays are used for pilot duty in control circuits because their contacts are intended to interrupt only small currents usually rated at 15 A or less.

The National Electrical Manufacturers Association (NEMA) defines a *magnetic contactor* as a magnetically actuated device for repeatedly establishing or interrupting an electric power circuit. Unlike relays, contactors are designed to make and break electric power circuits without being damaged. Such loads include lights, heaters, transformers, capacitors, and electric motors for which overload protection is separately provided or not required (Fig. 8-1 on page 218).

The electromagnetically operated contactor is one of the most useful mechanisms ever devised for closing and opening electric circuits. Figure 8-2 on page 219 shows three of its applications. Contactors are used to switch power ON and OFF to a pump, as shown in Fig. 8-2(*a*). They are also used to switch power ON and OFF to a distribution panel, as shown in Fig. 8-2(*b*). They are also used with pilot devices to control the temperature and liquid level of a tank, as illustrated in Fig. 8-2(*c*).

The advantages of using magnetic contactors instead of manually operated control equipment include the following:

- Where large currents or high voltages have to be handled, it is difficult to build a suitable manual apparatus. Furthermore, such an apparatus is large and hard to operate. On the other hand, it is a relatively simple matter to build a magnetic contactor that will handle large currents or high voltages,

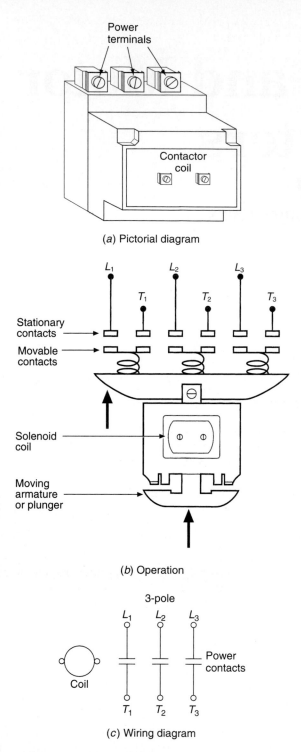

(a) Pictorial diagram

(b) Operation

(c) Wiring diagram

Fig. 8-1   Magnetic contactor.

and the manual apparatus must control only the coil of the contactor.

- Contactors allow multiple operations to be performed from one operator (one location) and interlocked to prevent false and dangerous operations.
- Where the operation must be repeated many times an hour, a distinct saving in effort will

result if contactors are used. The operator simply has to push a button and the contactors will automatically initiate the proper sequence of events.

- Contactors can be automatically controlled by very sensitive pilot devices. Pilot devices of this nature are limited in power and size, and it would be difficult to design them to handle heavy current directly.
- High voltage may be handled by the contactor and kept entirely away from the operator, thus increasing the safety of an installation. The operator also will not be in the proximity of high-power arcs, which are always a source of danger from shocks, burns, or perhaps injury to the eyes.
- With contactors the control equipment may be mounted at a remote point. The only space required near the machine will be the space needed for the pushbutton. It is possible to control one contactor from as many different pushbuttons as are desired, with only the necessity of running a few light control wires between the stations.
- With contactors, automatic and semiautomatic control is possible with equipment such as programmable logic controllers (PLCs).

The principal parts of a magnetic contactor are the electromagnet and the contacts. Figure 8-3(a) on page 220 shows four different types of operating electromagnets: clapper, bell-crank, horizontal-action, and vertical-action. The magnetic circuit consists of soft steel with high permeability and low residual magnetism. The magnetic pull developed by the coil must be sufficient to close the armature against the forces of gravity and the contact spring. To prevent the armature from being held in by residual magnetism, a permanent air gap must be provided in the magnetic circuit [Fig. 8-3(b)]. This is generally accomplished by placing a shim of nonmagnetic material between the core and the supporting frame, under the core head or at the core face.

Contactor coils have a number of insulated turns of wire in order to give the necessary ampere-turns to operate on small currents. Coils are built to operate over a range of 80 to 110 percent of rated ac or dc voltage. Direct current contactor coils have a large number of turns and a *high* ohmic resistance. The current through them is limited by the resistance. The current through an ac coil is limited by the impedance of the circuit, and the reactance has a greater effect than the resistance. Consequently, the resistance of an ac contactor coil is

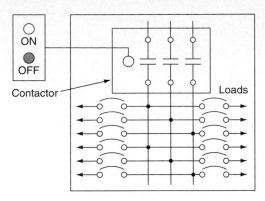

(*b*) Contactor used for a distribution panel

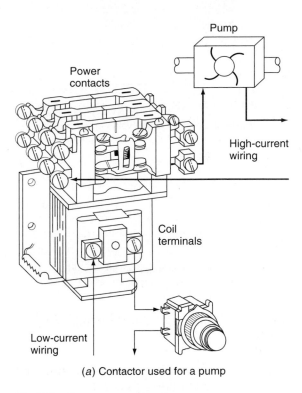

(*a*) Contactor used for a pump

Fig. 8-2   Contactor applications.

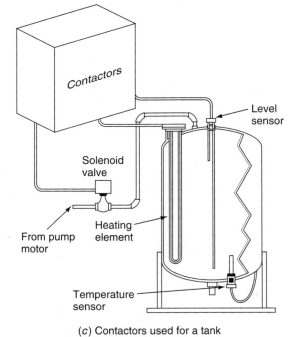

(*c*) Contactors used for a tank

low, and the number of turns is relatively small.

With a dc magnet, the current in the coil is the same whether the contactor is open or closed. With an ac magnet, however, the current in the coil is largely determined by the reactance of the circuit, which is lower when the contactor is open because of the air gap in the magnetic circuit. Therefore, there will be a high in-rush of current through the coil when the contactor is first connected to the supply line. The in-rush may be 5 to 20 times as high as the current that will flow through the coil when the contactor has closed. This fact must be taken into account when ac contactors and relays are used. Moreover, care must be taken that the pilot device handling the coil circuit has ample capacity to pass the in-rush current.

The attraction of an electromagnet operating on alternating current is pulsating and

equals zero twice during each cycle. As a result, the armature of an ac contactor coil will have a tendency to drop out or "chatter." This creates a noisy contactor and wear on its moving parts. The noise and wear can be eliminated by the use of *shading coils,* as shown in Fig. 8-4 on page 221. A shading coil is a single turn of conducting material (generally copper or aluminum) mounted on the face of the magnetic assembly. It sets up an auxiliary magnetic attraction that is out of phase with the main field and of sufficient strength to hold the armature tight to the core even though the main magnetic field has reached zero on the sine wave. With well-designed shading coils, ac contactors can be made to operate very quietly. A broken shading coil will make its presence known; the contactor will immediately become extremely noisy. Such a condition should be remedied at once because the

Clapper type

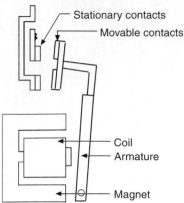

Stationary contacts
Movable contacts
Coil
Armature
Magnet

Bell-crank type

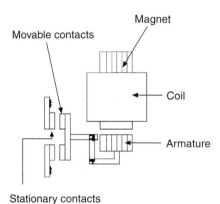

Magnet
Movable contacts
Coil
Armature
Stationary contacts

Horizontal-action type

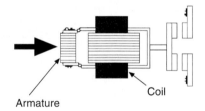

Armature
Coil

Vertical-action type

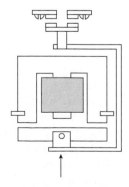

(a) Types of construction

Fig. 8-3 Contactor operating electromagnets.

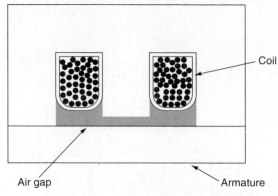

Coil
Air gap
Armature

(b) Permanent air gap prevents armature from being held in by residual magnetism

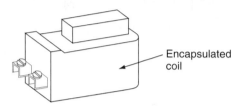

Encapsulated coil

(c) Contactor coil is molded into an epoxy resin to increase moisture resistance and coil life

Fig. 8-3 Contactor operating electromagnets (continued).

contactor will be subject to overheating and will soon cease to operate properly.

The core and armature of ac contactor assemblies are made of laminated steel, whereas dc assemblies are solid owing to the lack of eddy currents with continuous direct current. An ac contactor assembly may *hum* excessively because of improper alignment, foreign matter between contact surfaces, or loose laminations.

▐▶ **YOU MAY RECALL** that eddy currents are small amounts of current flow induced in the core and armature material by the magnetic flux lines.

An ac coil produces a counter electromotive force (cemf) that limits current flow through the coil when the contactor is energized. The amount of cemf depends on the proper alignment of the armature and core pole pieces and the resulting magnetic circuit. Misalignment or obstruction of the armature's ability to travel reduces cemf and causes an increased current flow in the coil. Depending on the amount of increased current, the coil may merely run hot, or it may burn out if the current increase is large enough and remains for a sufficient length of time.

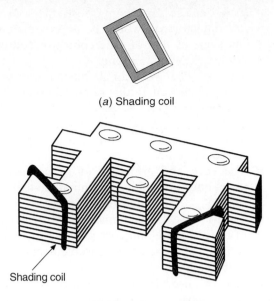

(a) Shading coil

Shading coil

(b) Magnetic assembly

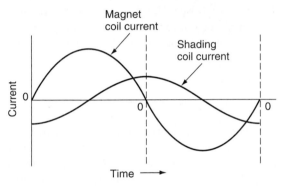

Magnet
coil current

Shading
coil current

Current

0

0

0

Time

(c) Magnet and shading coil currents are out-of-phase

Fig. 8-4   Shading coil.

The *main* contacts of a contactor act as a switch, opening and closing the circuit to the load. Generally the contactor is supplied in one-, two-, three-, or four-pole arrangements. Main contacts must carry their rated current without overheating, "make" current without bouncing or welding, and interrupt current without undue arcing.

One normally open (NO) *auxiliary* contact is generally supplied as standard on most contactors (Fig. 8-5). Additional NO and NC (normally closed) auxiliary contacts can be obtained as an option. The auxiliary contact has a much lower current rating and is used like a relay for interlocking or holding circuits. Figure 8-6 shows the schematic circuit for a three-phase heater circuit controlled by a magnetic contactor. The main power contacts CR-1,

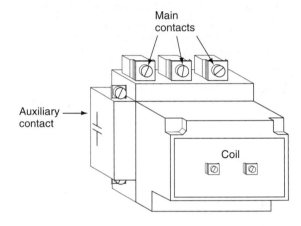

Main
contacts

Auxiliary
contact

Coil

Fig. 8-5   Main and auxiliary contacts.

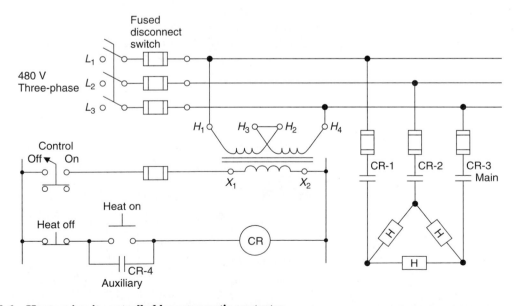

Fig. 8-6   Heater circuit controlled by a magnetic contactor.

CR-2, and CR-3 close to energize the heating elements at line voltage. The auxiliary contact CR-4 closes to seal in CR coil whenever the HEAT ON pushbutton is momentarily pressed.

Contacts are generally made of copper and/or silver. Silver contacts are not solid metal. They are silver-coated to reduce contact resistance. Contacts are subject to both electrical burning and mechanical wear. Copper contacts, during normal operation, should remain relatively clean owing to the rubbing action when the contacts open and close. This self-cleaning action removes dust and atmospheric coating from contacts *every* time they open or close. Copper contacts that seldom open and close, however, will discolor. The discoloration that forms is not a good conductor and may cause unnecessary heating if not removed. On larger contacts silver inserts are brazed or welded on copper contacts (on the heel), so silver carries the current and copper carries the arc on interruption. Most manufacturers recommend that silver contacts *never* be filed. Silver contacts need not be cleaned because the black discoloration that appears is silver oxide, which is a relatively good conductor of electricity.

Self-Test

Answer the following questions.

1. Unlike relays, contactors are designed to make and break electric _____ circuits without being damaged.
2. True or false? Where large currents or high voltages have to be handled, it is difficult to build suitable manual apparatus.
3. True or false? With contactors the control equipment may be mounted at a remote point.
4. The principal parts of a magnetic contactor are the _____ and the _____.

5. True or false? The ohmic resistance of an ac contactor coil would be approximately the same as that of a dc contactor coil of equivalent voltage rating.
6. True or false? With an ac magnet, the current in the coil is the same whether the contactor is open or closed.
7. A(n) _____ coil is installed in the magnetic assembly of an ac contactor coil to reduce chatter.
8. The core and armature of ac contactor assemblies are made of laminated steel to reduce _____ currents.
9. True or false? Misalignment or obstruction of the armature and core pole pieces of an ac coil will cause an increased current flow in the coil.
10. True or false? The main contacts of a contactor have a much lower current rating than the auxiliary contacts.
11. True or false? The discoloration that forms on silver contacts is not a good conductor and may cause unnecessary heating if not removed.

8-2 ARC SUPPRESSION

When large current switching contactors are utilized, some type of *arc suppression* is required to keep the contacts in the device from burning up. If a resistance load is interrupted (i.e., the contacts part), high current flows and a very low voltage across the arc is enough to sustain it [Fig. 8-7(a)]. As the distance between contacts increases, the resistance of the arc increases, the current decreases, and the voltage necessary to sustain the arc across the contacts increases [Fig. 8-7(b)]. Finally, a distance is reached at which full line voltage across the contacts is insufficient to maintain the arc [Fig. 8-7(c)]. The current drops to zero, and the arc extinguishes.

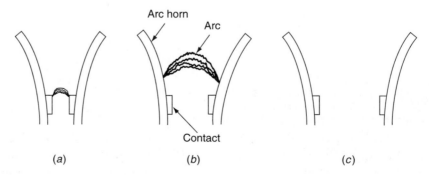

Fig. 8-7 Arc deflection between contacts.

With full-load current passing through this increasing resistance, a substantial temperature rise is created on the surface of the contacts. This temperature rise is often high enough to cause the contact surfaces to become molten and emit ions of vaporized metal into the gap between the contacts. Because this hot ionized vapor is electrically conductive, it will permit the current to continue to flow in the form of an arc. This arc can produce additional heat, which, if continued, can harm the contact surfaces.

Any arc can be extinguished by sufficient elongation, but to interrupt a current of 200 A at 230 V in a typical motor circuit, the required distance between contacts would have to be approximately 5 in. In larger-size contactors, over 400 A, *separate* arcing contacts are used. These are made of more durable material than the main contacts. Thus, the separate arcing contacts protect the main contacts so that they last longer. The arc must not be interrupted too quickly because it may restrike, owing to high rate of change and proportionally high voltage. Proper action is a sharp clear report, not a hissing tearing sound.

Direct current arcs are considerably more difficult to extinguish than ac arcs. A dc supply causes current to flow constantly and with great stability across a much wider gap than does an ac supply of equal voltage. An ac arc tends to be self-extinguishing when a set of contacts is being opened. In contrast to a dc supply of constant voltage, an ac sine wave supply voltage varies in magnitude and polarity passing through zero twice a cycle. As a result, the arc will have a maximum duration of no more than half a cycle. Also, during any half-cycle, the maximum arcing current is reached only once in that half-cycle.

To combat prolonged arcing in dc circuits, the contactor switching mechanism is constructed so that the contacts will separate rapidly and with enough air gap to extinguish the arc as soon as possible on opening. It is also necessary in closing dc contacts to move the contacts together as quickly as possible to avoid some of the same problems encountered in opening them. As a result, dc contactors tend to be faster-acting and somewhat larger than ac contactors in order to allow for the additional air gap.

Arcing may also occur on contactors when they are closing. One way this can occur is if the contacts come close enough together that a voltage breakdown occurs and the arc is able to bridge the open space between the contacts.

Another way this can occur is if a rough edge of one contact touches the other first and melts, causing an ionized path that allows current to flow. In either case, the arc lasts until the contact surfaces are fully closed.

Arc chutes are used on each set of contacts to help confine, divide, and extinguish arcs and the gases created by them (Fig. 8-8). Made of ceramic or asbestos compositions to withstand the high temperatures, they are designed to:

- Confine the arc to prevent it from striking other parts of the contactor.
- Cool the arc to reduce ionization and extinguish the arc more quickly.

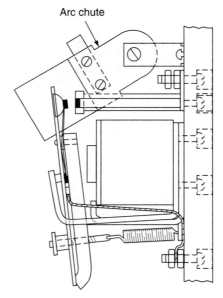

Arc chute

(a) Installed on a large contactor

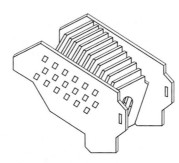

(b) Vertical barriers plus arc covers confine arcs to separate chambers and quickly quench them

Fig. 8-8 Arc chutes.

Arc chutes often contain heavy copper coils, called *blowout coils,* mounted above the contacts in series with the load to provide better arc suppression. These magnetic blowout coils help to extinguish an electric arc at contacts opening under ac and dc loads (Fig. 8-9). The blowout coil provides a magnetic field that blows out the arc in much the same way as you would blow out a match.

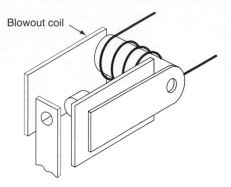

(a) Blowout coil enclosed in an arc chute

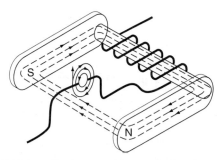

(b) Interaction between blowout field and arc current results in a force perpendicular to the direction of the blowout field

Fig. 8-9   Magnetic blowout coil.

The blowout coil is connected in series with the contacts, so load current is flowing through the coil as long as the contactor is closed or as long as there is an arc between the contacts. The current sets up a magnetic field through the core and pole pieces of the blowout structure and across the arcing tips of the contactor. When an arc is formed, the arc sets up a magnetic field around itself. The two magnetic fields repel each other, and the arc is forced upward and away from the contacts. It thus becomes longer and longer until it breaks and is extinguished. The extinguishing action is extremely rapid. It not only speeds up the operation of the contactor, but it also greatly reduces the wear on and burning of the contacts. Proper blowout design, therefore, is an important factor in contact life.

Contacts should be kept clean and periodically checked for wear. It is recommended to change a contact when worn 50 percent. If contacts are worn or burned to the point that replacement is necessary, they should always

be replaced in pairs to ensure that complete and proper surface contact is maintained. As part of a preventive maintenance program, large contactors should be checked periodically for contact wear, contact wipe, shunt terminal connections, free movement of the armature, blowout structure, blowout coil connections, coil structure, correct contact spring tension, and correct air gap.

Self-Test

Answer the following questions.

12. When switching high currents, some type of _____ suppression is normally required.
13. True or false? Direct current arcs are considerably more difficult to extinguish than ac arcs.
14. True or false? Arcing can occur only when contacts are open.
15. Arc _____ are used on large contactors to help confine the arc.
16. A blowout coil provides a(n) _____ that blows out the arc.
17. True or false? Worn contacts should always be replaced in pairs.

8-3   CONTACTOR SIZES AND RATINGS

Magnetic contactors are rated by NEMA according to the contactors' ability to carry rated current for 8 h without overheating according to size and type of load to be controlled. Tables 8-1 and 8-2 indicate the number size designation—00, 1, 2, 9, etc.—for general purpose ac and dc contactors and establish the current load carried by *each* contact in the contactor. As the NEMA size number classification increases, so does the current capacity and physical size of the contactor. Larger contacts are needed to carry and break the higher currents, and heavier mechanisms are required to open and close the contacts.

Magnetic contactors are also rated for the type of load to be utilized or actual applications. Load utilization categories include:

- *Nonlinear loads* such as tungsten lamps for lighting (large hot-to-cold resistance ratio—typically 10:1 or higher; current and voltage in phase).
- *Resistive loads* such as heating elements for furnaces and ovens (constant resistance; current and voltage in phase).

Table 8-1 Standard NEMA Ratings of AC Contactors, 60 Hz

Size	8-h Open Rating (A)	Power (hp)				
		Three-Phase			Single-Phase	
		200 V	230 V	230/460 V	115 V	230 V
00	9	1½	1½	2	⅓	1
0	18	3	3	5	1	2
1	27	7½	7½	10	2	3
2	45	10	15	25	3	7½
3	90	25	30	50	—	—
4	135	40	50	100	—	—
5	270	75	100	200	—	—
6	540	150	200	400	—	—
7	810	—	300	600	—	—
8	1215	—	450	900	—	—
9	2250	—	800	1600	—	—

Table 8-2 Standard NEMA Ratings of DC Contactors

Size	8-h Open Rating (A)	Power Rating (hp)		
		115 V	230 V	550 V
1	25	3	5	—
2	50	5	10	20
3	100	10	25	50
4	150	20	40	75
5	300	40	75	150
6	600	75	150	300
7	900	110	225	450
8	1350	175	350	700
9	2500	300	600	1200

- *Inductive loads* such as industrial motors and transformers (low initial resistance until the transformer becomes magnetized or the motor reaches full speed; current lags behind voltage).
- *Capacitive loads* such as industrial capacitors for power-factor correction (low initial resistance unit capacitor charges; current leads voltage).

Enclosed magnetic contactors must be housed in an approved enclosure based on the environment in which they must operate to provide mechanical and electrical protection. Electrical codes mandate the type of enclosure to use. More severe environments require more substantial enclosures. Severe environmental factors to be considered include:

- Exposure to damaging fumes.
- Operation in damp places.
- Exposure to excessive dust.
- Subject to vibration, shocks, and tilting.
- Subject to high ambient air temperature.

There are two general types of enclosures: *nonhazardous-location enclosures* and *hazardous-location enclosures*. Nonhazardous-location enclosures are further subdivided into the following categories:

- General-purpose (least costly)
- Watertight
- Oiltight
- Dusttight

Hazardous-location enclosures are *extremely costly,* but they are necessary in some applications. Hazardous-location, explosion-proof enclosures involve forged or cast material and special seals with precision-fit tolerances. The explosion-proof enclosures are constructed so that an explosion inside will not escape the enclosure. If an internal explosion were to blow open the enclosure, a general-area explosion and fire could ensue. Hazardous-location enclosures are classified into two categories:

- Gaseous vapors (acetylene, hydrogen, gasoline, etc.).

Inductive loads

Capacitive loads

Nonhazardous-location enclosures

Hazardous-location enclosures

■ Combustible dusts (metal dust, coal dust, grain dust, etc.).

All industrial electrical and electronic enclosures must conform to standards published by NEMA to meet the needs of location conditions. Although the enclosures are designed to provide protection in a variety of situations, the internal wiring and physical construction of the device remains the same. Consult the National Electrical Code (NEC) and local codes to determine the proper selection of an enclosure for a particular application.

Self-Test

Answer the following questions.

18. True or false? As the NEMA size number classification increases, the physical size of the contactor decreases.
19. True or false? A motor would be classified as a resistive load.
20. The two general types of enclosures are _____-location enclosures and _____-location enclosures.

8-4 MAGNETIC MOTOR STARTERS

The basic use for the magnetic contactor is for switching power in resistance heating elements, lighting, magnetic brakes, or heavy industrial solenoids. Contactors can also be used to switch motors if separate overload protection is supplied. A *magnetic starter* (Fig. 8-10) is a contactor with an overload relay physically and electrically attached.

The magnetic motor starter is similar to the contactor in design and operation. Both have one important feature in common: Contacts operate when the coil is energized. The important difference is the use of overload relays on the motor starter.

In their simplest and most widely used form, magnetic motor starters consist of a three- or four-pole magnetic contactor and an overload relay. These devices are mounted in a suitable enclosing case, which may be of general-purpose sheet-metal construction; dust-tight, water-tight, or explosion-resisting; or whatever may be required by the installation conditions. START and STOP pushbuttons may be mounted in the cover of the case. A separately mounted START-STOP pushbutton may also be used, in which case only the reset button would be mounted in the cover (Fig. 8-11). The starters

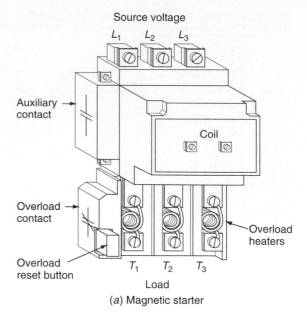

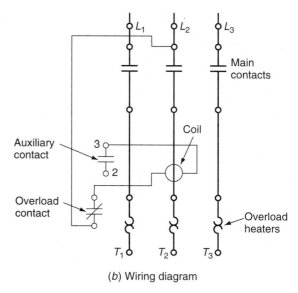

(a) Magnetic starter

(b) Wiring diagram

Fig. 8-10   Magnetic motor starter.

are also built in skeleton form, without enclosure, for mounting in a motor control center or control panel on a machine.

The control circuit of a magnetic motor starter is very simple. It involves only energizing the starter coil when the START button is pressed and deenergizing it when the STOP button is pressed or when the overload relay trips.

The overload relay incorporated into the motor starter distinguishes motor starter from a contactor. Contactor use is restricted to fixed lighting loads, electric furnaces, and other resistive loads that have set current values. Motors are subject to high starting currents and periods of load, no-load, short duration overload, and so on. They must have protec-

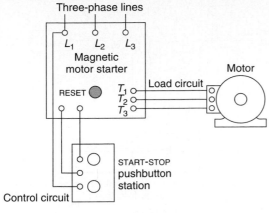

Three-phase lines

(a) Pictorial diagram

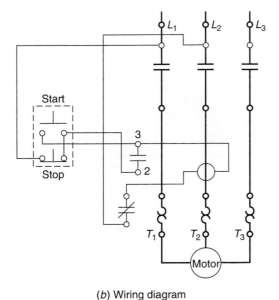

(b) Wiring diagram

(c) Schematic diagram of control circuit

Fig. 8-11   Magnetic motor starter with separately mounted START-STOP pushbutton.

tive devices with the flexibility required of the motor and driven equipment. The purpose of overload protection is to protect the motor windings from excessive heat resulting from motor overloading. The motor windings will not be damaged when overloaded for a short period of time. If the overload should persist, however, the sustained increase in current should cause the overload relay to operate, shutting off the motor.

External-overload protection devices, which are mounted in the starter, attempt to simulate the heating and cooling of a motor by sensing the current flowing to it. The goal is to protect the motor from overheating. The current drawn by the motor is a reasonably accurate measure of the load on the motor and thus of its heating. Four common forms are magnetic-overload relays, thermal-overload relays, electronic-overload relays, and fuses.

Magnetic-overload relays (Fig. 8-12) operate on the magnetic action of the load current that is flowing through a coil. When the load current becomes too high, a plunger is pulled up into the coil, interrupting the circuit. The tripping current is adjusted by altering the initial position of the plunger with respect to the coil.

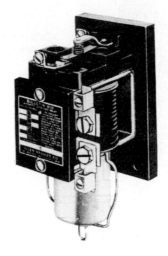

(Courtesy of Allen-Bradley Canada)

Fig. 8-12   Magnetic-overload relay.

A *thermal-overload relay* uses a heater connected in series with the motor supply (Fig. 8-13 on page 228). The amount of heat produced increases with supply current. If an overload occurs, the heat produced causes a set of contacts to open, interrupting the circuit. The tripping current is changed by installing a different heater for the required trip point. This type of protection is very effective because the heater closely approximates the actual heating within the windings of the motor and has a thermal "memory" to prevent immediate reset and restarting. There are two common types: the *bimetallic* type, which utilizes a bimetallic strip, and the *melting alloy* type, which utilizes the principle of heating solder to its melting point.

With a thermal-overload relay the same current that goes to the motor coils (causing the

(a) Thermal-overload relay symbol

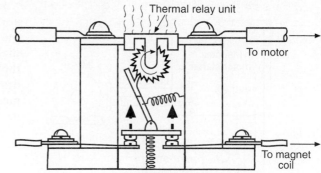

Drawing shows operation of melting alloy overload relay. As heat melts alloy, rachet wheel is free to turn. The spring then pushes contacts open.

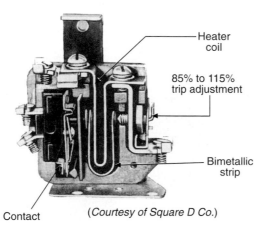

(Courtesy of Square D Co.)

(b) Bimetallic thermal-overload relay

One-piece thermal unit

Solder pot (heat-sensitive element) is an integral part of the thermal unit. It provides accurate response to overload current, yet prevents nuisance tripping.

Heating winding (heat-producing element) is permanently joined to the solder pot, so proper heat transfer is always ensured. No chance of misalignment in the field.

(Courtesy of Square D Co.)

(c) Melting alloy thermal-overload relay

Fig. 8-13   Thermal-overload relay.

motor to heat) also passes through the thermal elements of the overload relays. The thermal element is connected mechanically to an NC overload (OL) contact. When an excessive current flows through the thermal element for a long enough time period, the contact is tripped open. This contact is connected in series with the control coil of the starter. When the contact opens, the starter coil is deenergized. In turn, the starter's main power contacts open to disconnect the motor from the line. Some overload relays have a *trip indicator* built into the unit to indicate that an overload has taken place within the device. After a relay has tripped, the cause of the overload should be investigated. The problem should be solved before the RESET button is depressed to put the starter back into operation.

Selecting the *proper heater size* when installing magnetic motor starters is most important. All starter manufacturers list the size of overload heaters for a given starter application. The lists show the range of motor currents with which they should be used. These may be in increments of from 3 to 15 percent of full-load current (FLC). The smaller the increment, the closer the selection can be to match the motor to its actual work. Full-load current, operating temperature, service factor, and ambient temperature determine whether the size of the heater unit selected will actually protect the motor from an overload condition. An overload condition does *not* include short-circuit conditions; it is considered to be an excessive current condition for a long enough period of time to damage the motor. Heater units are selected from the manufacturer's tables by matching the FLC with the heater unit number. This assumes several things. First, the motor is operating at a maximum allowable temperature rise of 40°C and a service factor of 1.15. Many times the starter unit is not located near the motor. If this is the case and the temperature at the starter is higher than that of the motor, then a next higher rating is selected; if the temperature is lower, then the next lower size rating should be used.

An *electronic-overload relay* uses a current transformer and electronic circuitry. The transformer senses the current flowing to the motor, the electronic circuitry senses the changes in the motor current, then trips when the current reaches full-load (Fig. 8-14). If an overload condition exists, the sensing circuit interrupts the power circuit. The tripping current can be easily adjusted to suit a particular application. Electronic overloads often perform additional protective functions such as ground-fault and phase-loss protection.

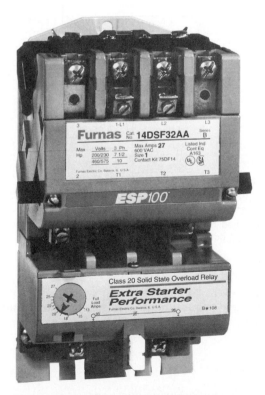

Fig. 8-14 Electronic overload relay mounted within a starter. (*Courtesy Furnas Electric Company*)

Some advantages of solid-state electronic-overload relays over thermal-overload types are the following:

- No buying, stocking, installing, or replacing of heater coils.
- Reduction in the heat generated by the starter.
- Energy savings (up to 24 W per starter) through the elimination of heater coils.
- Insensitivity to temperature changes in the surrounding environment.
- High repeat trip accuracy (±2 percent).
- Easily adjustable to a wide range of full-load motor currents.

Dual-element or *time-delay fuses* also can be used to provide overload protection. They have the disadvantage of being nonrenewable and must be replaced each time they operate.

Like the power contactor, magnetic motor starters are available in many sizes and types of enclosures to meet different requirements. The *International Electrotechnical Commission* (*IEC*) is European based and primarily used to rate electrical equipment used in Europe. The National Electrical Manufacturers Association is based in the United States and primarily used to rate electrical equipment in North America. Both IEC and NEMA rate contactors and motor starters. Both of their rating systems can be used to select motor control devices for maximum performance and high productivity, but it is important that the user understand the differences between IEC's and NEMA's standards in order to achieve the desired results.

Contactors and motor starters designed to NEMA standards are rated in *horsepower* and assigned a NEMA number. The size number may be selected by applying the appropriate horsepower, voltage, frequency, and phase to an easy-to-use chart (Fig. 8-15 on page 230).

Rather than define the specific ratings and sizes of control devices, the IEC works under the philosophy that performance is an integral part of the selection procedure. The user must first identify the utilization category of the specific application, then choose a product that is capable of handling the intended load in that utilization category (Fig. 8-16 on page 231). IEC contactors and motor starters are smaller per horsepower rating than those of NEMA. Also, IEC devices are built with materials required for average applications. NEMA devices, however, are built for a higher level of performance, and their electrical life is typi-

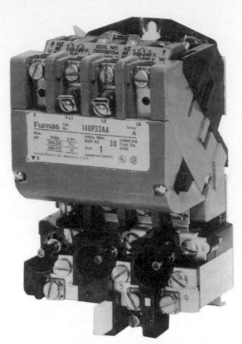

(*Courtesy of Furnas Electric Company*)

(*a*) Starter designed to NEMA standards

Size of Controller	Continuous Current Rating in Amperes	Horsepower at			
		60 Hz		50 Hz	60 Hz
		200 V	230 V	380 V	460 or 575 V
00	9	1 1/2	1 1/2	1 1/2	2
0	18	3	3	5	5
1	27	7 1/2	7 1/2	10	10
2	45	10	15	25	25
3	90	25	30	50	50
4	135	40	50	75	100
5	270	75	100	150	200
6	540	150	200	300	400
7	810	—	300	—	600
8	1215	—	450	—	900
9	2250	—	800	—	1600

(*b*) NEMA ratings for three-phase, single-speed, full-voltage magnetic controllers for nonplugging and nonjogging duty

Fig. 8-15   NEMA-rated contactors and starters.

cally 2.5 to 4 times higher for the same equivalent IEC device. IEC devices are less expensive, but they are more application sensitive; therefore, greater knowledge and care are needed during the selection process. IEC devices are commonly used in original equipment manufacturer machines, where machine specifications are known and will not change. NEMA devices are commonly used where

machine requirements and specifications could vary.

Self-Test

Answer the following questions.

21. A magnetic motor starter is a contactor with a(n) _____ added.

(Courtesy Furnas Electric Company)

(a) Starter designed to IEC standards

Utilization Category	Typical Duty
AC1	Noninductive or slightly inductive loads
AC2	Starting of slip-ring motors
AC3	Starting of squirrel-cage motors and switching off only after the motor is up to speed
AC4	Starting of squirrel-cage motors with inching and plugging duty

(b) Common IEC utilization categories for AC contactors

Fig. 8-16 IEC-rated contactors and starters.

22. True or false? A magnetic motor starter may have case-mounted or remote START-STOP pushbuttons.
23. True or false? Motors have set current values that do not fluctuate much.
24. True or false? The current drawn by a motor is a measure of its heating.
25. Two types of thermal-overload relays are the _____ and the _____.
26. True or false? Thermal-overload relays use NO contacts.
27. True or false? Overload relays are designed to provide short-circuit protection.
28. True or false? An electronic overload relay uses heater coils to sense the current flowing to the motor.
29. Motor starters designed to NEMA standards are rated in _____ and assigned a NEMA _____.
30. True or false? IEC motor starters are smaller per horsepower rating than those of NEMA.

8-5 SOLID-STATE CONTACTOR

Solid-state switching means interruption of power by nonmechanical electronic means. A solid-state contactor is a power-switching device designed to replace magnetic contactors for applications involving both resistive and inductive loads (Fig. 8-17).

Solid-state three-phase contactors, such as the one shown in Fig. 8-17(*a*), are especially suited to high-cycling applications owing to the absence of arc-producing air gap contacts. Sizes range from 10 to 600 A, with input volt-

(Courtesy Payne Engineering)
(a) Solid-state three-phase contactor

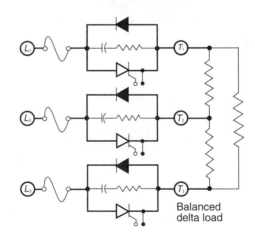

Balanced delta load

(b) Typical circuit schematic

Fig. 8-17 Solid-state power contactor.

ages from 240 to 550 Vac. Solid-state contactors now replace three-pole electromechanical contactors in industrial furnaces and ovens, mining and materials handling, and other industrial heating applications.

Silicon-controlled rectifiers (SCRs) enable reliable control of electric power from 1 kW to 1000 kW for most types of resistance heaters, motors, and other inductive loads. As the most common high-power semiconductor device, it consists of a specially treated silicon disc enclosed in a plastic or ceramic housing with metal power leads arranged for the anode and cathode connections and a smaller tab or wire for the gate connection. The SCR, like a contact, is in either the ON state (closed contact) or the OFF state (open contact). The SCR is analogous to a "latched relay" circuit—once the SCR is triggered, it will stay ON until its current decreases to zero. When current through the SCR stops, the "SCR switch" will open and stay open until retriggered (Fig. 8-18).

Since an SCR passes current in one direction only, two SCRs are necessary to switch ac power. The two SCRs are connected in inverse-parallel (back-to-back), as shown in Fig. 8-19, so that current can flow in both directions. Half the current is carried by each SCR, and a sinusoidal ac current flows through the resistive load R when gates G_1 and G_2 are fired at 0° and 180°, respectively. By varying the time interval between trigger pulses, as shown in Fig. 8-19(b), the output voltage is varied by blocking part of the input so that voltage is applied to the load for only part of each half-cycle.

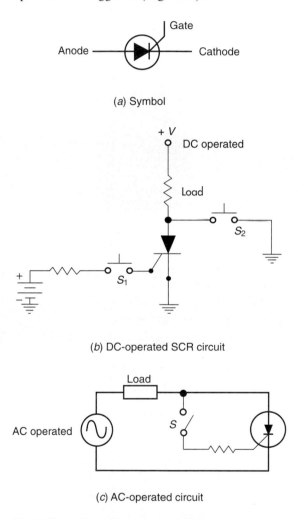

(a) Symbol

(b) DC-operated SCR circuit

(c) AC-operated circuit

Fig. 8-18   Silicon-controlled rectifier (SCR).

The dc-operated SCR circuit shown in Fig. 8-18(b) permits current to be switched to a load by the momentary closure of switch S_1 and removed from the load by the momentary closure of switch S_2. In ac-operated circuits, such as the one in Fig. 8-18(c), the SCR turn-off is automatic, because the current returns to zero twice each cycle.

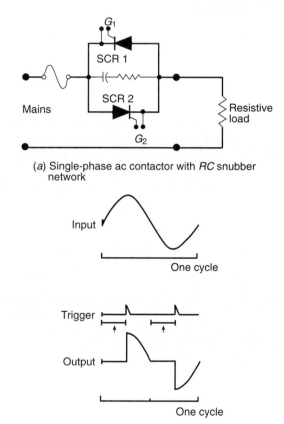

(a) Single-phase ac contactor with *RC* snubber network

(b) Output voltage is varied by blocking part of the input

Fig. 8-19   Solid-state single-phase ac contactor.

In contrast to a magnetic contactor, an electronic contactor is absolutely silent, and its "contacts" never wear out. Inductive loads and voltage transients are both seen as problem areas in solid-state ac control. Proper design of any solid-state ac control includes a resistor and capacitor in series connected in parallel to each power pole. These *"RC"* or "snubber" networks divert the charging current from the SCRs and help prevent unwanted turn-on.

All silicon semiconductors, when in the ON state, still have a small voltage drop across the junction of 1 to 2 V. The resultant 1 to 2 W of heat per conducted ampere through the device must be removed to the outside environment. Properly designed heat sinks accomplish this normally by conduction or convection heat transfer to maintain the silicon below its maximum temperature level.

The abrupt switching of the SCRs from the blocking state to the conducting state, particularly at higher current levels, may sometimes cause objectionable transients on the power line and create *radio frequency interference* (*RFI*). *Zero-fired control* refers to turning ON the SCRs at the zero voltage crossing for full cycles, applying full power or no power with the proportion of full cycles, or determining the resulting power to the load. This is sometimes called *integral cycle mode* or "burst fir-ing." The result is elimination of power line disturbances and RFI.

Self-Test

Answer the following questions.

31. True or false? Solid-state switching involves the interruption of power by non-mechanical means.
32. The two heavy power leads of a silicon-controlled rectifier (SCR) are known as the _____ and _____.
33. True or false? Two SCRs are necessary to switch ac power.
34. True or false? SCRs are normally mounted in heat sinks.
35. True or false? Zero-fired control refers to turning ON the SCR at the zero voltage crossing for full cycles.

Radio frequency interference (RFI)

Zero-fired control

Integral cycle mode (burst firing)

SUMMARY

1. A magnetic contactor is a magnetically actuated device for repeatedly establishing or interrupting an electric power circuit.
2. Advantages gained by the use of magnetic contactors include safer and simpler control of large currents or high voltages, accurate execution of repetitive machine operations, automatic control by pilot devices, and remote control of equipment.
3. The principal parts of a magnetic contactor are the electromagnet and the contacts.
4. Contactor coils are designed to operate over a range of 80 to 110 percent of rated ac or dc voltage.
5. Direct current contactor coils have high ohmic resistance; current is limited by resistance only.
6. Alternating current contactor coils have low ohmic resistance; current is limited mainly by the reactance of the coil.
7. Shading coils are used with AC contactor coils to prevent chatter of the armature assembly.
8. Alternating current contactor assemblies are made of laminated steel to reduce eddy current flow.
9. The main contacts of a contactor must carry their rated current without overheating, make current without bouncing or welding, and interrupt current without undue arcing.
10. The auxiliary contacts of a contactor have a much lower current rating and are used like a relay to interlock or hold circuits.
11. Contacts are generally made of copper and/or silver.
12. The discoloration that forms on copper contacts is not a good conductor.
13. The discoloration that forms on silver contacts is a relatively good conductor.
14. When large currents are switched, some type of arc suppression is required to keep the contacts from burning.
15. As the distance between contacts increases, the resistance of the arc increases, the current decreases, and, the voltage necessary to sustain the arc across the contacts increases.
16. A dc supply causes current to flow constantly. Because of this, dc arcs are considerably more difficult to extinguish than ac arcs.
17. Arc chutes are used on large contactors to help confine, divide, and extinguish arcs.
18. Blowout coils, connected in series with the load, provide a magnetic field that blows out the arc.
19. Worn contacts should always be replaced in pairs.
20. Magnetic contactors are NEMA rated by their ability to carry rated current for 8 h without overheating.

21. As the contactor NEMA size number classification increases, so does the current capacity and physical size of the contactor.
22. Load utilization categories for magnetic contactors include nonlinear loads, resistive loads, inductive loads, and capacitive loads.
23. Magnetic contactors are housed in enclosures based on the environment in which they must operate.
24. There are two general types of enclosures: nonhazardous-location enclosures and hazardous-location enclosures.
25. A magnetic motor starter is a contactor with an overload relay added.
26. The purpose of overload protection is to protect the motor windings from excessive heat resulting from motor overloading.
27. Overload protection devices, mounted in the starter, attempt to simulate the heating and cooling of a motor by sensing the current flowing to it.
28. A thermal-overload relay consists of a heater electrically connected in series with the motor supply and mechanically connected to an NC contact.
29. There are two common types of thermal-overload relays: the bimetallic type and the melting alloy type.
30. The thermal-overload heater size is selected according to the full-load current of the motor, operating temperature, service factor, and ambient temperature.
31. An electronic-overload relay uses a current transformer that senses current flow and electronic circuitry that senses changes in the motor current.
32. Motor starters designed to NEMA standards are rated in horsepower and assigned a NEMA number.
33. With motor starters designed to IEC standards, performance is an integral part of the selection procedure.
34. IEC contactors and motor starters are smaller per horsepower rating than those designed to NEMA standards.
35. The silicon-controlled rectifier (SCR) is the most common high-power switching semiconductor.
36. Two SCRs connected in inverse-parallel are necessary to switch ac power.
37. Zero-fired control refers to turning ON the SCR at the zero voltage crossing of the applied ac waveform.

CHAPTER REVIEW QUESTIONS

Answer the following questions.

8-1. What is a magnetic contactor?
8-2. List five advantages gained by the use of contactors instead of manually operated control equipment.
8-3. List four different types of operating electromagnets.
8-4. Compare ac and dc contactor coils with regard to:
    a. Operating voltage range.
    b. Ohmic resistance.
    c. Current flow with the contactor open and closed.
8-5. What function does the shading coil perform?
8-6. Why are ac contactor assemblies made of laminated steel?
8-7. What function do auxiliary contacts perform?
8-8. **a.** Why are some contacts silver-coated?
    b. Why do manufacturers recommend that discolored silver contacts not be filed?
8-9. Why is arc suppression needed?
8-10. Why is it harder to extinguish an arc on contacts passing direct current than on contacts passing alternating current?
8-11. Describe two methods used to reduce an arc across a contact.
8-12. List six things to check as part of a routine preventative program for large contactors.

8-13. How does NEMA rate contactors?

8-14. List four types of load utilization categories.

8-15. Why are contactors and starters placed in an enclosure?

8-16. What are the two major classes of electrical equipment enclosures?

8-17. What is a magnetic motor starter?

8-18. Explain the purpose of motor overload protection.

8-19. List four types of motor overload protective devices.

8-20. Explain how a thermal-overload relay operates.

8-21. List four major factors to be considered when selecting the proper heater size for a motor thermal-overload relay.

8-22. How does an electronic-overload relay sense current flowing to the motor?

8-23. Compare NEMA- and IEC-rated contactors and motor starters with regard to:
   a. Selection for a particular application.
   b. Size per horsepower rating.
   c. Electrical life.
   d. Cost.

8-24. What is a solid-state contactor?

8-25. What is the major advantage of an electronic contactor over the magnetic type?

8-26. Explain the necessity of using heat sinks when switching high power circuits using SCRs.

8-27. What is the major advantage of zero-fired SCR control?

CRITICAL THINKING QUESTIONS

8-1. To play it safe, why not use explosion-proof enclosures in all applications?

8-2. What type of enclosures (hazardous-location or nonhazardous-location) would you use for each of the following environments?
   a. In a paint booth.
   b. Outside, in the weather.
   c. At a feed mill.
   d. Inside a plant, for lathe controls.

8-3. Identify several possible causes or things to investigate for each of the following reported problems with a contactor or motor starter:
   a. Noisy magnet.
   b. Coil failure.
   c. Wear on magnet.
   d. The overheating of a blowout coil.
   e. Pitted, worn, or broken arc chutes.
   f. Failure to pick up.
   g. Short contact life.
   h. Broken flexible shunt.
   i. Failure to drop out.
   j. Insulation failure.
   k. Failure to break arc.
   l. An overload that trips on low current.
   m. Failure to trip (motor burnout).
   n. Failure to reset.

Answers to Self-Tests

1. power
2. true
3. true
4. electromagnet; contacts
5. false
6. false
7. shading
8. eddy
9. true
10. false
11. false
12. arc

13. true
14. false
15. chutes
16. magnetic field
17. true
18. false
19. false
20. nonhazardous; hazardous
21. overload relay
22. true
23. false
24. true

25. bimetallic; melting alloy
26. false
27. false
28. false
29. horsepower; number
30. true
31. true
32. anode; cathode
33. true
34. true
35. true

CHAPTER 9

Motor Control Circuits

∎

CHAPTER OBJECTIVES

This chapter will help you to:

1. *Explain* the principles and applications of motor protection circuits.
2. *List* and *describe* the methods by which a motor can be started.
3. *Describe* the operation of reversing and jogging motor control circuits.
4. *List* and *describe* the methods of stopping a motor.
5. *Explain* the operating principles of variable-speed motor drives.

Motor control circuits play an important part in industry today. This chapter is intended to give students an insight into properly designed, coordinated, and installed motor control circuits. The four major motor control topics covered in this chapter are motor protection, starting and stopping, reversing, and speed control.

∎

9-1 MOTOR PROTECTION AND INSTALLATION

Motor protection safeguards the motor, the supply system, and personnel from various upset conditions of the driven load, the supply system, or the motor itself. Article 430 of the National Electrical Code (NEC) covers the installation requirements for motors and motor circuits. In addition, Underwriters Laboratory (UL) product listings and certifications often dictate the choice of components for motor branch circuits. A suitable *disconnecting means* of sufficient capacity is required, as part of a motor branch circuit, within sight from the motor. The purpose of the disconnect device is to open the supply conductors to the motor, allowing personnel to work safely on the installation. The NEC defines "within sight from" to mean that the disconnecting means is visible and not more than 50 ft from the equipment being controlled. When the source for the motor branch circuit is within sight from the motor, the same disconnecting means can be used to satisfy all the code requirements. Where the source for the motor branch circuit is not within sight from the motor, another disconnecting means is required, as shown in Fig. 9-1 on page 238. If the source, controller, and motor are not within sight from each other, three separate disconnecting means are needed.

The disconnecting means shall have an ampacity rating of at least 115 percent of the full-load current rating of the motor. The disconnecting means shall open all the ungrounded supply conductors and shall be gang operated so that one operating mechanism opens all poles simultaneously. There shall be a clear indication as to whether the disconnect is in the open (OFF) or closed (ON) position. The disconnect must be capable of being locked in the open position.

The requirements for disconnect switches used in motor branch circuits rated 600 V and less are defined in Article 430, Part H, of the NEC. The disconnect switch must have a horsepower rating equal to, or greater than, the horsepower rating of the motor at the applicable voltage. This horsepower rating is required in the event that a motor stalls and the motor controller fails to properly open the circuit. When the disconnect switch is operated, it then has to interrupt the locked-rotor current of the motor, which is typically 600 percent of the motor full-load current. Opening the switch of an underrated disconnect that is not able to extinguish the arc could cause the arc to jump to the metal enclosure or jump line to line, causing a short circuit that could blow a hole in the cover and cause possible injury to the operator. An exception to the horsepower rating of the switches applies to ac

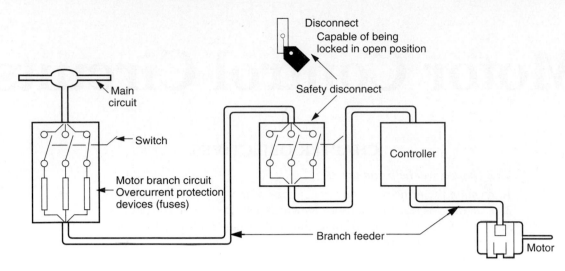

Fig. 9-1   Disconnecting means for motor branch circuits.

motors that are rated 100 hp or more where the disconnecting means permitted is a general-use *isolation switch* that is plainly marked "DO NOT OPERATE UNDER LOAD."

An ordinary disconnect switch is not to be used for normal motor starting and stopping. This is because the motor contains considerable *inductance*. When the switch is opened under load, a spark will jump across the switch as it opens. This spark can burn the switch blades. Never stand in front of a switch when it

is being operated. It is *safest* to keep your face and body off to one side and use your left hand to operate the switch. Although severe arcing does not happen often, the danger of serious damage to the switch and injury to the operator is always there.

Motors typically have starting currents that are *six times* their full-load running current (Fig. 9-2). Unless the design of the circuit feeding the motor allows for this large starting current, the motor will be unnecessarily shut

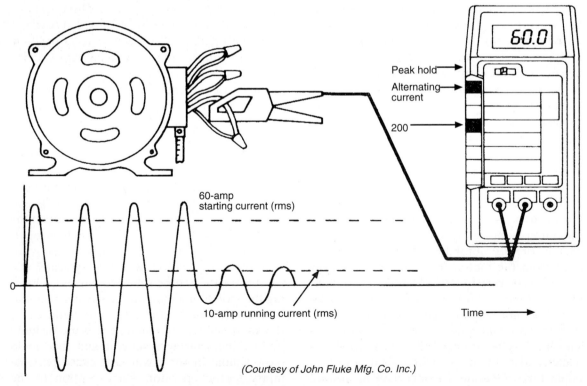

(Courtesy of John Fluke Mfg. Co. Inc.)

Fig. 9-2   Motor starting current is much higher than running current.

down. Once the motor is properly started and running, it must then be protected from overheating as a result of mechanical overloading of the motor.

The basic components required for a motor installation are shown in Fig. 9-3. The *motor controller* is the device that does the actual ON/OFF switching of the motor. Section G of Article 430 governs the selection of the controller. Basically, the controller must have a horsepower rating not lower than the horsepower rating of the motor it is controlling.

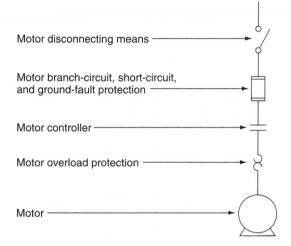

Motor disconnecting means

Motor branch-circuit, short-circuit, and ground-fault protection

Motor controller

Motor overload protection

Motor

Fig. 9-3   NEC requirements for each motor installation.

Often a motor is loaded *beyond* its designed capacity. For example, when using a saw, if the board is damp or the cut is too deep, the motor will be overloaded and slow down. The current flow in the windings will increase and heat the motor beyond its design temperature. A jammed pump or an extra heavy load on a hoist will have the same effect on a motor. If nothing is done, there may be permanent damage to the motor's windings, and expensive repairs will be needed.

A *motor-overload protection device* is installed in *each* power line of the motor to protect the motor, the motor controller, and the branch circuit conductors against excessive heating resulting from motor overloads or the motor's failure to start. In some instances the thermal protection is integral to the motor. In the majority of applications, however, a separate overload protection device is required. Articles 430-32 and 430-34 specify the guidelines and maximum values for the selection of the overload device.

The current rating of a thermal overload relay is chosen to protect the motor against sustained overloads. The overload (OL) contact opens after a period of time, which depends upon the magnitude of the overload current. Figure 9-4 shows the tripping time as a multiple of the rated overload relay current. At rated current (multiple 1), the relay does not trip, but at twice rated current, it trips after an interval of 40 s.

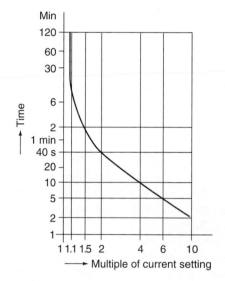

Fig. 9-4   Typical curve of a thermal overload relay.

Short-circuit protection (including ground fault protection) is intended to protect motor circuit components, such as the conductors, switches, controllers, overload relays, motor, and so on, against short-circuit currents or grounds (Fig. 9-5 on page 240). This protection is provided by fuses or circuit breakers. Short-circuit protection devices are sized based on a percentage of full-load amperes (FLA) of the motor. A careless selection of a fuse or circuit breaker may result in damage to the wiring system and possibly injury to personnel. The electrical code is very specific in regard to the selection of protection devices. Paragraph 430-52 and Table 430-152 of Article 430 outline the maximum sizing restrictions of the short-circuit protection device. They allow the following types of devices to be utilized as the short-circuit protection device:

- Nontime-delay fuse.
- Dual element fuse (time-delay fuse).
- Instantaneous trip breaker (only if adjustable and part of a combination controller having motor overload and also short-circuit and ground-fault protection in each conductor).
- Inverse time breaker.

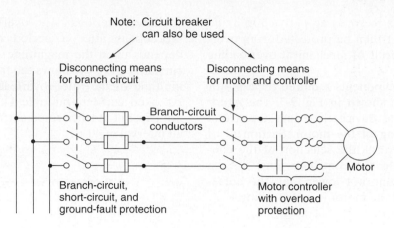

Fig. 9-5   Two levels of protection (overload and short-circuit) are required for motor branch circuits.

Figure 9-6 shows a wiring diagram of a 230-V, 10-hp motor with proper sized No. 8 TW motor *power circuit conductors*. The size conductor needed for motor connection is determined by the full-load running current of the motor. Section 430-6a of the NEC states that the conductor size shall be determined by Tables 430-147, 430-148, 430-149, and 430-150, *instead of motor nameplate current*. Section 430-22 states that the conductors supplying a single motor shall have an ampacity of not less than 125 percent of the motor's full-load current. This, in effect, means that the conductors are loaded to only 80 percent of their rating under full-load conditions, which is standard for circuits with continuous loads. Note that it is not necessary to size the conductors to match the rating permitted for the branch-circuit protection device, which can be as high as 600 percent of the motor's full-load current. The conductors are adequately protected by the motor overload protection. Tables 310-16 through 310-31 are used to determine the conductor size after the ampacity has been determined. Where motors are connected to large-capacity systems, the short-circuit current ratings should be investigated. On long circuit runs, the conductor sizes must be investigated for excessive voltage drop.

When conductors are used to supply two or more motors on the same circuit, feeder conductor ampacity can be determined by adding the full-load currents of all the motors in the circuit; then 25 percent of the *largest* motor's full-load current is added to the total (Fig. 9-7).

Article 430, Part F, of the NEC covers the requirements for motor control circuit conductors. The *motor control circuit conductor* is defined as, "the circuit of the control apparatus

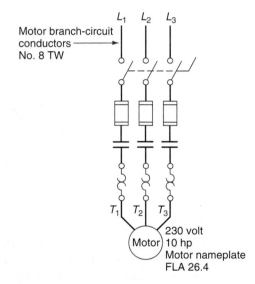

Fig. 9-6   Sizing motor branch-circuit power conductors.

or system that carries the electrical signals, directing the performance of the controller." For example, the conductors from the stop-start station to the controller itself are for control of the starter only and do not carry the main power or motor current. A control circuit tapped on the load side of the motor branch circuit fuse that controls the motor on that branch circuit shall be protected against overcurrent as in Section 430-72. Figure 9-8 shows a wiring diagram of a remote stop-start station with a supplemental 20-A fuse added to properly protect the motor control circuit conductors.

A motor can be protected from overload in two ways: by internal protection that is mounted in the motor, and by external protection that is mounted in, or near, the motor

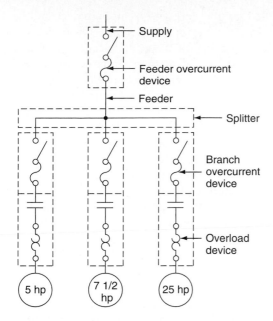

Fig. 9-7 Single-line diagram for a multiple motor installation.

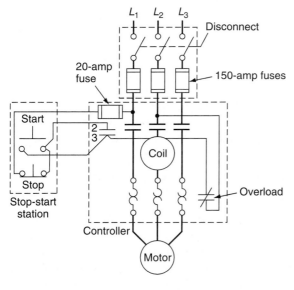

Fig. 9-8 Protection of control circuit conductor.

starter. *Internal overload protection* is usually a temperature detector in the motor windings. Common internal protection devices include:

- *Thermostatic devices.* These are usually embedded in the windings and directly connected to the control circuitry.
- *Resistance temperature detector.* An RTD is used to indicate accurately temperature in the windings.
- *Thermocouple.* This is a pair of dissimilar conductors, joined at the end.
- *Thermistor.* This is a semiconductor that changes resistance greatly with changes in temperature.

External overload protection, such as that provided by an overload relay in the motor starter, is only sensing the motor current in an attempt to determine the heating in the motor windings. This method of protection may be less effective if the motor is covered in dirt. External overload protection also is not sensitive to changes in ambient temperature around the motor. Thus, a motor can be overheated, and the external overload protection will not trip. Internal protectors are recommended for applications where the ambient (surrounding) temperature can change drastically, or where the cleanliness of the motor cannot be guaranteed. External overload relays are also less effective if the motor is started and stopped frequently, or if it is restarted too quickly after an overload trip. Internal thermal protection is the best for frequent starting applications.

Other types of motor protection circuits include:

- *Low-voltage protection.* This type of protection operates when the supply voltage drops below a set value to provide machine-operator protection. The motor must be *restarted* upon resumption of normal supply voltage.
- *Low-voltage release protection.* This method interrupts the circuit when the supply voltage drops below a set value, and reestablishes the circuit when the supply voltage returns to normal.
- *Phase-failure protection.* This form of protection interrupts the power in all phases of a polyphase circuit upon failure of any one phase. Normal fusing and overload protection may not adequately protect a polyphase motor from damaging single-phase operation. Without this protection, the motor will continue to operate if one phase is lost. Large negative-sequence currents are developed in the rotor circuit, causing excessive current and heating in the stator circuit, which will eventually burn out.
- *Phase reversal protection.* This method operates upon detection of a phase reversal in a polyphase circuit. It is used in applications, such as elevators, where it would be damaging or dangerous to have the motor inadvertently run in reverse.
- *Ground-fault protection.* This method of protection operates when one phase of a motor shorts to ground, thus preventing high currents from damaging the stator windings and the iron core.

Magnetism

Generator action

Counter emf (cemf)

Locked-rotor current

Full-voltage (across-the-line) starter

Manual starters

Self-Test

Answer the following questions.

1. The purpose of the _____ device is to open the supply conductors to the motor.
2. The locked-rotor current of a motor is typically _____ percent of the motor full-load current.
3. True or false? An ordinary disconnect switch is not to be used for normal motor starting and stopping.
4. True or false? The motor controller is the device that does the actual ON/OFF switching of the motor.
5. Motor overload protection is designed to protect the motor against _____ overloads.
6. Motor short-circuit protection is provided by _____ or _____.
7. The conductors supplying a single motor must have an ampacity of not less than _____ percent of the motor's full-load current.
8. True or false? The size of the motor control circuit conductors is determined by the full-load running current of the motor.
9. Internal overload protection usually consists of a(n) _____ detector in the motor windings.
10. True or false? External overload protection is the best for frequent starting applications.
11. True or false? A motor with low-voltage release protection must be restarted upon resumption of normal supply voltage.
12. True or false? Single-phase operation of a polyphase motor produces excessive current heating.

9-2 MOTOR STARTING

▐▐▐➡ **YOU MAY RECALL** that electric motors operate on *magnetism*. The amount of current needed to create the magnetism depends on the size and design of the motor. Motors are rated in horsepower or watts. The higher the rating of the motor, the higher the starting and running currents will be.

Every motor, when turning, tends to produce a voltage and current within its windings. This *generator action* produces a voltage *opposite* to the applied voltage. When a motor is running, a *counter emf (cemf)* or *voltage* is generated by the rotor cutting through magnetic lines of force. This reduces the current supplied to the motor. Upon being started and before the motor has begun to turn, however, there is no cemf to limit the current, and initially there is a high in-rush of current. Because this current occurs before the rotor has moved, the locked-rotor current will be the same whether the motor is being started at no-load or at full-load. The term *locked-rotor current* is derived from the fact that it is determined by locking the motor shaft so that it cannot turn, then applying rated voltage to the motor, and measuring the current. Although the locked-rotor current may be up to six times the normal running current, it normally lasts for only a fraction of a second.

The main factor in determining the amount of counter generated voltage and current in the motor is its *speed*. Therefore, all motors tend to draw much more current during the *starting* period (*starting* current) than when rotating at *operating* speed (*running* current). If the load placed on a motor reduces the speed, less generated current will be developed and more applied current will flow. That is, the *greater* the load on the motor, the *slower* the motor will rotate and the *more* applied current will flow through its windings. If the motor is jammed or prevented from rotating in any way, a *locked-rotor* condition is created, and the applied current becomes very high. This high current will cause the motor to burn out quickly.

FULL VOLTAGE STARTING OF AC INDUCTION MOTORS

A *full voltage*, or *across-the-line*, *starter* is designed to apply full line voltage to the motor upon starting. Full voltage starters may be either manual or magnetic. *Manual starters* are often used for small motors, those up to about 10 hp. They consist of a switch with one set of contacts for each phase and a thermal overload device. The starter contacts remain closed when power is removed from the circuit; the motor restarts when power is reapplied. Manual starters for fractional-horsepower, single-phase motors are popular because they serve as both motor protectors and branch feeder disconnects. Figure 9-9 shows a single-phase starter having one overload relay and a toggle switch. Because the current required by single-phase motors is usually small, the single-phase starter is not needed as an operational disconnect. Ordinary thermostats, relays, and the like can normally serve to connect and disconnect such motors when automatic operation is

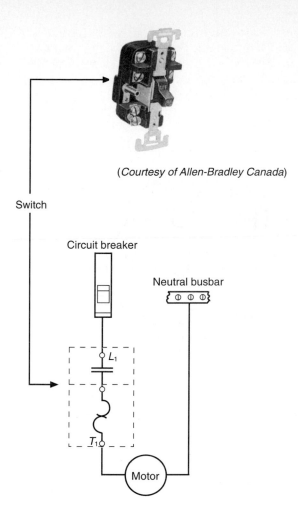

(Courtesy of Allen-Bradley Canada)

Switch

Circuit breaker

Neutral busbar

L_1

T_1

Motor

Fig. 9-9   Fractional horsepower manual starting switch.

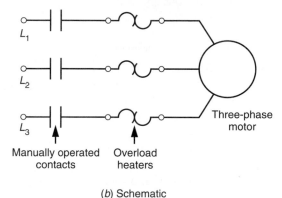

(Courtesy of Allen-Bradley Canada)

(*a*) Manual starter

L_1

L_2

L_3

Three-phase motor

Manually operated contacts   Overload heaters

(*b*) Schematic

Fig. 9-10   Manual three-phase starter.

desired. When an overload occurs, the starter handle automatically moves to the center position to signify that the contacts have opened because of overload and the motor is no longer operating. The starter contacts cannot be reclosed until the overload relay is reset manually. The relay is reset by moving the handle to the full OFF position after allowing about 2 min for the heater to cool.

The manual three-phase starter shown in Fig. 9-10 is operated by pushing a button on the starter enclosure cover that *mechanically* operates the starter. This type of starter provides overload protection, but it cannot be used where low or undervoltage protection is required or for remote or automatic operation. When an overload relay trips, the starter mechanism unlatches, opening the contacts to stop the motor. The contacts cannot be reclosed until the starter mechanism has been reset by pressing the STOP button; first, however, the thermal unit needs time to cool.

These starters are designed for infrequent starting of small ac motors (10 hp or less) at voltages ranging from 120 to 600 V.

Larger motors require the use of contactors to handle their load current, and it is generally required that starters operate automatically under control of remotely located devices such as pushbuttons, switches, thermostats, relays, or solid-state controllers. Therefore, the most common type of motor starter is the *magnetic across-the-line starter,* which is operated by an electromagnet or a solenoid. A manual starter must be mounted so that it is easily within reach of the machine operator. With magnetic control, pushbutton stations are mounted nearby, but automatic control pilot devices can be mounted almost anywhere on the machine.

The operating principle that makes a magnetic starter different from a manual starter is the use of an electromagnet. Figure 9-11 on page 244 shows a typical magnetic across-the-line ac starter and its associated connection diagram. The starter circuit has three main components: a magnetic contactor, an overload relay, and a control station. When the START

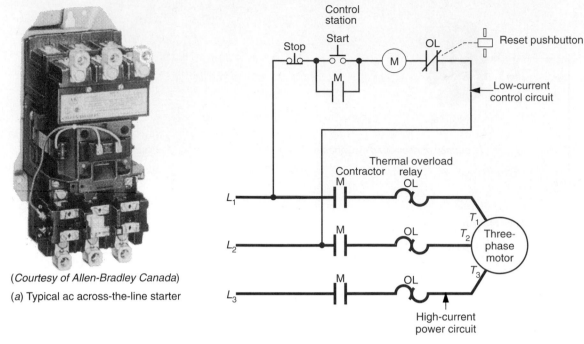

(*Courtesy of Allen-Bradley Canada*)

(*a*) Typical ac across-the-line starter

(*b*) Schematic

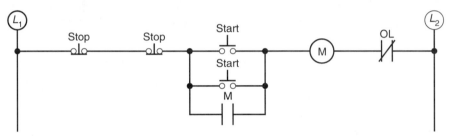

(*c*) Motor can be started or stopped from a number of locations by connecting additional start buttons in parallel with one another and stop buttons in series with one another

Fig. 9-11   Three-phase magnetic starter.

button is pressed, coil M is energized. When coil M is energized, it closes *all* M contacts. The M contacts in series with the motor close to complete the current path to the motor. These contacts are part of the *power circuit* and must be designed to handle the full-load current of the motor. Control contact M (across from the START button) closes to *seal in* the coil circuit when the START button is released. This contact is part of the *control circuit*; as such, it is required to handle the small amount of current needed to energize the coil. The starter has three overload heaters, one in each phase. The normally closed (NC) relay contact OL opens automatically when an overload current is sensed on any phase to deenergize the M coil and stop the motor.

A *combination starter* (Fig. 9-12) consists of an across-the-line starter and a disconnect means wired together in a common enclosure.

The cover of the enclosure is interlocked with the external operating handle of the disconnect means. The door cannot be opened while the disconnect means is closed. When the disconnect means is open, all parts of the starter are accessible; however, the hazard is reduced because the readily accessible parts of the starter are not connected to the power line.

REDUCED VOLTAGE STARTERS

There are two primary reasons for using a reduced voltage when starting a motor:

- It limits line disturbances.
- It reduces excessive torque to the driven equipment.

When starting a motor at full voltage, the current drawn from the power line is typically

Fig. 9-12 Combination starter. *(Courtesy Furnas Electric Co.)*

motor and the torque produced by the motor are reduced. Table 9-1 shows the typical relationship of voltage, current, and torque for a NEMA design B motor.

Utility current restrictions, as well as in-plant bus capacity, may require motors above a certain horsepower to be started with reduced voltage. High inertia loads may require control of the acceleration of the motor and load. If the driven load or the power distribution system cannot accept a full voltage start, some type of reduced voltage or *soft starting* scheme must be used. Typical reduced voltage starters include primary-resistance starters, autotransformers, wye-delta starters, part-winding starters, and solid-state starters. These devices can be used only where low starting torque is acceptable.

Figure 9-13 shows the general arrangement of a *primary-resistance starter,* which is no longer in common use today. The ac motor is often compared to a transformer in which the stator winding is the primary. The primary-resistance starter adds resistance to the stator circuit during the starting period, thus reducing the current drawn from the line. Closing the

600 percent of normal full-load current. The large starting in-rush current of a big motor could cause line voltage dips and brownouts. In addition to high starting currents, the motor also produces starting torques that are higher than full-load torque. In many applications this starting torque can cause excessive mechanical damage such as belt, chain, or coupling breakage. When a reduced voltage is applied to a motor at rest, both the current drawn by the

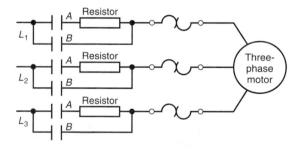

Fig. 9-13 Primary-resistance starter.

Table 9-1 Typical Voltage, Current, and Torque Characteristics for NEMA Design B Motors							
		Motor Starting Current as a % of:		Line Current as a % of:		Motor Starting Torque as a % of:	
Starting Method	% Voltage at Motor Terminals	Locked Rotor Current	Full-Load Current	Locked Rotor Current	Full-Load Current	Locked Rotor Torque	Full-Load Torque
Full voltage	100	100	600	100	600	100	180
Autotrans.							
80% tap	80	80	480	64	307	64	115
65% tap	65	65	390	42	164	42	76
50% tap	50	50	300	25	75	25	45
Part-winding	100	65	390	65	390	50	90
Star delta	100	33	198	33	198	33	60
Solid-state	0–100	0–100	0–600	0–100	0–600	0–100	0–180

contacts at *A* connects the motor to the supply via resistors that provide a voltage drop to reduce the starting voltage available to the motor. The resistors' value is chosen to provide adequate starting torque while minimizing starting current. Motor in-rush current declines during acceleration, thus reducing the voltage drop across the resistors and providing more motor torque. This results in smooth acceleration. After a set period of time, contacts *A* open and the resistors are shorted out by contacts *B*, applying full voltage to the motor.

One type of *autotransformer starter* is shown in Fig. 9-14. This type of starter uses autotransformers to reduce the voltage at start-up. When the motor approaches full speed, the autotransformers are bypassed. An autotransformer is a single-winding transformer on a laminated core with taps at various points on the winding. The taps are usually expressed as a percentage of the total number of turns and, thus, a percentage of applied voltage output. The three autotransformers are connected in a wye configuration, with taps selected to provide adequate starting current. The motor is first energized at a reduced voltage by closing contacts *A*. After a short time, the autotransformers are switched out of the circuit by opening contacts *A* and closing contacts *B*, thus applying full voltage to the motor. The autotransformers need not have high capacity because they are only used for a very short period of time.

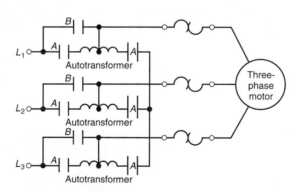

Fig. 9-14   Autotransformer starter.

Wye-delta starters (Fig. 9-15) can be used with three-phase ac motors where all six leads of the stator windings are available (on some motors only three leads are accessible). The major advantage of the wye-delta starter is the absence of starting resistors and transformers. By first closing contacts *A* and *B*, the windings are connected in a wye configuration, which

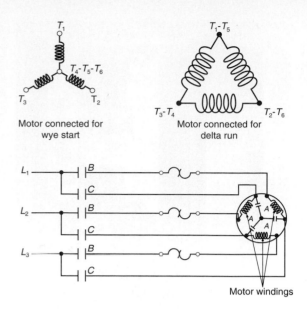

Fig. 9-15   Wye-delta starter.

presents only 58 percent of rated voltage of the motor. Full voltage is then applied by reconnecting the motor in a delta configuration by closing contacts *C* and opening those at *A*. The starting current and torque are 33 percent of their full voltage ratings, which limits applications to loads that require very low starting torque.

Part-winding starters are sometimes used on motors wound for dual voltage operation, such as a 230/460-V motor. These motors have two sets of windings connected in parallel for low-voltage and connected in series for high-voltage operation. When used on the *lower voltage*, they can be started by first energizing only *one* winding, limiting starting current and torque to approximately one-half of the full voltage values. The second winding is then connected normally, once the motor nears operating speed about 2 s later (Fig. 9-16).

Electromechanical reduced voltage starters must make a transition from reduced voltage to full voltage at some point in the starting cycle. At this point there is normally a line current surge. The amount of surge depends upon the:

- Type of transition used.
- Speed of the motor at the transition point.

There are two methods of transition from reduced voltage to full voltage, namely, *open circuit transition* and *closed circuit transition*. Open transition means that the motor is actually disconnected from the line for a brief

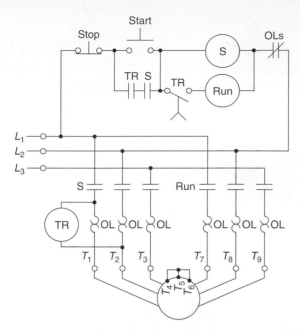

Fig. 9-16   Two-start part-winding starter.

to full speed as possible. This also minimizes the amount of surge on the line (Fig. 9-17).

Solid-state starters (Fig. 9-18 on page 248) provide stepless reduced voltage starting of ac motors. In this process, known in industry as *soft start,* the voltage is increased gradually as the motor starts. This provides smooth, stepless, torque-controlled acceleration. The same principles of current and torque apply to both electromechanical reduced voltage starters and solid-state controllers.

In solid-state starters, high-power semiconductors such as SCRs (silicon controlled rectifiers) are used to control the voltage to the motor (Fig. 9-19 on page 249). An SCR allows current to flow in one direction only. The amount of conduction of an SCR is controlled by the pulses received at the gate of the SCR. When two SCRs are connected back to back, the ac power to a load can be controlled by changing the firing angle of the line voltage during each half-cycle. By changing the angle, it is possible to increase or decrease the voltage and current to the motor. The starter uses a *micro-computer* to control the firing of the SCRs. Six SCRs are used in the power section to provide full cycle control of the voltage and current.

Some solid-state starters offer the choice of three starting modes—soft start, current limit, or full voltage—in the same device. In addi-

period of time when the transition takes place. With closed transition, the motor remains connected to the line during transition. Open transition will produce a higher surge of current because the motor is momentarily disconnected from the line. Transfer from reduced voltage to full voltage should occur at as close

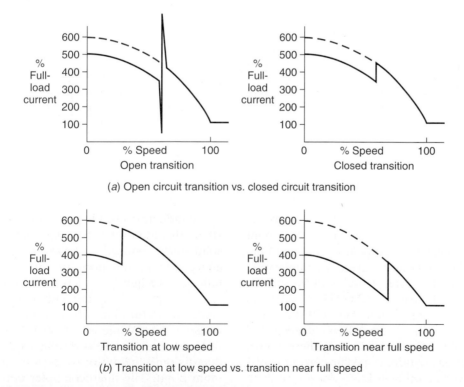

(a) Open circuit transition vs. closed circuit transition

(b) Transition at low speed vs. transition near full speed

Fig. 9-17   Transition from reduced voltage to full voltage.

(*Courtesy of Allen-Bradley Canada*)

(*a*) Starter module

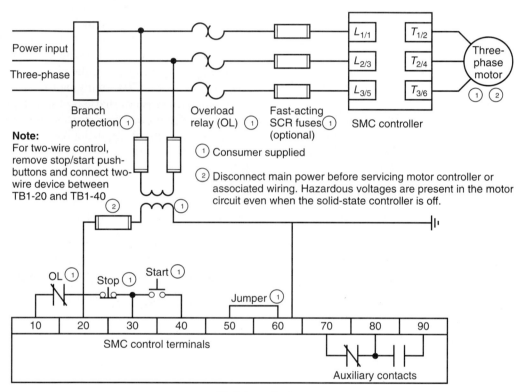

Power input

Three-phase

Branch protection ① Overload relay (OL) ① Fast-acting SCR fuses ① (optional) SMC controller

$L_{1/1}$ $T_{1/2}$
$L_{2/3}$ $T_{2/4}$
$L_{3/5}$ $T_{3/6}$

Three-phase motor ① ②

Note:
For two-wire control, remove stop/start push-buttons and connect two-wire device between TB1-20 and TB1-40 ②

① Consumer supplied

② Disconnect main power before servicing motor controller or associated wiring. Hazardous voltages are present in the motor circuit even when the solid-state controller is off.

OL ① Stop ① Start ① Jumper ①

| 10 | 20 | 30 | 40 | 50 | 60 | 70 | 80 | 90 |

SMC control terminals

Auxiliary contacts

(*b*) Schematic diagram

Fig. 9-18 Typical solid-state starter.

tion, the controller allows adjustment of the time for the *soft start ramp* or current limit maximum value, which enables selection of the starting characteristic to meet the application (Fig. 9-20). The time to full voltage can be adjusted usually from 2 to 30 s. The result is no large current surge when the solid-state controller is set up and correctly matched to the load. Current limit can be used where power line limitations require a specific current load.

Solid-state controllers have control over the voltage applied to the motor owing to the

semiconductors used in the power circuit, even when the motor is up to speed. This allows solid-state reduced voltage to provide an energy-saving function for motors that run unloaded or lightly loaded for long periods of time. The microcomputer within the solid-state controller determines when the motor is lightly loaded. The voltage to the motor can then be reduced by properly controlling the semiconductors until the motor is operating at an optimum point. The microcomputer detects when a load is reapplied and increases the voltage to

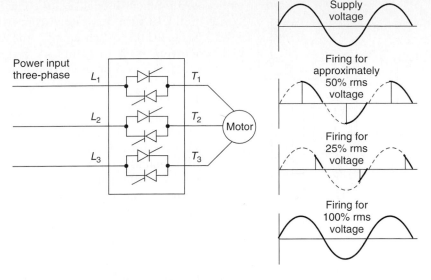

Fig. 9-19 Using SCRs, the voltage and current can be slowly and steplessly increased to the motor.

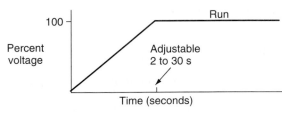

(a) Soft start allows voltage to be gradually increased to the motor (acceleration ramp period)

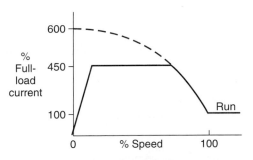

(b) Current limit is used to limit maximum starting current and is adjustable from 200% to 450% of full-load amps

Fig. 9-20 Soft start and current limit control.

prevent stalling. Other features that are available with solid-state controllers include additional protection to the motor and controller and diagnostics that aid in setup and troubleshooting. Protection typically provided includes shorted SCR, phase loss, open load lead, SCR overtemperature, and stalled motor. Appropriate LEDs illuminate to aid in troubleshooting when one of these faults trip out the solid-state reduced voltage controller.

DC MOTOR CONTROL

Direct current motors are used far less than ac units (they need special starting equipment). Like ac motors, small dc motors can be connected directly across the line for starting because a little friction and inertia are overcome quickly in gaining full speed and cemf. Fractional horsepower manual starters or magnetic contactors and starters are used for across-the-line starting of small dc motors.

With a dc magnetic motor starter, it is important to realize that the breaking of the power circuit produces an arc, which will burn the power contacts if not extinguished quickly. To help extinguish the arc, the starter is equipped with three power contacts connected in series. Figure 9-21 on page 250 shows the wiring diagram for a dc across-the-line starter.

As with large ac motors, large dc motors must be provided with a means for limiting the starting current to reasonable values. One solution is to connect a tapped resistor in series with the armature, as shown in the constant-speed dc starter circuit of Fig. 9-22 on page 250. When the M power contact closes, full line voltage is applied to the shunt field while the resistor is connected in series with the armature. After the first time delay, A_1 contact closes, bypassing a section of the resistance. Following the second time delay period, A_2 contact closes, bypassing the resistor and allowing the motor to operate at base speed. The starting method is closed transition. Today electronic methods are often used to limit starting current and provide speed control of dc motors.

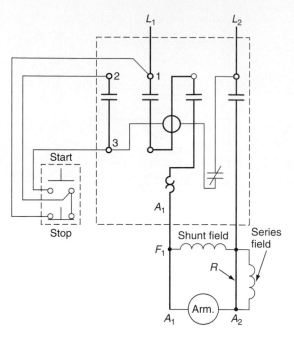

Fig. 9-21   Direct current across-the-line starter.

ELECTRIC CLUTCH

Electric clutches are used to:

- Engage very large motors to their loads after the motors have reached running speeds.
- Provide smooth starts for operations in which the material being processed might be damaged by abrupt starts.
- Start high-inertia loads when starting may be difficult for a motor that is sized to handle only the running load.

In general, an electric clutch consists of an electromagnet disc and an armature disc (Fig.

9-23). When disengaged, the clutch allows the motor to turn separately from its load. When engaged, the clutch couples the two together, allowing both to rotate. A magnetic field is generated as soon as current flows through the electromagnet disc. This draws the armature disc into direct contact with the magnet. The electromagnet disc is faced with friction material to provide positive engagement between the two discs. Engage and disengage motions are signaled by a switch or sensor. As soon as current to the coil is removed, a spring action separates the faces to provide a small clearance between the discs.

Certain electric clutches are controllable; that is, the torque produced is directly proportional to the amount of voltage and current applied to the coil. As a result, the shift from positive, instantaneous engagement to soft, cushioned starts can be accomplished by simply varying the coil current.

Self-Test

Answer the following questions.

13. Electric motors operate on _____.
14. Generator action within a motor produces a counter voltage and current that flows _____ to the applied current.
15. The greater the load on the motor, the _____ it will rotate and _____ applied current will flow through its windings.
16. True or false? The starter contacts of a manual starter remain closed when power is removed from the circuit.

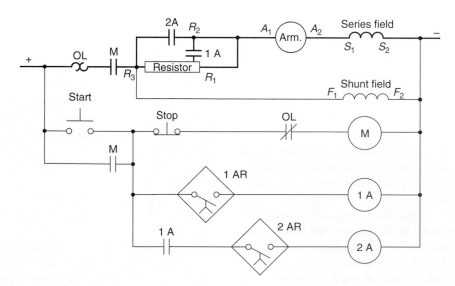

Fig. 9-22   Constant-speed dc starter.

(a) Shaft-mounted electro clutch

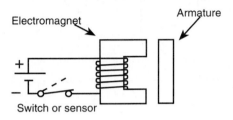

(b) ON/OFF clutch control provides fast and accurate actuation

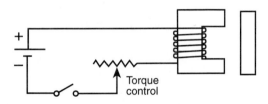

(c) Controllable clutch control provides adjustable control from zero to maximum torque

Fig. 9-23 Electric clutch. (*Courtesy Warner Electric*)

17. True or false? A three-phase manual starter can be operated remotely by a start-stop pushbutton station.
18. The operating principle that makes a magnetic starter different from a manual starter is the use of a(n) _____.
19. A combination starter consists of an across-the-line starter and a(n) _____ means wired together in a common enclosure.
20. When a reduced voltage is applied to a motor at rest, both the _____ and _____ produced by the motor are reduced.
21. Primary-resistance starters add _____ to the motor stator circuit during the starting period.
22. The transformer used in an autotransformer starter is a(n) _____-winding transformer with taps.

23. With a wye-delta reduced voltage starter, the motor windings are connected in a(n) _____ configuration on starting.
24. Part-winding reduced voltage starters are used on motors that are wound for dual-_____ operation.
25. The two methods of transition from reduced voltage to full voltage are known as _____ circuit transition and _____ circuit transition.
26. True or false? Solid-state soft start starters make a number of open transitions from reduced voltage to full voltage.
27. The amount of conduction of an SCR is controlled by the pulses received at the _____ of the SCR.
28. True or false? The ramp time for a solid-state starter refers to the time it takes to reach full voltage.
29. True or false? Direct current motors are used more often than ac units.
30. Starting current to dc motors can be limited by connecting a tapped resistor in series with the motor _____.
31. In general, an electric clutch consists of a(n) _____ disc and a(n) _____ disc.

9-3 MOTOR REVERSING AND JOGGING

Interchanging any two leads to a three-phase induction motor will cause it to run in the *reverse* direction. A three-phase reversing starter (Fig. 9-24 on page 252) consists of *two contactors* enclosed in the same cabinet. As seen in the power circuit, the contacts (*F*) of the forward contactor, when closed, connect L_1, L_2, and L_3 to motor terminals T_1, T_2, and T_3, respectively. The contacts (*R*) of the reverse contactor, when closed, connect L_1 to motor terminal T_3 and connect L_3 to motor terminal T_1, causing the motor to run in the opposite direction. Whether operating through either the forward or reverse contactor, the power connections are run through the same set of overload relays.

Mechanical and electrical *interlocks* are used to prevent the forward and reverse contactors from being activated at the same time, which would cause a short circuit (Fig. 9-25 on page 252). With the *mechanical interlock,* the first coil to close moves a lever to a position that prevents the other coil from closing its contacts when it is energized. *Electrical pushbutton interlocks* use double-contact (NO and NC)

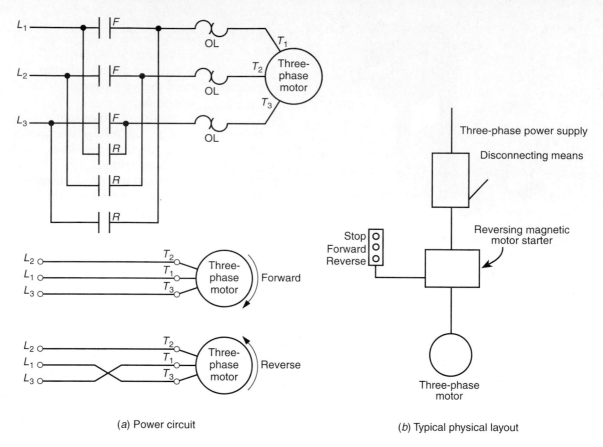

(a) Power circuit

(b) Typical physical layout

Fig. 9-24 Three-phase reversing starter.

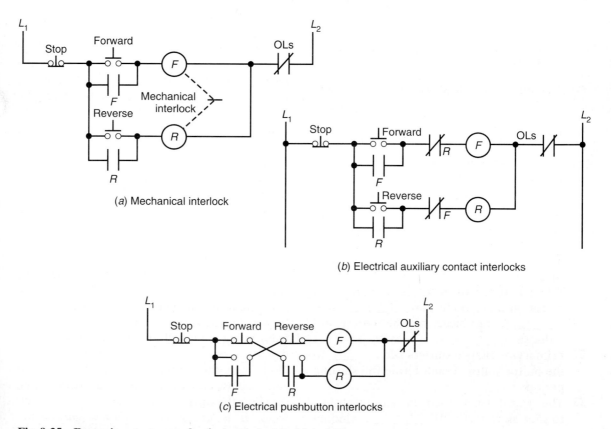

(a) Mechanical interlock

(b) Electrical auxiliary contact interlocks

(c) Electrical pushbutton interlocks

Fig. 9-25 Reversing starter mechanical and electrical interlocks.

pushbuttons. When the FORWARD pushbutton is pressed, the NC contacts open the reverse-coil circuit. There is no need to press the STOP button before changing the direction of rotation. If the forward button is pressed while the motor is running in the reverse direction, the reverse control circuit is deenergized and the forward contactor is energized and held closed.

The reversal of a dc motor can be accomplished in two ways:

- Reversing the direction of the armature current and leaving the field current the same.
- Reversing the direction of the field current and leaving the armature current the same.

Most dc motors are reversed by switching the direction of current flow through the armature. The switching action generally takes place in the armature because the armature has a much lower inductance than the field. The lower inductance causes less arcing of the switching contacts when the motor reverses its direction. Figure 9-26 shows the power circuits for dc motor reversing using electromechanical and solid-state control. In Fig. 9-26(a) the forward (FOR) contactor causes current to flow through the armature in one direction, and the reverse (REV) contactor, when closed, causes current to flow through the armature in the

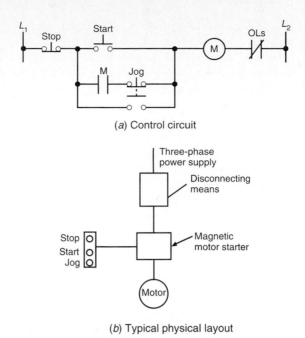

(a) Control circuit

(b) Typical physical layout

Fig. 9-27 JOG/START/STOP **pushbutton control circuit.**

opposite direction. In Fig. 9-26(b) two sets of SCRs are provided. One set is used for current flow in one direction through the armature, and the second set is used for current flow in the opposite direction.

Jogging (sometimes called *inching*) is the momentary operation of a motor for the purpose of accomplishing small movement of the driven machine. It involves an operation in which the motor runs when the pushbutton is pressed and will stop when the pushbutton is released. Jogging is used when motors must be operated momentarily (e.g., for machine tool setup). Repeatedly high starting currents created by jogging causes excessive heating of the power circuit contacts; therefore, if a motor is expected to be jogged more than five times per minute, the rating of the motor starter must be *derated* (the motor requires a larger normal-rated starter).

Figure 9-27 shows a jog control circuit requiring a double contact jog pushbutton: one NC contact and one NO contact. When the JOG button is pressed, the seal in circuit to the starter coil is opened by the NC contacts of the JOG pushbutton. As a result, the starter coil will not lock in; instead, it can only stay energized as long as the JOG button is fully depressed. As a result, a jogging action can be obtained.

On quick release of the JOG pushbutton, in the jog control circuit of Fig. 9-27, should its

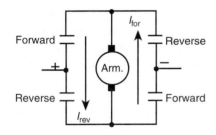

(a) Motor reversing using electromechanical control

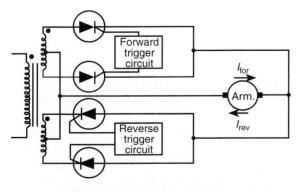

(b) Motor reversing using solid-state circuitry

Fig. 9-26 DC motor reversing-power circuits.

CHAPTER 9 MOTOR CONTROL CIRCUITS **253**

NC contacts reclose *before* the starter maintaining contacts (M) open, the motor would continue to run. *This could be hazardous to workers and machinery.* A jogging attachment can be used to prevent the reclosing of the NC contacts of the JOG button. This device ensures that the starter holding circuit is not reestablished if the JOG button is released too rapidly. Jogging can be repeated by reclosing the JOG button; it can be continued until the jogging attachment is removed.

The control relay jogging circuit shown in Fig. 9-28 is much safer than the previous circuit. A single contact JOG pushbutton is used; in addition, the circuit incorporates a *jog control relay* (CR). Pressing the start pushbutton completes a circuit for the CR coil, closing the CR1 and CR2 contacts. The CR1 contact completes the circuit for the M coil, starting the motor. The M-maintaining contact closes; this maintains the circuit for the M coil. Pressing the JOG button energizes the M coil only, starting the motor. Both CR contacts remain open, and the CR coil is deenergized. The M coil will not remain energized when the JOG pushbutton is released.

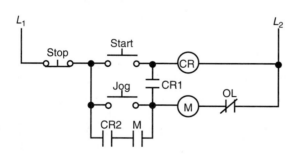

Fig. 9-28 JOG/START/STOP pushbutton control using a jog relay.

Figure 9-29 shows the use of a *selector switch* in the control circuit to obtain jogging. The START button doubles as a JOG button. When the selector switch is placed in the *run* position, the maintaining circuit is not broken. If the START button is pressed, the M coil circuit

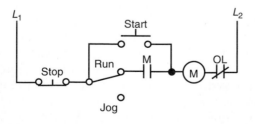

Fig. 9-29 Selector switch jog control circuit.

is completed and maintained. Turning the selector switch to the *jog* position opens the maintaining circuit. Pressing the START button completes the circuit for the M coil, but the maintaining circuit is open. When the START button is released, the M coil is deenergized.

Self-Test

Answer the following questions.

32. A three-phase reversing starter consists of _____ contactors enclosed in the same cabinet.
33. True or false? Closing the forward and reverse contactors at the same time would cause a short circuit.
34. True or false? Most dc motors are reversed by switching the direction of current flow through the armature.
35. Jogging is used where motors must be operated _____.
36. True or false? When operated in the jog mode, the sealed-in circuit to the starter coil must be opened.
37. True or false? Use of a jog control relay increases the reliability of the jogging control function.

9-4 MOTOR STOPPING

The most common method of stopping a motor is to remove the supply voltage and allow the motor and load to coast to a stop. In some applications, however, the motor must be stopped more quickly or held in position by some sort of braking device. Electric braking uses the windings of the motor to produce a retarding torque. The kinetic energy of the rotor and the load is dissipated as heat in the rotor bars of the motor. There are two different means of electric braking: plugging and dynamic braking.

Plugging stops a polyphase motor quickly, by momentarily connecting the motor for reverse rotation while the motor is still running in the forward direction. Plugging a motor more than five times a minute requires the motor starter to be derated.

A *zero-speed switch* (also known as a plugging switch) is coupled to a moving shaft on the machinery whose motor is to be plugged. The zero-speed switch prevents the motor from reversing after it has come to a stop. As the zero-speed switch rotates, centrifugal force or a magnetic clutch causes the contacts to open or close, depending on the intended use. Each zero-speed switch has a rated operating speed range within which the contacts will be switched; for example, 50 to 200 rpm.

The control schematic of Fig. 9-30 shows one method of plugging a motor to stop from one direction only. Pushing the START button closes the forward contactor. As a result, the motor runs forward. The NC contact F opens the circuit to the reverse contactor. The forward contact on the speed switch closes. Pushing the STOP button drops out the forward contactor. The reverse contactor is energized, and the motor is plugged. The motor speed decreases to the setting of the speed switch, at which point its contact opens and drops out the reverse contactor. This contactor is used only to stop the motor using the plugging operation; it is not used to run the motor in reverse.

Many machines, large and small, require that the motor be able to reverse. Most small machines are not adversely affected by reversing the motor before coming to a stop. The same is *not* true of larger pieces of equipment. The sudden reversing torque applied when a large motor is reversed (without slowing the motor speed) could damage the driven machinery, and the extremely high current could af-

fect the distribution system. *Antiplugging protection*, according to NEMA, is obtained when a device prevents the application of a counter torque until the motor speed is reduced to an acceptable value.

In the antiplugging circuit shown in Fig. 9-31 on page 256, the motor can be reversed but not plugged. Pressing the FORWARD button completes the circuit for the "F" coil, closing the "F" power contacts and causing the motor to run in the forward rotation. The "F" zero-speed switch contact opens because of the forward rotation of the motor. Pressing the STOP button deenergizes the "F" coil, which opens the "F" power contacts, causing the motor to slow down. Pressing the REVERSE button will not complete a circuit for the "R" coil until the "F" zero-speed switch contact recloses. Therefore, when the rotating equipment reaches near zero speed, the reverse circuit may be energized, and the motor will run in the reverse rotation.

Dynamic braking is a method of braking that uses the motor as a *generator* during the braking period immediately after the motor is turned OFF. Connecting the motor in this way makes the motor act like a loaded generator that develops a retarding torque, which rapidly stops the motor. The generator action converts

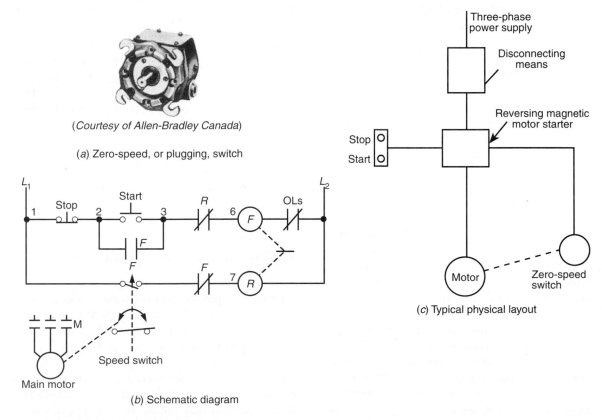

(*Courtesy of Allen-Bradley Canada*)

(*a*) Zero-speed, or plugging, switch

(*b*) Schematic diagram

(*c*) Typical physical layout

Fig. 9-30 Plugging a motor to a stop.

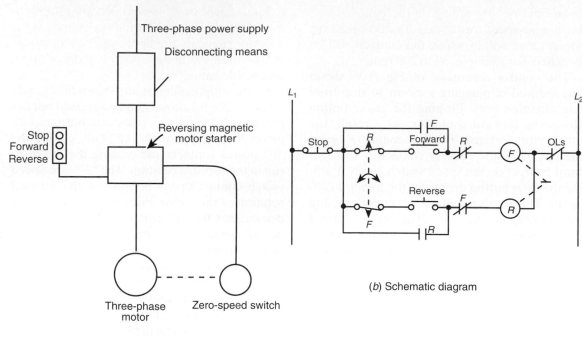

(a) Typical physical layout

(b) Schematic diagram

Fig. 9-31 Antiplugging protection circuit.

the mechanical energy of rotation to electric energy that can then be dissipated as heat in a resistor.

The circuit shown in Fig. 9-32 illustrates how dynamic braking is applied to a dc motor. When the STOP button is depressed, NC contact M completes the braking circuit through the braking resistor, which acts like a load. The shunt field windings of the dc motor are left connected to the power supply. The armature generates a cemf voltage. This cemf causes current to flow through the resistor and armature. The smaller the ohmic value of the braking resistor, the greater the rate at which energy is dissipated and the faster the motor comes to rest. Neither plugging nor dynamic braking can hold the motor stationary after the motor has stopped.

Electric braking can be achieved with a three-phase induction motor by removing the ac power supply from the motor and applying direct current to one of the stator phases. Figure 9-33 illustrates one method used to apply direct current to the motor after the alternating current is removed. This circuit uses a bridge rectifier circuit to change the alternating current into direct current. An OFF-delay timer is connected in parallel with the motor-starter coil. This OFF-delay timer controls an NO contact that is used to apply power to the braking contactor for a short period of time af-

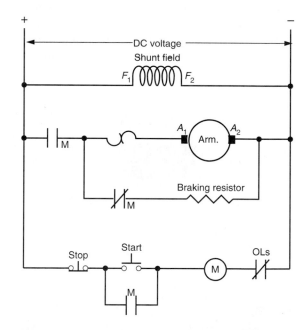

Fig. 9-32 Dynamic braking applied to a dc motor.

ter the STOP pushbutton is pressed. The timing contact is adjusted to remain closed until the motor comes to a complete stop. A transformer with tapped windings is used in this circuit to adjust the amount of braking torque applied to the motor. The motor starter (M) and braking contactor (B) are mechanically and electrically interlocked so that the ac and dc supplies are not connected to the motor at the same time.

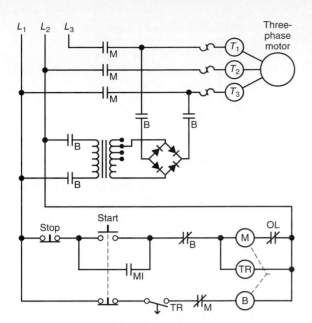

Fig. 9-33 Electric braking applied to a three-phase motor.

The term *electromechanical friction brake* refers to a device external to the motor that provides retarding torque. Figure 9-34 shows a typical friction brake. Most rely on friction in a drum or a disc brake arrangement, and they are set with a spring and released by a solenoid. When the motor is running, the solenoid is energized, which keeps the brake shoes from touching the drum mounted on the motor shaft. When the motor is turned OFF, the solenoid is deenergized, and braking occurs as a

result of the friction between the shoes and the wheel.

An advantage of using dynamic braking is that motors can be stopped rapidly without causing brake linings or drums to wear. Dynamic braking, however, cannot be used to hold a suspended load. Electromechanical friction brakes have the ability to hold a motor stationary and are used in applications such as cranes that require the load to be held. Electrically released brakes are fail-safe in that they automatically stop and hold loads indefinitely when power is lost.

Friction brakes require more maintenance than other braking methods. This maintenance consists of adjusting brakes and replacing worn shoes. The braking torque developed is directly proportional to the braking surface area and spring pressure. Spring pressure is normally adjustable on almost all friction brakes. Low pressure, equally distributed over a large braking surface, will result in even wear and braking torque.

The *electric load brake* (also known as an *eddy current brake*) is a simple, rugged device that consists of an iron rotor mounted inside a stationary field assembly (Fig. 9-35 on page 258). The field assembly consists of a coil and an iron structure designed in such a way that, when direct current flows through the coil, alternate magnetic poles are produced in the iron—that is, a north pole next to a south pole and so on. When the rotor iron moves past the stator poles, alternate fields are induced, caus-

(a) Solenoid-operated brake used on machine tools, conveyors, and small hoists

Fig. 9-34 Electromechanical friction brake.

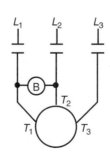

(b) Typical ac shunt brake coil connection

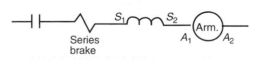

(c) Typical dc series brake coil connection

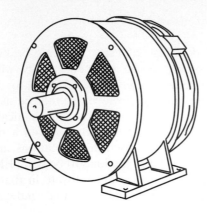

Fig. 9-35   Electric load brake.

ing eddy currents to flow in the rotor. The interaction of the stationary fields and the fields induced by the eddy currents in the rotor produces retarding torque. The braking energy produces heat in the rotor; this heat is dissipated as air is drawn past the rotor fins. If a wound rotor is used, the amount of braking force can be controlled by altering the rotor current. Eddy current brakes cannot hold the motor stationary.

Self-Test

Answer the following questions.

38. Plugging is a method of stopping a motor by momentarily _____ its direction of rotation.
39. Antiplugging protection prevents the motor from _____ until the motor speed is reduced.
40. Dynamic braking is a method of braking that uses the motor as a(n) _____ during the braking period.
41. Electric braking of a three-phase induction motor can be achieved by applying _____ to one of the stator phases.
42. True or false? An electromechanical friction brake cannot be used to hold a suspended load.
43. Usually, electromechanical friction brakes are set with a(n) _____ and released by a(n) _____.
44. Owing to its principle of operation, an electric load brake is also known as a(n) _____ current brake.

9-5 MOTOR SPEED CONTROL

Often motor speed must be changed to meet load demand. In general, motor speed control can be classified into four areas:

- Multispeed motors.
- Variable speed drives for induction and synchronous motors.
- Wound rotor induction motor control.
- DC motor controllers.

MULTISPEED MOTORS

Induction motors with *multiple speed windings* are suitable for applications requiring up to four discrete speeds. The speed is selected by connecting the windings in different configurations and is essentially constant at each setting. Multispeed motors are often found in applications such as ventilating fans and pumps. There are two main types: the separate winding motor and the consequent pole motor.

The *separate winding motor,* as the name implies, uses two or more windings that are electrically separate from each other. Each winding can deliver the motor's horsepower at the rated speed. The mechanical arrangement of the windings determines the number of magnetic poles per phase built into the motor, and thus the different speeds. The more poles per phase, the slower the operating revolutions per minute of the motor when that set of poles is being used. Because the windings are independent of one another, the speeds designed into the motor can be quite varied, such as 3600 rpm/600 rpm or 900 rpm/720 rpm.

The *consequent pole motor* uses a special winding that can be reconnected, using contactors, to obtain different speeds. Two-speed consequent pole motors always have a speed ratio of 2:1. There are three types of consequent pole motors: constant horsepower, constant torque, and variable torque. Their names indicate the output characteristics of the motors.

Figure 9-36 shows the wiring and line diagram of a two-speed separate winding motor starter. The starter consists of two contactors, mechanically and electrically designed not to be activated at the same time (interlocking). The control station is a three-element, HIGH/LOW/STOP station connected for starting at either the HIGH or LOW speed. The change from LOW to HIGH can be made without first pressing the STOP button. When changing from HIGH to LOW, however, the STOP button must be pressed between speeds. As with all multispeed starters, overload relays are provided for both the high- and low-speed circuits to ensure adequate protection on each speed range.

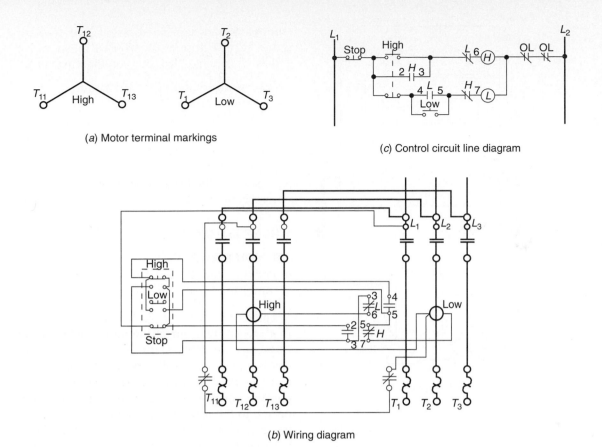

(a) Motor terminal markings

(c) Control circuit line diagram

(b) Wiring diagram

Fig. 9-36 Two-speed separate winding motor starter.

VARIABLE SPEED DRIVES

A *variable speed drive* is used to provide continuous range process speed control (as compared to discrete speed control as in multispeed or pole changing motors). Variable speed drives may be referred to by a variety of names, such as adjustable speed drives, adjustable frequency drives, and variable frequency inverters.

An electrical variable speed drive is an electrical system that is composed of the *motor, drive controller,* and *operator's controls* (either manual or automatic). It is capable of adjusting both speed and torque of a constant speed electric motor. Electrical drive systems—alternating current or direct current—are used in any application where simple starter control of a motor is inadequate.

Selecting the proper drive system is dependent on the application at hand. This could involve, as an example, operator-controlled variance of speed, or the use of control-feedback systems to maintain steady motor speed in spite of load fluctuations or other disturbances. Speed control can be *open loop,* where no feedback of actual motor speed is used, or *closed*

loop, where feedback is used for more accurate speed regulation (Fig. 9-37 on page 260). Often a tachometer is included to achieve good speed regulation. The tachometer is mounted on the motor and produces a speed feedback signal that is used within the controller. In closed-loop control, a change in demand is compensated by a change in the power supplied to the motor, which acts to maintain a constant speed (within regulation capability).

The *drive controller* is an electronic device that can control the speed, torque, horsepower, and direction of an ac or dc motor. Common control functions associated with adjustable speed drives include:

- *Preset speed.* Preset speed refers to one or more fixed speeds at which the drive will operate.
- *Base speed.* Base speed is the manufacturer's nameplate rating where the motor will develop rated horsepower at rated load and voltage. With dc drives, it is commonly the point where full armature voltage is applied with full rated field excitation. With ac systems, it is commonly the point where 60 Hz is applied to the induction motor.

Variable speed drive

Open-loop speed control

Closed-loop speed control

Drive controller

Preset speed

Base speed

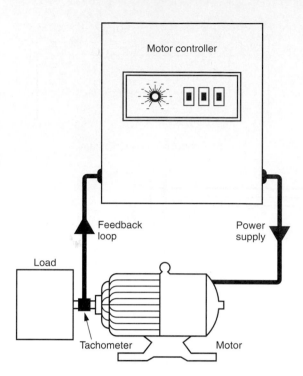

Fig. 9-37   Electrical variable speed drive.

- *Speed range.* The speed range extends from the speed minimum to maximum at which a motor must operate under constant or variable torque load conditions. A 50:1 speed range for a motor with a top speed of 1800 rpm means the motor must operate as low as 36 rpm and still remain within regulation specification. Controllers are capable of wider controllable speed ranges than motors because there is no thermal limitation, only electrical. Controllable speed range of a motor is limited by the motor's ability to deliver 100 percent torque below base speed without additional cooling.

- *Speed regulation.* Speed regulation is the numerical measure, in percent, of how accurately the motor speed can be maintained. It is the percentage of change in speed between full-load and no-load—the ability of a drive to operate a motor at constant speed (under varying load) without "hunting" (alternately speeding up and slowing down). It is related to both the characteristics of the load being driven and the electrical time constants in the drive regulator circuits.

- *Regenerative control.* A regenerative drive contains the inherent capability and/or power semiconductors to control the flow of power to and from the motor.

- *Four-quadrant operation.* Four-quadrant operation refers to the four combinations of

forward and reverse rotation and forward and reverse torque of which a regenerative drive is capable. The four combinations are:

- Forward rotation/forward torque (motoring)
- Forward rotation/reverse torque (regeneration)
- Reverse rotation/reverse torque (motoring)
- Reverse rotation/forward torque (regeneration)

WOUND ROTOR AC MOTOR DRIVES

➡ **YOU MAY RECALL**   that a *wound rotor motor* is a specially constructed motor that can accomplish speed control.

The motor rotor is constructed with windings that are brought out of the motor through slip rings on the motor shaft. These windings are connected to a controller, which places variable resistors in series with the windings. By changing the amount of external resistance connected to the rotor circuit, the motor speed can be varied (the lower the resistance, the higher the speed). Wound rotor motors are most common in the range of 300 hp and above.

Figure 9-38 shows the power circuit for a typical wound rotor motor controller. It consists of a magnetic starter (M), which connects the primary circuit to the line, and two secondary accelerating contactors (S and H), which control the speed. When operating at

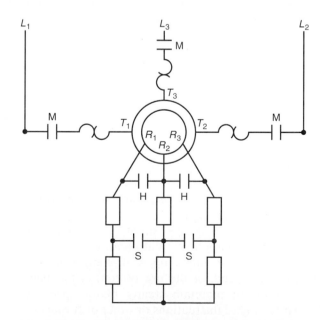

Fig. 9-38   Speed control of a wound rotor induction motor using resistance.

low speed, contactors S and H are both open, and full resistance is inserted in the rotor's secondary circuit. When contactor S closes, it shunts out part of the total resistance in the rotor circuit; as a result, the speed increases. When contactor H closes, all resistance in the secondary circuit of the motor is bypassed; thus, the motor runs at maximum speed.

One disadvantage of using resistance to control the speed of a wound rotor induction motor is that a lot of heat is dissipated in the resistors; the efficiency, therefore, is low. Also, speed regulation is poor; for a given amount of resistance, the speed varies considerably if the mechanical load varies.

Figure 9-39 shows *saturable reactor* (also known as *magnetic amplifier*) *speed control* of a wound rotor induction motor drive. The saturable reactor is a device that is similar in appearance and reliability to a transformer. It consists of ac windings whose impedance varies *inversely* with the amount of excitation control current flowing through the dc control winding. The speed is changed by varying the current flow through the dc control winding. Increasing the dc control current causes speed to increase. In Fig. 9-39(*a*), the flux due to current in the ac load coils flows through the outer legs but cancels out in the center leg, and hence no ac voltage is induced in the control coil center leg. The dc flux produced by current in the control coil flows through both iron cores. The current which flows in the ac load coils is directly proportional to the dc control coil current.

Figure 9-40 on page 262 shows *chopper speed control* of a wound rotor induction motor. The three-phase bridge rectifier feeds the rectified power to the capacitor, which, in turn, supplies the high current pulses drawn by the chopper. By varying the chopper on-time, the apparent resistance across the bridge recti-

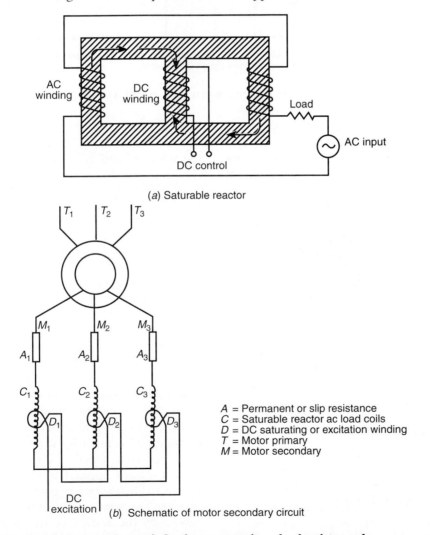

(a) Saturable reactor

A = Permanent or slip resistance
C = Saturable reactor ac load coils
D = DC saturating or excitation winding
T = Motor primary
M = Motor secondary

(b) Schematic of motor secondary circuit

Fig. 9-39 Speed control of a wound rotor induction motor using a load resistor and saturable reactor.

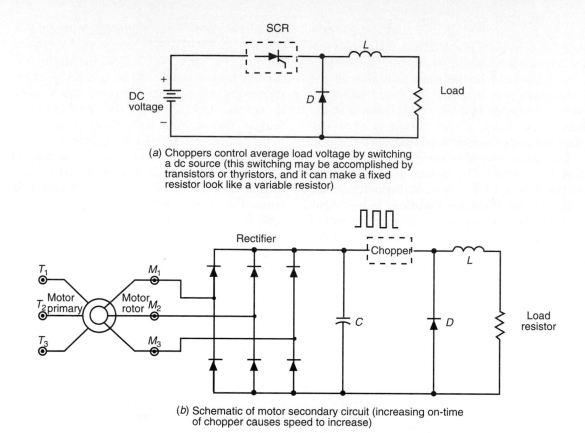

(a) Choppers control average load voltage by switching a dc source (this switching may be accomplished by transistors or thyristors, and it can make a fixed resistor look like a variable resistor)

(b) Schematic of motor secondary circuit (increasing on-time of chopper causes speed to increase)

Fig. 9-40 **Speed control of a wound rotor induction motor using a load resistor and chopper.**

fier can be made either HIGH or LOW. Increasing the on-time of the chopper lowers the apparent resistance and increases motor speed.

One of two things can happen to power extracted from the rotor circuit: It is either wasted as heat, or it can be recovered and converted into useful electric or mechanical energy. Figure 9-41 shows a wound rotor induction motor speed control that returns power to the ac source. The ac rotor voltage is rectified and fed to an inverter. The inverter changes the dc power back into ac power, which matches that of the line. To achieve this, the SCRs have to be triggered within a precisely defined range. A transformer is used to couple the two circuits together.

DIRECT CURRENT DRIVES

The dc drive technology is the oldest form of electrical speed control. The speed of a dc motor is the simplest to control, and it can be varied over a very wide range.

▐▐▶ **YOU MAY RECALL** that the speed of a dc motor is controlled by varying the applied

voltage across the armature, the field, or both the armature and the field.

The drive system consists of a dc motor and a controller. The motor is constructed with armature and field windings. Both of these windings require a dc excitation for motor operation. Usually the field winding is excited with a constant-level voltage from the controller. Then, to operate the motor a dc voltage from the controller is applied to the armature of the motor. The controller is a phase-controlled bridge rectifier with logic circuits to control the dc voltage delivered to the motor armature. *Speed control* is achieved by regulating the armature voltage to the motor.

Armature voltage control of a dc motor is the usual method of changing the speed of a dc motor. Figure 9-42 illustrates methods of electronic speed control by variation of armature voltage. Motor speed is directly proportional to the voltage applied to the armature. The SCR is the major power control element of the circuits. The conduction of the SCR is controlled by the setting of the speed-reference potentiometer, which varies the ON time of the SCR per each positive half-cycle and so varies

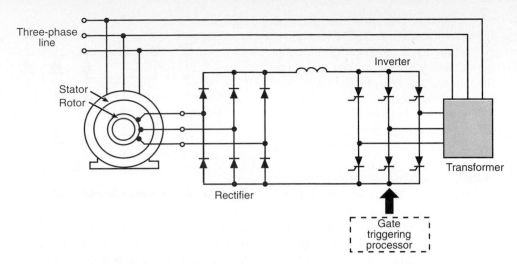

Fig. 9-41 Speed control of a wound rotor motor that returns power to the ac source.

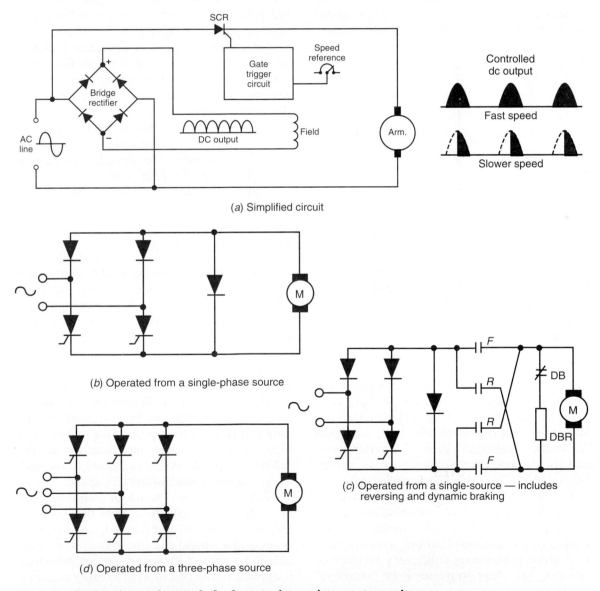

(a) Simplified circuit

(b) Operated from a single-phase source

(c) Operated from a single-source — includes reversing and dynamic braking

(d) Operated from a three-phase source

Fig. 9-42 Electronic speed control of a dc motor by varying armature voltage.

the voltage applied to the armature. The ac input is applied directly to the SCR because it will rectify (change to direct current) as well as control the voltage. A bridge rectifier is used to convert the alternating current to direct current that is required for the field circuit to operate. Armature voltage–controlled dc drives are constant-torque drives, capable of rated motor torque at any speed up to rated motor base speed.

Field voltage–controlled dc drives provide constant horsepower and variable torque. *Field weakening* is the act of reducing the current applied to a dc motor shunt field. This action weakens the strength of the magnetic field and thereby increases the motor speed. The weakened field reduces the cemf generated in the armature; therefore the armature current and the speed increase. Figure 9-43 illustrates a typical dc motor field regulator used for voltage-controlled field applications.

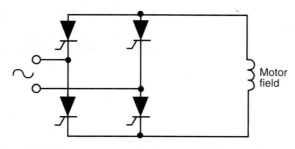

Fig. 9-43 DC motor field regulator.

Nonregenerative dc drives are those where the dc motor rotates in only one direction, supplying torque in high friction loads such as mixers and extruders. The load exerts a strong natural brake.

Regenerative dc drives can invert the dc electric energy produced by the generator's or motor's rotational mechanical energy. All dc motors are dc generators as well. Cranes and hoists use dc regenerative drives to hold back "overhauling loads." Flywheel applications, such as stamping presses, also have overhauling loads. Regenerative drives are better speed control devices than nonregenerative drives, but are more expensive and complicated.

Figure 9-44 illustrates a regenerative dc motor drive that uses two identical converters connected in reverse parallel. Both are connected to the armature, but only *one* operates at a given time, acting either as a rectifier or inverter. The other converter is on "standby," ready to take over whenever power to the

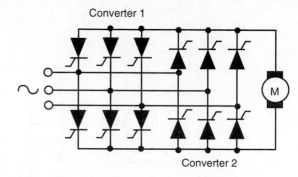

Fig. 9-44 Regenerative dc motor drive.

armature has to be reversed. During normal steady-state operation, converter 1 acts as a rectifier, supplying power to the motor. During this period gate pulses are withheld from converter 2 so that it is inactive. When motor speed is reduced, the control circuit withholds the pulses to converter 1 and simultaneously applies pulses to converter 2. During this period, the motor acts as a generator, and converter 2 conducts current through the armature in the reverse direction. This current reverses the torque, and the motor speed decreases rapidly.

Another method for producing regeneration in dc motors is *field reversing*. It is accomplished by changing the direction of current through the motor field, which reverses the polarity of the motor cemf to account for generator action.

EDDY CURRENT DRIVES

Eddy current drives can be used to control the speed of standard ac squirrel-cage induction motors. The eddy current drive consists of two distinct parts. One part, the mechanical unit, consists of the eddy current clutch and an induction motor. The motor runs at constant speed and provides a source of energy for the clutch. By controlling the excitation to the clutch, the amount of slip between the motor and the output shaft can be regulated and the output speed can be varied. If the excitation is high, the output speed increases toward the full motor speed. As the excitation is lowered, the speed decreases toward zero speed.

Figure 9-45 shows the major components of an eddy current clutch: a steel drum that is directly driven by an ac motor, a rotor with poles, and a wound coil that provides the variable flux required for speed control. A voltage is applied to the coil of wire in order to establish a flux. The magnetic flux crosses the air

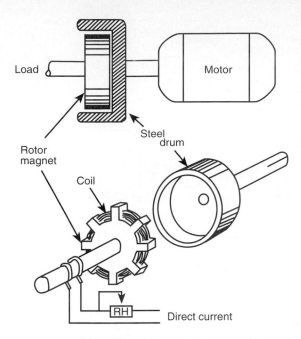

Fig. 9-45 Eddy current drive—clutch assembly.

gap and penetrates the steel drum. The rotation of the drum with relation to the magnet generates eddy currents and magnetic fields in the drum. Magnetic interaction between the two units transmits torque from the motor to the load. By varying the applied voltage, the amount of torque transmitted and, therefore, the speed can be varied.

The clutch excitation is controlled from the eddy current controller. The controller uses high gain amplifiers and a closed loop speed control circuit to sense the need for clutch excitation. The control circuitry is less sophisticated and more complex than that found in other systems.

VARIABLE FREQUENCY AC DRIVES

Speed control of ac squirrel-cage motors can be accomplished if the *frequency* of the voltage applied to the stator is varied to change the synchronous speed.

▐▮➡ **YOU MAY RECALL** that the speed of a squirrel-cage induction motor is directly proportional to the frequency of the ac input voltage.

A typical variable frequency drive can produce output frequencies from 2 to 90 Hz. Therefore, a standard squirrel-cage induction motor rated for 1725 rpm at 60 Hz can be made to run at speeds from about 60 rpm (2 Hz) to about 2700 rpm (90 Hz).

High-power, solid-state electronics have made efficient and accurate *variable frequency drives* (commonly called *inverters*) possible. The basic drive consists of the inverter itself, which converts the 60 Hz of incoming power to a variable frequency and variable voltage. The variable frequency is the actual requirement that will control the motor speed.

The inverter has a large portion of sophisticated circuits that require skilled technicians for service. The use of large-scale integrated circuits and microprocessor circuits, however, permits self-diagnostics that aid in troubleshooting.

To keep variable frequency motors running efficiently and to prevent them from overheating, the ratio of voltage to frequency (V/Hz) must be maintained. As the frequency is reduced, the voltage must also be reduced to limit motor current. Inductive reactance decreases with frequency; therefore; the motor would take excess current at lower frequencies without voltage control. This type of speed control, when used with induction motors, is becoming the most cost-effective and popular system.

Figure 9-46 on page 266 shows the general arrangement of a typical variable frequency adjustable-speed ac drive. The circuit has two states of power conversion: a *rectifier* and an *inverter*. The rectifier changes the incoming three-phase ac power to dc power and delivers this power to the inverter circuit. The inverter circuit changes the dc power back to an adjustable frequency ac output that controls the speed of the motor. The inverter is composed of electronic switches (thyristors or transistors) that switch the dc power ON and OFF to produce a controllable ac power output at the desired frequency and voltage. A *regulator* modifies the inverter switching characteristics so that the output frequency can be controlled. Its inputs include sensors to measure the control variables.

There are three major types of inverter designs available today. These are known as *current source inverters* (*CSI*), *variable voltage inverters* (*VVI*), and *pulse-width-modulated* (*PWM*) *inverters.*

The VVI (Fig. 9-47 on page 266) controls the voltage and frequency to the motor to produce variable speed operation. The distinguishing characteristic between this type of inverter and the PWM inverter is the scheme used to control the voltage. Variable voltage inverters control the voltage in a separate section from the output section used for frequency generation. A controlled rectifier transforms supply alternating current to vari-

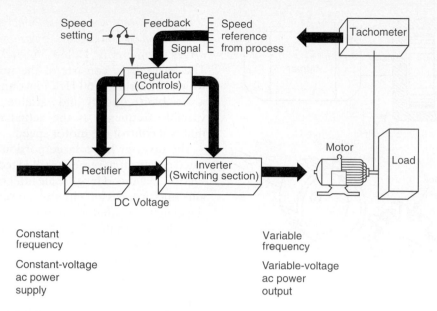

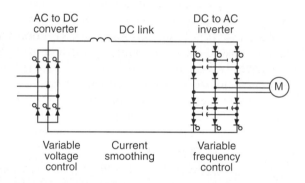

Fig. 9-46 Variable frequency adjustable-speed ac drive.

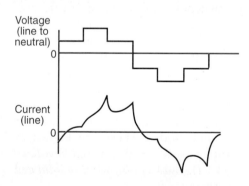

(a) Typical schematic of power circuit

(b) VVI—output waveforms

Fig. 9-47 Variable voltage inverter (VVI).

able voltage direct current. The converter can be an SCR bridge or a diode bridge rectifier with a dc chopper. The voltage regulator presets dc bus voltage to motor requirements.

The frequency controls are accomplished by an output bridge circuit that switches the variable voltage to the motor at the desired frequency. The VVI is the simplest adjustable frequency drive; however, it has the poorest output waveform, and it requires the most filtering to the inverter. These drives are available from fractional horsepowers to about 500 hp.

The CSI (Fig. 9-48) controls the current output to the motor. The actual speed of the motor is sensed and then compared to the reference speed. An error is used to generate a demand for more or less current to the motor. Alternating current transformers are used to adjust the controlled rectifier. The input converter is similar to the VVI drive. A current regulator is used to preset the dc bus current. The output inverter SCRs deliver a six-step current frequency pulse, which the voltage waveform follows. The voltage exhibits commutation spikes when the SCRs fire. Braking power is returned to the distribution system.

The main advantage of using CSI drives lies in their ability to control current and, therefore, torque. This applies in variable torque applications. Current source inverters are available in a wide range of horsepowers, but most often 50 hp and above.

The PWM inverter (Fig. 9-49 on page 268) accomplishes both frequency and voltage control at the output section of the drive. Diode rectifiers provide constant dc voltage. Because the inverter receives a fixed voltage, amplitude of the output waveform is always constant. The inverter adjusts the width of output voltage

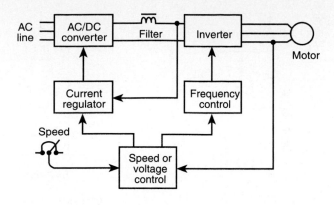

(a) Block diagram of typical CSI drive

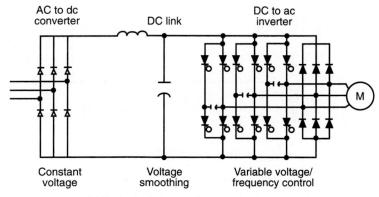

(b) Typical schematic of power circuit

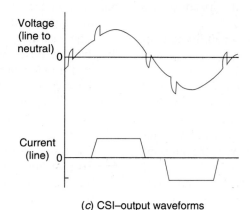

(c) CSI–output waveforms

Fig. 9-48   Current source inverter (CSI).

is approximately sinusoidal. The better wave-form requires less filtering. PWM inverters are the most complex and costly of the three main ac variable speed drive types. PWM drives are available from 1 to 1000 hp.

Self-Test

Answer the following questions.

45. True or false? The speed of multispeed induction motors is selected by connecting the windings in different configurations.

46. The two main types of multispeed induction motors are the _____ motor and the _____ motor.

47. An electrical speed drive is capable of adjusting both _____ and _____ of a constant speed electric motor.

48. True or false? Open-loop speed control uses feedback for more accurate speed regulation.

49. A 20:1 speed range for a motor with a top speed of 1800 rpm means that the motor must operate as low as _____ rpm

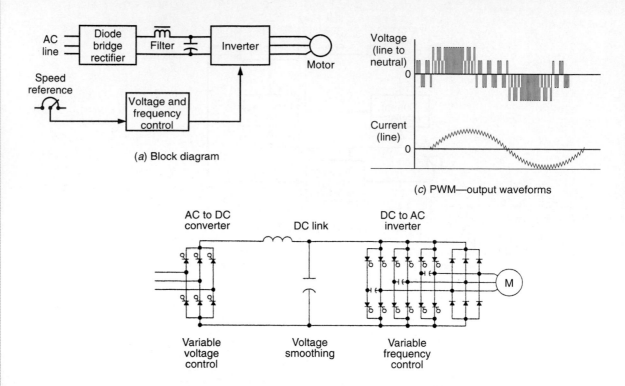

(a) Block diagram

(c) PWM—output waveforms

(b) Typical schematic of power circuit

Fig. 9-49 Pulse-width-modulated (PWM) inverter.

and still remain within specifications required by regulation.

50. A regenerative drive has the ability to control the flow of power _____ and _____ the motor.

51. True or false? The higher the external rotor resistance of a wound rotor induction motor, the higher the speed.

52. A saturable reactor consists of ac windings whose _____ varies with the amount of dc control current.

53. Chopper speed control of a wound rotor motor involves varying the chopper _____.

54. True or false? Power extracted from the rotor circuit of a wound rotor motor can be recovered and converted into useful electric energy.

55. True or false? Armature voltage control is the usual method of changing the speed of a dc motor.

56. True or false? Weakening the field of a dc motor will cause its speed to decrease.

57. True or false? Nonregenerative dc drives are those where the dc motor rotates in only one direction.

58. Eddy current drives vary speed by controlling the excitation to the _____.

59. True or false? The speed of a squirrel-cage induction motor is directly proportional to the frequency of the ac input voltage.

60. The _____ circuit of a variable frequency ac drive outputs ac power at the desired frequency to control motor speed.

61. List the three major types of inverter designs available.

SUMMARY

1. Motor protection safeguards the motor, the supply system, and personnel from various upset conditions of the driven load, the supply system, and the motor itself.

2. A suitable disconnecting means is required as part of the motor branch circuit to allow personnel to work safely on the installation.

3. An ordinary disconnect switch should not be used for normal motor starting and stopping.

4. The motor controller is the device that does the actual ON/OFF switching of the motor.

5. Motor overload protection protects against excessive heating resulting from motor overloads or the motor's failure to start.

6. Motor short-circuit protection protects against short-circuit currents or grounds.

7. Conductors supplying current to a motor must have an ampacity of not less than 125 percent of the motor's full-load current.

8. External overload protection is usually provided by an overload relay in the motor starter.

9. Internal overload protection is usually a temperature detector in the motor winding.

10. Other types of motor protection circuits include the following: low-voltage protection, low-voltage release protection, phase-failure protection, phase-reversal protection, and ground-fault protection.

11. When a motor is running, a cemf or voltage is generated by the rotor's cutting through magnetic lines of force; this reduces the current supplied to the motor.

12. Although the locked-rotor current may be perhaps six times the normal running current, it normally lasts for only a fraction of a second.

13. If the load placed on a motor reduces the speed, less generated current will be developed and more applied current will flow.

14. A full voltage, or across-the-line, starter is designed to apply full line voltage to the motor upon starting.

15. The manual three-phase starter is operated by pushing a button on the starter enclosure cover that mechanically operates the starter.

16. The operating principle that makes a magnetic starter different from a manual starter is the use of an electromagnet.

17. A combination starter consists of an across-the-line starter and a disconnect means wired together in a common enclosure.

18. There are two primary reasons for using a reduced voltage when starting a motor:
    a. To limit line disturbances.
    b. To reduce excessive torque to the driven equipment.

19. The primary-resistance starter adds resistance to the stator circuit during the starting period, thus reducing the current drawn from the line.

20. An autotransformer starter uses autotransformers to reduce the voltage at start-up; when the motor approaches full speed, the autotransformers are bypassed.

21. The wye-delta starter starts a motor by first connecting in a wye configuration, which presents only 58 percent of rated voltage of the motor. Full voltage is then applied by reconnecting the motor in a delta configuration.

22. Part-winding starters are sometimes used on motors wound for dual voltage operation. When used on the lower voltage, they can be started by first energizing only *one* winding, limiting starting current and torque to approximately one-half of the full voltage values. The second winding is then connected normally once the motor nears operating speed.

23. With open-circuit transition, the motor is actually disconnected from the line for a brief period of time when the transition takes place.

24. With closed-circuit transition the motor remains connected to the line during transition.

25. Open transition will produce a higher surge of current because the motor is momentarily disconnected from the line.

26. Solid-state starters provide stepless reduced voltage starting of motors.

27. In solid-state starters, high-power semiconductors such as silicon controlled rectifiers (SCRs) are used to control the voltage to the motor.

28. Features available with solid-state controllers include additional protection to the motor and controller and diagnostics that aid in setup and troubleshooting.

29. With a dc magnetic motor starter, it is important to realize that the breaking of the power circuit produces an arc, which will burn the power contacts if not extinguished quickly.

30. In general, an electric clutch consists of an electromagnet disc and an armature disc. When disengaged, the clutch will allow a motor to turn separately from its load, but when engaged, the clutch will couple the two, allowing both to rotate.

31. Interchanging any two leads to a three-phase induction motor will cause the motor to run in the reverse direction.

32. Interlocking of reversing starters is designed to prevent the forward and reverse contactors from being activated at the same time, which would cause a short circuit.

33. Most dc motors are reversed by switching the direction of current flow through the armature.

34. Jogging is the momentary operation of a motor for the purpose of accomplishing small movement of the driven machine.

35. When the JOG button is pressed in jogging circuits, the sealed-in circuit to the starter coil is opened.

36. Plugging can stop a polyphase motor very quickly. This is accomplished by momentarily connecting the motor for reverse rotation while the motor is still running in the forward direction.

37. As the zero-speed switch rotates, centrifugal force or a magnetic clutch causes the contacts to open or close, depending on the intended use.

38. Antiplugging protection is obtained when a device prevents the application of a counter torque until the motor speed is reduced to an acceptable value.

39. Dynamic braking is a method of braking that uses the motor as a generator during the braking period immediately after the motor is turned OFF.

40. Neither plugging nor dynamic braking can hold the motor stationary after it has stopped.

41. Electric braking can be achieved with a three-phase induction motor by removing the ac power supply from the motor and applying direct current to one of the stator phases.

42. Electromechanical friction brakes rely on friction in a drum or a disc brake arrangement. They are set with a spring and released by a solenoid.

43. Electromechanical friction brakes have the ability to hold a motor stationary.

44. In an eddy current brake, the interaction of the stationary fields and the fields induced by the eddy currents in the rotor produces retarding torque.

45. The speed of multispeed induction motors is selected by connecting the windings in different configurations.

46. A variable speed drive is used to provide continuous range process speed control.

47. An electrical variable speed drive is an electrical system that is composed of the motor, drive controller, and operator's controls (either manual or automatic).

48. Speed control can be open-loop, where no feedback of actual motor speed is used, or closed-loop, where feedback is used for more accurate speed regulation.

49. Common control functions associated with adjustable speed drives include preset speed, base speed, speed range, speed regulation, and regenerative control.

50. The speed of a wound rotor induction motor is changed by changing the amount of external resistance connected to the rotor circuit (the lower the resistance, the higher the speed).

51. With electronic dc drives, the field winding is usually excited with a constant level voltage. The controller is a phase-controlled bridge rectifier with logic circuits to control the dc voltage delivered to the motor armature. Speed control is achieved by regulating the armature voltage to the motor.

52. Nonregenerative dc drives are those where the dc motor rotates in one direction only.

53. Regenerative dc drives can invert the dc electric energy produced by the generator's or motor's rotational mechanical energy.

54. With an eddy current drive, the motor runs at constant speed and provides a source of energy for the clutch. By controlling the excitation to the clutch, the amount of slip between the motor and the output shaft can be regulated and the output speed varied.

55. The speed of a squirrel-cage induction motor is directly proportional to the frequency of the ac input voltage.

56. A typical variable frequency, adjustable-speed ac drive has two states of power conversion: a rectifier and an inverter. The rectifier changes the incoming ac power to dc power and delivers this power to the inverter circuit. The inverter circuit changes the dc power back to an adjustable frequency ac output that controls the speed of the motor.

57. The three major types of inverter designs available are known as current source inverters (CSI), variable voltage inverters (VVI), and pulse-width-modulated (PWM) inverters.

58. PWM inverters are the most complex type; they produce an approximately sinusoidal output waveform that requires less filtering.

Answer the following questions.

9-1. Explain the purpose of motor protection.

9-2. What is the purpose of the motor disconnecting means?

9-3. Describe the safest way to operate a disconnect switch.

9-4. What is a motor controller?

9-5. Specifically, what does the motor overload protection device protect against?

9-6. What is used to provide motor short-circuit protection?

9-7. When conductors are used to supply two or more motors on the same circuit, how is the feeder conductor ampacity determined?

9-8. Which conductors are considered to be motor control circuit conductors?

9-9. List four types of internal overload protection devices.

9-10. What type of applications may require internal overload protection?

9-11. State the condition under which each of the following motor protection circuits will operate:
    a. Low-voltage protection.
    b. Low-voltage release protection.
    c. Phase-failure protection.
    d. Ground-fault protection.

9-12. **a.** Why is there initially a high in-rush of current while the motor is being started and before the motor has begun to turn?
    b. How do the motor starting current value and the normal running current compare?

9-13. What will create a motor locked rotor condition?

9-14. How is a full voltage starter designed to start a motor?

9-15. Compare the way contacts of a manual and magnetic starter are operated.

9-16. Of what does a combination starter consist?

9-17. State two reasons for the use of reduced voltage starting.

9-18. Name four typical types of reduced voltage starters.

9-19. Explain the difference between open-circuit and closed-circuit transition.

9-20. Explain how an electric clutch is engaged.

9-21. How can the direction of rotation of a three-phase induction motor be reversed?

9-22. Why are mechanical and electrical interlocks used with three-phase reversing starters?

9-23. Why are most dc motors reversed by switching the direction of current flow through the armature?

9-24. Describe the jogging operation of a motor.

9-25. **a.** Explain the principle of operation of electric braking.
    b. State two means of electric braking.

9-26. Explain the principle of operation of an electromechanical friction brake.

9-27. What produces the retarding torque in an electric load brake?

9-28. List the four general areas that can be used to classify motor speed control.

9-29. Compare the winding connections of a separate winding and consequent pole multispeed motor.

9-30. Of what is an electrical variable speed drive composed?

9-31. Compare open-loop and closed-loop speed control.

9-32. Define the term *base speed* as it commonly applies to:
    a. DC drives.   **b.** AC drives.

9-33. In what four quadrants is a regenerative drive capable of operating?

9-34. Explain how the speed of a wound rotor induction motor is changed.

9-35. Explain the principle of operation of an electronic dc drive system.

9-36. Explain the principle of operation of an eddy current drive.

9-37. In a variable frequency drive, both the output frequency and voltage applied to the motor are varied. Why?

CRITICAL THINKING QUESTIONS

9-1. Refer to Fig. 9-4. How long would it take for the thermal overload relay to trip if the fault current was four times the rated current?

9-2. Refer to Table 9-1. Which starting method can provide the lowest motor starting current?

9-3. Refer to the illustrations listed and identify the type of transition (open circuit or closed circuit) from reduced voltage to full voltage.
   a. Figure 9-13.   **d.** Figure 9-16.
   b. Figure 9-14.   **e.** Figure 9-19.
   c. Figure 9-15.

9-4. Refer to Fig. 9-21. Under what condition would jumper R be removed?

9-5. Refer to Fig. 9-22. Assume that a field loss relay is to be connected into the circuit that will sense shunt field current and automatically stop the motor if current to the shunt field is lost while the motor is running.
   a. What additional protection would this circuit provide?
   b. At what point could the coil of the field loss relay be connected into the circuit?
   c. At what point could the NO contact of field loss relay be connected into the circuit?

9-6. Refer to Fig. 9-25(*c*). Assume the NC contact on the FORWARD pushbutton is defective and *open* at all times. How would the forward and reverse operation of the motor be affected?

9-7. Refer to Fig. 9-26(*a*). Assume that the forward and reverse contactors are both closed at the same time. What would this condition cause?

9-8. Refer to Fig. 9-28. Assume that relay coil CR becomes open-circuited. How would the run and jog operation of the motor be affected?

9-9. Refer to Fig. 9-30(*b*). Assume that the speed switch contacts fail to open when the motor reverses its direction of rotation. What would happen when the STOP button is momentarily pressed with the motor running in the forward direction?

9-10. Refer to Fig. 9-32. Assume that the shunt field is reconnected directly across A_1 and A_2 of the armature. How would this affect the dynamic braking? Why?

9-11. Refer to Fig. 9-33. Assume that the braking torque applied to the motor is to be reduced. What change must be made to the circuit?

9-12. Refer to Fig. 9-36(*c*). Assume that the operation of the control circuit is to be such that the change from one speed to the other cannot be made without first pressing the STOP button. What modification would you make to the circuit?

9-13. Refer to Fig. 9-39(*b*). Assume that one of the dc excitation windings becomes open. How would this affect the operation of motor speed control?

9-14. Refer to Fig. 9-40(*b*). Assume that the ohmic size of the load resistor is increased. What effect would this have on the top speed range of the motor?

9-15. Refer to Fig. 9-42(*a*). Assume that the SCR becomes shorted from anode to cathode. How would this affect the operation of motor speed control?

Answers to Self-Tests

1. disconnect
2. 600
3. true
4. true
5. sustained
6. fuses; circuit breakers
7. 125
8. false
9. temperature
10. false
11. false
12. true
13. magnetism
14. opposite
15. slower; more
16. true
17. false
18. electromagnet
19. disconnect
20. current; torque
21. resistance
22. single

23. wye
24. voltage
25. open; closed
26. false
27. gate
28. true
29. false
30. armature
31. electromagnet; armature
32. two
33. true
34. true
35. momentarily
36. true
37. true
38. reversing
39. reversing
40. generator
41. direct current
42. false
43. spring; solenoid
44. eddy

45. true
46. separate winding; consequent pole
47. speed; torque
48. false
49. 90
50. to; from
51. false
52. impedance
53. on-time
54. true
55. true
56. false
57. true
58. clutch
59. true
60. inverter
61. variable voltage inverters (VVIs); current source inverters (CSIs); pulse-width-modulated (PWM) inverters

CHAPTER 10

Types of Control

■

CHAPTER OBJECTIVES

This chapter will help you to:

1. *Describe* different methods for achieving motion control.
2. *Explain* common terms referred to in conjunction with programmable motion control.
3. *Discuss* the operation of pressure control systems.
4. *Identify* the basic components of a temperature control system.
5. *Discuss* different methods of heating.
6. *Explain* how time-delay relays and timers work.
7. *Discuss* terms associated with the selection and operation of timers.
8. *Explain* the operation of counters used in counting, length, and position measurement.
9. *Describe* different methods for achieving sequence control.

This chapter is intended to present an overview of commonly used industrial types of control functions, which include motion control, pressure control, temperature control, time control, count control, and sequence control. Typical applications are illustrated and discussed.

■

10-1 MOTION CONTROL

Position indication and control play an important part in the control of machines. Most often, position information is provided by means of a pilot device which provides an adequate electrical signal. This information may be used for indication and control. In many cases the relative position of a machine part or process product may not be too critical. However, in some cases position information must be reliable to 0.001 in. or less.

Many machine tool operations require a repeated forward and reverse action in their operation. Figure 10-1 illustrates a reciprocating motion machine process that uses two *limit switches* to provide automatic control of the motor. Each limit switch (LS1 and LS2) has two sets of contact, one normally open and the other normally closed. The operation of the circuit can be summarized as follows:

- The START and STOP pushbuttons are used to initiate and terminate the automatic control of the motor by limit switches.

- Contact CR1 is used to maintain the circuit to the control relay during the running operation of the circuit.
- Contact CR2 is used to make and break the line circuit to the forward and reverse control circuit.
- Using the control relay and its START and STOP buttons also provides low-voltage protection.
- The normally closed contact of limit switch LS2 acts as the stop for the forward controller, and the normally open contact of limit switch LS1 acts as the start contact for the forward controller. The auxiliary contact of the forward starter is connected in parallel with the normally open contact of limit switch LS1 to maintain the circuit during the running of the motor in the forward direction.
- The normally closed contact of limit switch LS1 is wired as a stop contact for the reverse starter, and the normally open contact of limit switch LS2 is wired as a start contact for the reverse starter. The auxiliary contact on the reverse starter is wired in parallel

From page 274:

Limit switches

On this page:

Proximity switches

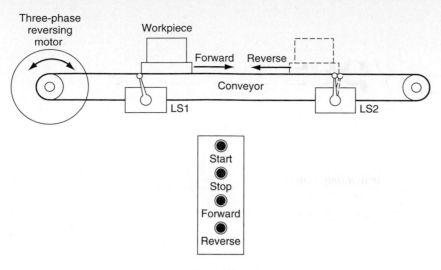

(a) Typical layout

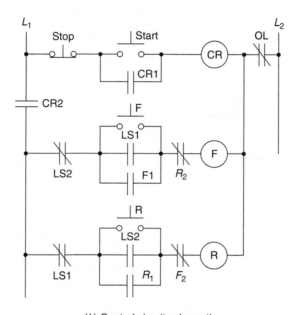

(b) Control circuit schematic

Fig. 10-1 Reciprocating motion machine process.

with the normally open contacts of limit switch LS2 to maintain the circuit while the motor is running in reverse.

- Electrical interlocking is accomplished by the addition of a normally closed contact in series with each starter and operated by the starter for the opposite direction of rotation of the motor.
- Reversal of the direction of rotation of the motor is provided by the action of the limit switches. When limit switch LS1 is moved from its normal position, the normally open contact closes energizing coil F and the normally closed contact opens and drops out coil R. The reverse action is performed by

limit switch LS2, thus providing reversing in either direction.

- The forward and reverse pushbuttons provide a means of starting the motor in either forward or reverse in order that the limit switches can take over automatic control.

Figure 10-2 on page 276 shows how two *proximity switches* can be used to automatically control the motion of a piston from position A to position B. The proximity switch functions in much the same way as the limit switch. It is positioned at the cylinder ends and changes state (opens or closes) once the piston approaches the switch. Proximity switch A (Prox-A) con-

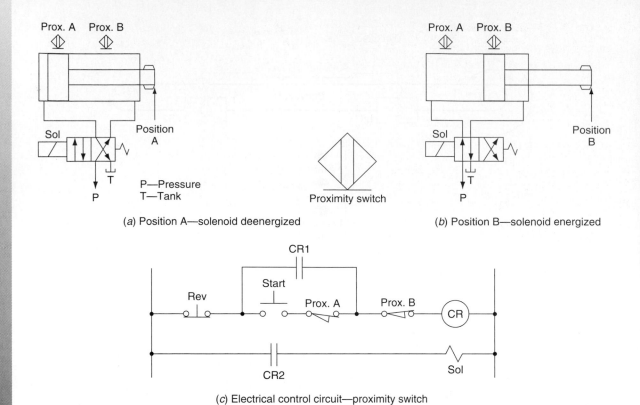

(a) Position A—solenoid deenergized

(b) Position B—solenoid energized

(c) Electrical control circuit—proximity switch positions shown as they would be with piston at position A

Fig. 10-2 Cylinder-piston motion control.

tact is *normally open* and proximity switch B (Prox-B) contact is *normally closed*. The operation of the circuit can be summarized as follows:

- At position A, Prox-A and Prox-B switches are both closed.
- START pushbutton switch is momentarily pressed closed.
- Relay coil CR is energized.
- Relay contact CR1 closes, completing the circuit around the start pushbutton switch and Prox-A switch.
- Relay contact CR2 closes, energizing the solenoid valve.
- The valve spool shifts and the piston starts to move from its start position at A to position B.
- Prox-A operates opening its normally open contact.
- At position B, Prox-B operates, opening its normally closed contact.
- Relay coil CR is deenergized.
- Relay contact CR1 opens, to open the seal-in circuit.
- Relay contact CR2 opens, deenergizing the solenoid.

- The valve spring returns the valve spool to its initial position, returning the piston to position A to complete the cycle. Prox-A operates, closing its normally open contact and thus ensuring that the piston is at position A before another cycle can be initiated.

Electrical overhead traveling cranes and hoists are a basic part of many industrial plants. Their purpose is to lift heavy loads. Electrical drive systems used for cranes and hoists include:

- Variable frequency ac motor controllers for squirrel-cage motors.
- DC motor controllers for dc motors.
- Eddy current clutches for ac motors.
- Wound rotor motor controllers for wound rotor ac motors.

Bridge cranes allow loads to be moved from one point to another, anywhere within the rectangular coverage of the bridge and runway. A rigid mounted arrangement with the bridge running perpendicular to the runway allows the load to be moved in any direction without hesitation and with the ultimate in ease of movement.

Monorail crane and hoist systems allow for movement along fixed routes. They are designed to operate overhead and not interfere with ground level activities. Curves, switches, turntables, brake devices, and drop/lifts can be added for increased flexibility.

Jib cranes are used for dedicated work stations and can be a useful addition to an overhead handling system. Freestanding units can be used in open areas without having to be tied to the building structure. Wall-mounted units are ideal for mounting on columns or walls.

Basically, crane motion control involves the control of hoist and traverse (bridge and trolley) motors on electric overhead traveling cranes. The control system that is normally applied to the hoist drive differs somewhat from that applied to the traverse drive. Figure 10-3 shows some typical crane control panels.

The hoist control requirements for the hoisting direction differ somewhat from those of the lowering direction. On lowering, when the load is heavy enough to overcome the friction of the drive, it begins to overhaul the motor. The ultimate goal of a hoist control system, from both performance and safety aspects, is to provide smooth, precise, stepless speed control, totally independent of load. You should be able to move at extremely slow speeds with no load or full load. Likewise, you should not exceed 100 percent speed with any load whether you are hoisting or lowering a load. The drive must also be able to come to a prompt safe stop in case of power loss or an emergency stop.

Programmable motion controllers are microprocessor- or computer-based systems used to implement precise motion control. Components of a typical programmable motion control system (Fig. 10-4) include:

- *Programmable motion controller*
  - Determines the actual task to be performed by the motor
  - Sets things like speed, distance, direction, and acceleration rate
- *Drive*
  - Is an electronic power amplifier which delivers the power to operate the motor in response to low-level control signals
  - In general, will be specifically designed to operate with a particular type of motor (for example, a stepper drive cannot be used to operate a dc brush motor)
- *Motor*
  - May be a stepper motor (either rotary or linear), a dc brush motor or a brushless servo motor

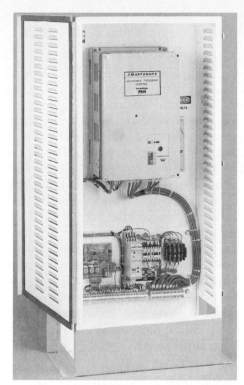

(*Courtesy Harnischfeger Corporation*)

(*a*) AC adjustable frequency drive enclosure

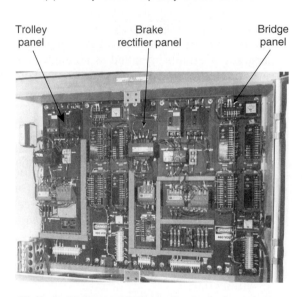

(*b*) Typical bridge and trolley motion in single enclosure

Fig. 10-3   Crane control panels.

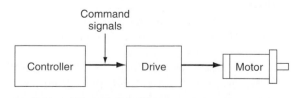

Fig. 10-4   Components in a typical programmable motion control system.

– Needs to be fitted with some kind of feedback device unless it is a stepper motor

Figure 10-5 shows typical programmable motion control applications. In Fig. 10-5(*a*), all three axes are automated, allowing flexibility to machine various parts. Microstepping drives provide plenty of power, resolution, and accuracy without the need to resort to more expensive closed-loop servo systems. The operator utilizes the jog function to position the grinding head at the proper "spark-off" height. From this point, the automatic control system takes over and finishes the part. In Fig. 10-5(*b*), the motor accelerates the material web to a continuous speed until a registration mark is sensed. The controller then signals a predefined registration move, activates the cutting mechanism, and repeats the process.

Common terms and parameters often referred to in conjunction with programmable motion control include:

- *Acceleration.* The change in velocity as a function of time. Acceleration usually refers to increasing velocity and deceleration describes decreasing velocity.
- *Accuracy.* A measure of the difference between expected position and actual position of a motor or mechanical system. Motor accuracy is usually specified as an angle representing the maximum deviation from expected position.
- *Closed loop.* A broadly applied term relating to any system in which output is measured and compared to input. The output is then adjusted to reach the desired condition. In motion control, the term is used to describe a system wherein a velocity or position (or both) transducer is used to generate correction signals by comparison to desired parameters.
- *Damping.* An indication of the rate of decay of a signal to its steady state value. Related to settling time.
- *Duty cycle.* For a repetitive cycle, the ratio of on time to total cycle time. Operating a motor beyond its recommended duty cycle results in excessive heat in the motor and drive.
- *Encoder.* A device which translates mechanical motion into electronic signals used for monitoring position or velocity.
- *Friction.* All mechanical systems exhibit some frictional force, and this should be taken into account when sizing the motor, as the motor must provide torque to overcome any system friction.

- *Holding torque.* Sometimes called *static torque,* it specifies the maximum external force or torque that can be applied to a stopped, energized motor without causing the rotor to rotate continuously.
- *Home.* A reference position in a motion control system derived from a mechanical datum or switch. Often designed as the "zero" position.
- *Hysteresis.* The tendency of a motor to resist a change in direction. This is a magnetic characteristic of the motor; it is not due to friction or other external characteristics.
- *Incremental motion.* A motion-control term used to describe a device that produces one step of motion for each step command (usually a pulse) received.
- *Inertia.* A measure of an object's resistance to change in velocity. The larger the inertial load, the longer it takes a motor to accelerate or decelerate that load.
- *I/O.* Abbreviation of input/output; refers to input signals from switches or sensors and output signals to relays, solenoids, and so on.
- *Leadscrew drives.* Lead screws convert rotary motion to linear motion.
- *Limits.* Properly designed motion-control systems have sensors called *limits* that alert the control electronics that the physical end of travel is being approached and that motion should stop.
- *Microstepping.* An electronic control technique that proportions the current in a step motor's windings to provide additional intermediate positions between poles. Produces smooth rotation over a wide speed range and high positional resolution.
- *Open loop.* Refers to a motion-control system in which no external sensors are used to provide position or velocity correction signals.
- *Pulse rate.* The frequency of the step pulses applied to a motor driver. The pulse rate multiplied by the resolution of the motor-drive combination (in steps per revolution) yields the rotational speed in revolutions per second.
- *Ramping.* The acceleration and deceleration of a motor. May also refer to the change in frequency of the applied step pulse train.
- *Regeneration.* Usually refers to a circuit in a drive amplifier which accepts and drains energy produced by a rotating motor either during deceleration or free-wheel shutdown.
- *Repeatability.* The degree to which the positioning accuracy for a given move performed repetitively can be duplicated.

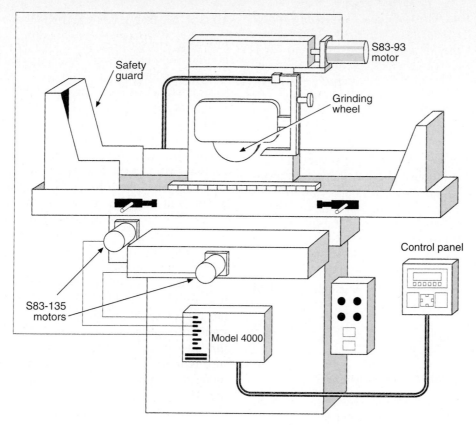

(a) Surface grinding machine linear motion feed control

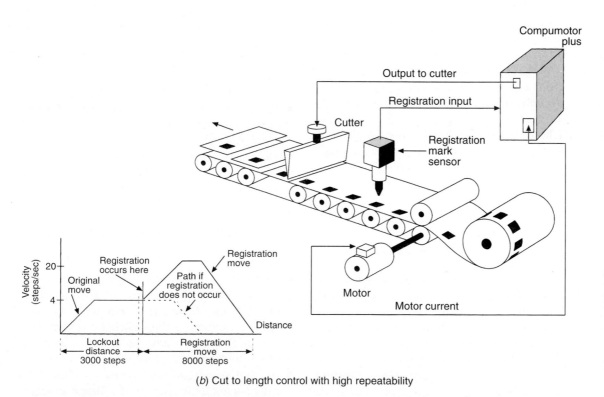

(b) Cut to length control with high repeatability

Fig. 10-5 Programmable motion control applications. *(Courtesy Parker Hannifin Corporation)*

- *Resolution.* The smallest positioning increment that can be achieved. Frequently defined as the number of steps required for a motor's shaft to rotate one complete revolution. For instance, a leadscrew with four revolutions per inch and a 25,000 step per revolution motor/drive would give 100,000 steps per inch. Each step would then be 0.00001 in.

- *Servo.* A system consisting of several devices which continuously monitor actual information (position, velocity), compares those values to desired outcome, and makes necessary corrections to minimize that difference.

- *Torque.* When selecting a motor/drive, the torque capacity of the motor must exceed the load. The torque any motor can provide varies with its speed and is usually specified with a torque/speed curve.

Self-Test

Answer the following questions.

1. True or false? A limit switch can be used to provide position information.
2. Refer to Fig. 10-1. Assume that the STOP button is pressed when the workpiece is midway between the two limit switches. How can automatic control be resumed with the workpiece initially moving in the reverse direction?
3. Refer to Fig. 10-2. Assume that the piston is moving from position A to B when all electrical power to the circuit is lost. How would this affect the movement of the piston?
4. True or false? The hoist control system of a crane is basically the same as that applied to the bridge and trolley.
5. Programmable motion controllers are _____ based systems.
6. The accuracy of a motion-control system is a measure of the difference between _____ position and _____ position.
7. _____ motion control describes a device that produces one step of motion for each step command received.
8. True or false? Leadscrew drives convert linear motion to rotary motion.
9. True or false? Ramping refers to the smallest positioning increment that can be achieved.
10. True or false? A servo system would be classified as an open loop system.

10-2 PRESSURE CONTROL

Pressure controls may either (1) *control* (maintain the pressure of gas, liquid, or solid at some specified value) or (2) *limit* (sense that pressure has reached some preset limit or is moving out of some safe range).

Pressure is defined as the *force per unit area.* The most common units of pressure are pounds per square inch (psi), inches of water column (wc) in a manometer, or inches of mercury (Hg) in a manometer. The most popular metric unit of measurement is the kilopascal (kPa). Pressure must be measured with respect to a given reference pressure, but it is most commonly atmospheric pressure at sea level or absolute zero pressure.

Pressure is one of the most important industrial process variables. Figure 10-6 illustrates the use of pressure limit control in a simple stamping operation. A normally closed *pressure switch* contact that opens on rising pressure is used. This pressure switch operates to signal the return of the pneumatic (air) cylinder. The advantage of using a pressure switch over a limit switch is that the workpiece will always receive the same amount of pressure before the cylinder returns. The operation of the circuit can be summarized as follows:

- START pushbutton switch is momentarily pressed closed.
- Relay coil CR and solenoid valve energize.
- Seal-in contacts CR1 close.
- The valve spool shifts and the cylinder advances.
- The cylinder continues to advance until the pressure switch contacts open (at the preset pressure).
- The solenoid and relay are deenergized.
- The valve spring returns the valve spool to its initial position, thus returning the cylinder.

Important *pressure switch specifications* include:

- *Adjustable operating range.* This is the range of pressures within which the pressure sensing element of the switch can be set to actuate the contact on the switch. For example, a pressure switch may have an adjustable operating range of 20 to 100 psi.
- *Adjustable differential range.* Sometimes called *deadband.* This is the range of pressure between the higher pressure limit, which changes the electrical contacts, and

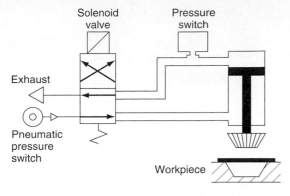

Exhaust

Pneumatic
pressure
switch

Workpiece

(a) Solenoid deenergized for start condition

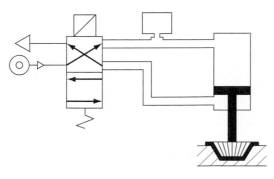

(b) Solenoid energized

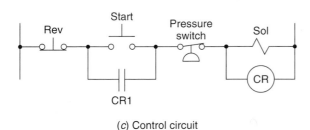

(c) Control circuit

Fig. 10-6   Pressure control of a stamping operation.

the lower pressure limit, which returns the electrical contacts to their normal condition. For example, a pressure switch may have an adjustable operating range of 20 to 100 psi with an adjustable differential of 5 to 15 psi.

- *Set point repeatability.* This is the ability of the switch to operate repetitively at its set point. This is typically ±1 percent of the maximum operating pressure.
- *Enclosure type.* This refers to the NEMA rating of the enclosure, for example, 4X watertight and corrosion resistance.
- *Electrical rating.* This refers to the current and voltage rating of the electrical contacts. For example, 10 A:125 Vac.
- *Switch arrangement.* This refers to the type of switch supplied. Two sets of contacts, one

NO (normally open) and one NC (normally closed), are standard on many pressure switches.

The pressure switch is used to transfer information concerning pressure to an electric circuit. Figure 10-7 on page 282 illustrates the electrical control of a pneumatically operated system. The pneumatic system contains two cylinder-piston assemblies. Each assembly is powered through a single-solenoid, spring-return operating valve. Both pressure switches are of the *normally closed* type. Limit switch LS1 uses one set of NO and one set of NC contacts. Limit switch LS2 uses one NO contact. The operation of one cycle of the circuit can be summarized as follows:

- START pushbutton switch is momentarily pressed closed.
- Relay coil CR1 energizes.
- Solenoid A energizes.
- Piston of cylinder A moves forward.
- The piston reaches the workpiece, builds pressure to a preset amount on PS1, operating the pressure switch contacts.
- Relay coil CR1 deenergizes.
- Solenoid valve A deenergizes.
- Piston A returns and on the return travel operates limit switch LS2 to energize relay coil CR2.
- Solenoid B energizes.
- Piston B now moves forward, meeting the workpiece and building pressure to an amount preset on PS2.
- Relay coil CR2 deenergizes.
- Piston B returns to its initial position.

Figure 10-8 on page 282 illustrates the application of a *pressure transducer* used with a solid-state starter to control an ac wound rotor motor. Unlike an ON/OFF pressure switch, the pressure transducer outputs a signal that is proportional to the pressure. This signal is compared to the pressure set point and determines whether the motor is switched on or off.

Self-Test

Answer the following questions.

11. Pressure controls may _____ or _____ pressure.

12. Refer to Fig. 10-6. Assume the start button is pressed, and the cylinder advances but returns as soon as the start button is released. What part of the circuit is at fault?

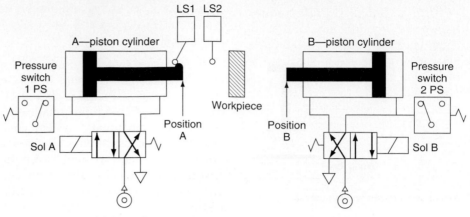

(a) Start of cycle— solenoids A and B are both deenergized

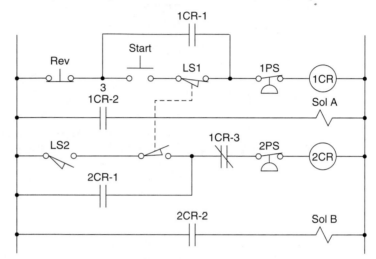

(b) Electrical control circuit—limit switch positions
shown as they would be with piston at position A

Fig. 10-7   Electrical control of a pneumatic-operated system.

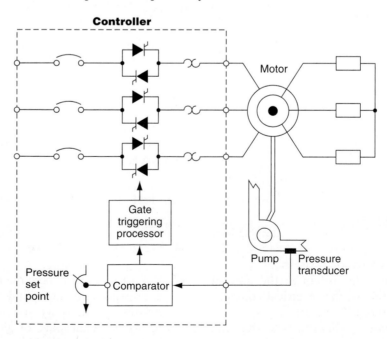

Fig. 10-8   Pressure transducer control of a motor.

13. True or false? Adjustable operating range and adjustable differential range refer to the same pressure switch specification.

14. The electrical rating of a pressure switch refers to the current and voltage rating of the electrical _____.

15. Refer to Fig. 10-7. Assume pressure switch PS2 is permanently open-circuited. How would this affect the movement of piston cylinder A and B?

16. Refer to Fig. 10-8. Assume the signal coming from the pressure transducer is lost. How would this affect the ON/OFF operation of the motor?

10-3 TEMPERATURE CONTROL

Temperature control may be used to maintain a specified temperature within a process or to protect against overtemperature conditions.

Figure 10-9 illustrates a temperature control system designed to maintain a preset temperature. In this application, a thermocouple is used to measure and transmit the temperature of the enclosure, while a solid-state contactor is used to turn the heater's coil on and off to provide the selected temperature.

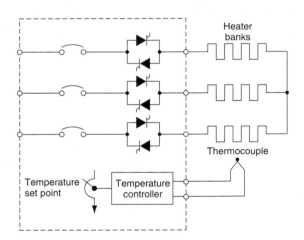

Fig. 10-9 Temperature control that uses a thermocouple as the temperature-sensing element.

▐▐▐➡ **YOU MAY RECALL** that a thermocouple is the junction of two dissimilar metals and has a voltage output proportional to the difference in temperature between the hot junction and the lead wires (cold junction).

A *temperature controller* is used to accurately control process temperature without extensive operator involvement. The controller accepts a temperature sensor such as a thermocouple or RTD as input and compares the actual temperature to the desired control temperature, or set point, and provides an output to a control element.

Figure 10-10 on page 284 shows the control panel of a typical temperature controller. The unit displays both the set and the process temperature and provides an accurate output control signal to maintain a process at the desired control point.

There are three basic types of temperature mode control, *ON/OFF*, *proportional*, and *PID*. Depending upon the system to be controlled, the operator can use one type or the other to control the process (Fig. 10-11 on page 285).

A variety of control outputs are available with temperature controllers. These include relay, solid-state relay (SSR), as well as voltage or current outputs. For example, relay contact output of 5 A at 250 Vac, 12 Vdc pulse for SSR drive, 4 to 10 mA, and 1 to 5 Vdc. Relay output capacity is normally enough to handle only a small heating element load. For large loads the relay is used to energize a contactor.

With *ON/OFF temperature control,* the output turns on when the temperature falls below the set point and turns off when the temperature reaches the set point. Control is simple, but overshoot and cycling about the set point can be disadvantageous in some processes. ON/OFF control is usually used where a precise control is not necessary, in systems which cannot handle the energy being turned on and off frequently, where the mass of the system is so great that temperatures change extremely slowly, or for a temperature alarm.

Proportional controls are designed to eliminate the cycling associated with ON/OFF control. A proportional controller decreases the average power being supplied to the heater as the temperature approaches set point. This has the effect of slowing down the heater, so that it will not overshoot the set point but will approach the set point and maintain a stable temperature. Depending upon the process and the precision required, either a simple proportional control or one with PID may be required.

A *PID,* or *three-mode controller,* combines the proportional, integral (reset) and derivative (rate) actions, and is usually required for tight control of temperature-sensitive applications. The outputs turn on and off in proportion to temperature deviation from the set point. The *rate function* (derivative action) shortens the time it takes the temperature to

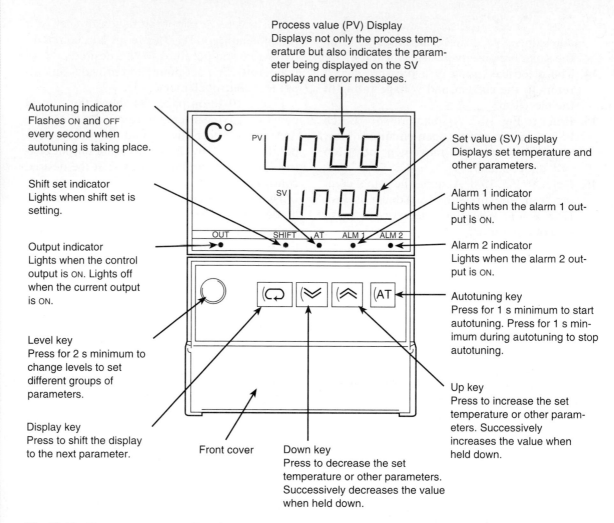

Process value (PV) Display
Displays not only the process temperature but also indicates the parameter being displayed on the SV display and error messages.

Autotuning indicator
Flashes ON and OFF every second when autotuning is taking place.

Shift set indicator
Lights when shift set is setting.

Output indicator
Lights when the control output is ON. Lights off when the current output is ON.

Level key
Press for 2 s minimum to change levels to set different groups of parameters.

Display key
Press to shift the display to the next parameter.

Set value (SV) display
Displays set temperature and other parameters.

Alarm 1 indicator
Lights when the alarm 1 output is ON.

Alarm 2 indicator
Lights when the alarm 2 output is ON.

Autotuning key
Press for 1 s minimum to start autotuning. Press for 1 s minimum during autotuning to stop autotuning.

Up key
Press to increase the set temperature or other parameters. Successively increases the value when held down.

Front cover

Down key
Press to decrease the set temperature or other parameters. Successively decreases the value when held down.

Fig. 10-10   Temperature controller. *(Courtesy Omron)*

stabilize near the set point. The *reset* function (integral action) eliminates any offset from the temperature set point.

Alarm outputs include ones for upper limit, lower limit, upper and lower limits, and upper and lower range limit. Also available is an alarm output and indication that a heater has burned out. This uses a current transformer to detect a drop in current consumption to trigger the alarm.

Dual-output temperature controllers are used in applications requiring two related control actions, such as cold start protection, quick start with auxiliary heaters, and alarms, or wherever an auxiliary-related control action is required. They can also be used for a heat-cool application as is shown in Fig. 10-12. Two controlled outputs are supplied, one for heating and another for cooling. This type of controller is required for processes which may require heat to start up but then generate excessive heat during operation.

A *temperature recorder* or *data logger* is sometimes permanently installed as part of the temperature-control system (Fig. 10-12). This instrument makes a permanent record of the process temperature by writing an analog trace or actual numeric values onto paper. There are two sets of units shown on most recording charts—one for the value of the measured temperature, the other for time. Temperature controllers are sometimes used to convert the sensor signal into a form that the recorder can accept.

Resistance heating uses current flowing through a conducting material to produce heat. When dc or ac voltage is connected directly across a piece of metal or other conducting material, current flows and generates heat at a rate equal to I^2R, where I is the current (in amperes) and R is the resistance (in ohms) of the material. Such metallic heaters may be part of a resistance-type furnace. The heating of materials by electricity is faster and cleaner and is

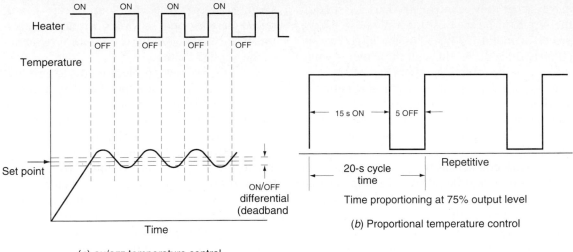

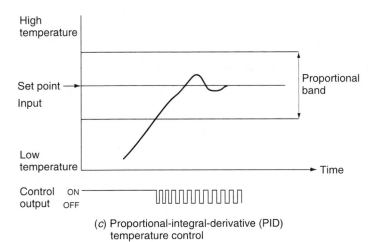

(a) ON/OFF temperature control

(b) Proportional temperature control

(c) Proportional-integral-derivative (PID) temperature control

Fig. 10-11   Types of temperature control.

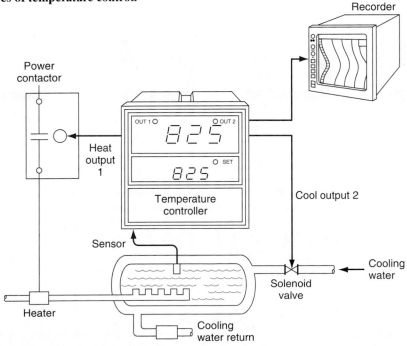

Fig. 10-12   Dual-output heat/cool control.

more accurately controlled than heating in fuel-fired furnaces. Resistance *welding* of two pieces of metal involves passing large current (1000 to 100,000 A) through these pieces while they are being forced together (Fig. 10-13).

Induction heating uses electromagnetic induction to produce heat (Fig. 10-14). A conducting material is placed within a work coil that is connected to a *high-frequency* ac supply.

The work coil acts as the primary winding of a transformer, and the workpiece being heated acts as a single-turn secondary winding that is short-circuited, so that current is forced to circulate within it. Cooling water often is circulated within the work coil. One advantage of induction is its ability to instantly produce heat. The amount of heat produced in the workpiece depends upon the frequency of the supply voltage and the amount of current flowing in the workpiece. The depth to which the heat penetrates depends upon the frequency of the supply voltage. The higher the frequency, the less the depth of heat penetration.

Dielectric heating is based on the principle of capacitance (Fig. 10-15). An insulating material is placed between two conducting plates connected to a high-frequency ac supply. High voltages at frequencies above 1 megahertz (MHz) are required. This high voltage and frequency cause the electrons in the insulating material to strain first in one direction and then in the other. This continuous and rapid strain reversal produces heat. Heat is produced instantly and uniformly within the material and is controlled by varying the frequency of the applied voltage.

Arc heating uses an electric arc to produce heat. The electric arc is produced by ionizing the air between two electrodes. The heat from the arc is used for welding and melting metal. In arc welding the material to be welded forms one electrode, and the welding rod the other. To form the arc, the welding rod is first touched to the material to be welded then pulled away. The ionized air conducts current, forming an arc. Once the arc is produced, the heat generated will cause the material and the rod to melt. The metal from the rod flows with that from the material to form a continuous seam.

Self-Test

Answer the following questions

17. Refer to Fig. 10-9. What solid-state switching device is used to turn the heater bank ON and OFF?

18. A temperature controller is used to accurately control process temperature without extensive _____ involvement.

19. True or false? The relay output capacity of a temperature controller is normally enough to handle only a small heating element load.

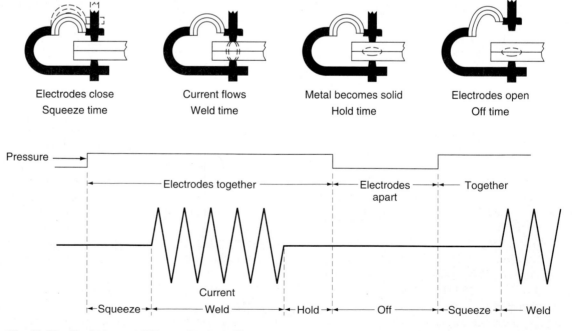

Electrodes close
Squeeze time

Current flows
Weld time

Metal becomes solid
Hold time

Electrodes open
Off time

Fig. 10-13   Resistance welding control profile.

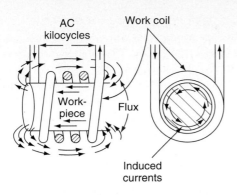

Fig. 10-14 Induction heating.

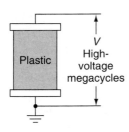

Fig. 10-15 Dielectric heating.

20. True or false? ON/OFF temperature control is usually used where a precise control is not necessary.
21. The two sets of variables shown on most temperature recording charts are _____ and _____.
22. _____ heating uses current flow through a conducting material to produce heat.
23. _____ heating uses electromagnetic induction to produce heat.
24. Dielectric heat is based on the principle of _____.
25. With arc heating, the arc used to produce the heat is created by _____ the air between two electrodes.

10-4 TIME CONTROL

The main function of a *timer* is to place information about elapsed time into a control circuit. Timing control can be accomplished using pneumatic, electromechanical, or electronic components.

A distinction can be made between timers and time-delay relays. Generally, *time-delay relays* are devices having a timing function after the timer coil has been energized or deenergized.

When reference is made to *timers,* the time function may start on one or more of the con-

tacts upon energization, or at any time after energization during a preset time cycle. Likewise, the timing function may stop on one contact or more during the cycle after timing has been started on the particular contact(s). In general, a timer opens or closes electrical circuits to selected operations according to a timed program. Instantaneous contacts are also available on both time-delay relays and timers.

▥▶ **YOU MAY RECALL**   that instantaneous contacts operate immediately, independent of the timing mechanism.

Terms associated with the selection and operation of timers and time-delay relays include:

- *Analog timer.* A timer that has a dial or knob which is used to select the set time.
- *Automatic reset.* Automatically returns the timer to the 0 state after the lapse of a given time.
- *Delayed contacts.* The switch contacts in a timer that switch according to the timing circuits.
- *Digital setting.* The use of pushbutton switches to select a timer preset value.
- *Electrical service life.* The service life of a timer when the control output is operated to switch the specified voltage/current load connected to the control output.
- *External/electrical reset.* Resets the timer by a required signal applied from an external source to the reset input signal terminals.
- *Inhibit function.* This is accomplished by applying a voltage to specific terminals. The timing cycle then stops and holds its outputs in the last state without resetting the circuit. Timing will continue and the outputs will be allowed to change according to their programmed operating modes when voltage is removed from the above-mentioned specific terminals.
- *Instantaneous contacts.* In timers, the relay or switch contacts that are operated instantaneously upon power up.
- *Load rating.* The maximum current and voltage of the load-circuit energy that may be switched by a timer for normal life expectancy.
- *Maintained start.* A constant closure of a contact to start and complete the control function.
- *Manual reset.* Mechanically resets the timer by manual operation.
- *Memory protective function during power failure.* The function by which the elapsed

time at the time of a power failure is retained until power is applied again to the timer.

- *Minimum setting.* The shortest cycle that you can set on a unit. If it is set below the minimum, it will probably not time at all, or operate erratically.
- *Momentary start.* An initiate signal that is usually shorter than the control function.
- *Operating voltage range.* The range of voltage over which a timer will perform to specifications.
- *Power reset.* Resets the timer by interrupting the operating supply voltage.
- *Repeat accuracy.* Difference of operating times measured when the timer repeats operation under the same condition with a given setting time. Formula for calculation (with operating time measured more than five times):

Percent repeat accuracy ± 1/2

$$\times \frac{T_{max} - T_{min}}{TMs} \times 100\%$$

where T_{max} = Maximum value of operating times measured at the same set time

T_{min} = Minimum value of operating times measured at the same set time

TMs = Maximum scale time

Since the repeat accuracy is expressed in terms of percentage against the maximum setting time, the absolute value of the repeat accuracy does not change even if the setting time is changed. Accordingly, the time specification should be taken into account as much as possible, so that the timer may be used in the vicinity of full scale.

- *Repeat cycle operation.* The operation to repeat ON/OFF at each given operating time.
- *Reset.* To restore the timing, display, and output sections of the timer to their initial states before the start of timing.
- *Timing chart.* A graphic representation of two or more sequences of events, all drawn to the same horizontal time scale, so that any point in one sequence occurs at the same time as any point directly above or below it in another sequence.

Timing functions will start from an electrical signal, initiated through any one of several components. This may be the pushbutton switch, relay contact, temperature switch, pres-

sure switch, limit switch, etc. Figure 10-16 shows an application for a time-delay relay in which the timing is started by pressing a pushbutton. The control circuit uses an on-delay relay with a set of normally closed contacts that are timed open. This circuit is used as a warning signal when moving equipment, such as a conveyor motor, is about to be started. The operation of the circuit can be summarized as follows:

- Coil CR is energized when the START pushbutton PB1 is momentarily actuated.
- Contact CR1 closes to seal in relay coil CR.
- Contact CR2 closes to energize timer coil TD.
- Contact CR3 closes to sound the horn.

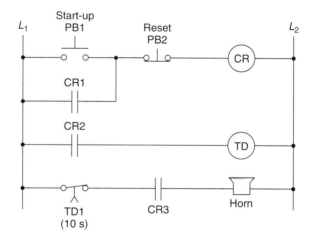

Fig. 10-16 Time-delay relay—warning signal circuit schematic.

- After a 10-s time-delay period, timer contact TD1 opens to automatically switch the horn off.
- The reset button is pressed to restore the relay to its original state before another alarm can be initiated.

Different types of timers are used for industrial applications such as heating, conveyor, and machine tool control. Timers may be classified according to their method of *timing* as being motor-driven or solid-state. They are also classified according to method of *operation* as being manually set timers, reset timers, and repeat cycle timers (Fig. 10-17).

A *manually set timer* is synchronous-motor-driven, adjustable to a selected time range by a large, easy-to-read set pointer. To operate, turn the set pointer to the required time interval. A load circuit is closed when the pointer is turned from zero, at which point the load circuit opens.

(a) Manually set timer

(b) Synchronous motor-driven analog reset timer

(c) Solid-state repeat cycle digital timer

Fig. 10-17   Types of timers. *(Courtesy Eagle Signal Controls)*

On a synchronous-motor-driven analog *reset timer,* the time range is knob-adjustable. The progress pointer is activated during the timing period. Reset timers measure a "piece of time" and are inactive both before and after timing. The period before the timing function is called *reset;* the period after the timing function is called *timed out.* The entire sequence of the reset timer is therefore reset–timing–timed out.

A *solid-state repeat cycle digital timer* will alternately turn an output ON and OFF as long as power is applied to the unit. It is microprocessor based for greater setting accuracy and fully programmable to provide a number of time ranges and operating modes in one unit. The time cycle is shown on the digital display. Annunciators at the right-hand side of the display indicate the operating mode. These annunciators flash to indicate when the timer is in the ON time cycle, and are constantly on to indicate when the timer is in the OFF time cycle.

Figure 10-18 on page 290 shows the schematic circuit for a motor-driven reset timer. It depends on the clutch and the synchronous motor for its operation. Control is achieved by using the limit switch as a sustained input. The limit switch must remain closed to energize the clutch solenoid and power the timing motor. The timer illustrated is connected for ON delay, requiring input power to close the clutch starting the timer. The contacts—both instantaneous and time delay—are connected to four loads marked A, B, C, and D. A code is added above each load to illustrate the sequence during reset, timing, and timed out. (0 denotes an OFF condition, and X denotes an ON switch condition.) Loads C and D are strictly relay type responses, using only the instantaneous contacts. Loads A and B utilize the combined action of the instantaneous and delay contacts to achieve the desired sequence. The limit switch is opened to reset the timer. Loss of plant power will also reset the timer, because the clutch will open when power is lost. The operation of the circuit can be summarized as follows, beginning with the timer in its reset mode and the switch open:

- Limit switch closes.
- Clutch and motor are energized to begin timing.
- NC instantaneous contacts 6–7 open to turn OFF load D.
- NO instantaneous contacts 6–8 close to turn ON load C.
- NO instantaneous contacts 9–10 close to turn ON load B.
- Load A remains OFF.
- Timer times out.
- NC timed contacts 4–5 open in order to turn OFF load B.
- NO timed contacts 4–3 close in order to turn ON load A.
- NC timed contacts 11–12 open to deenergize motor.
- Limit switch is opened to reset timer by deenergizing the clutch.

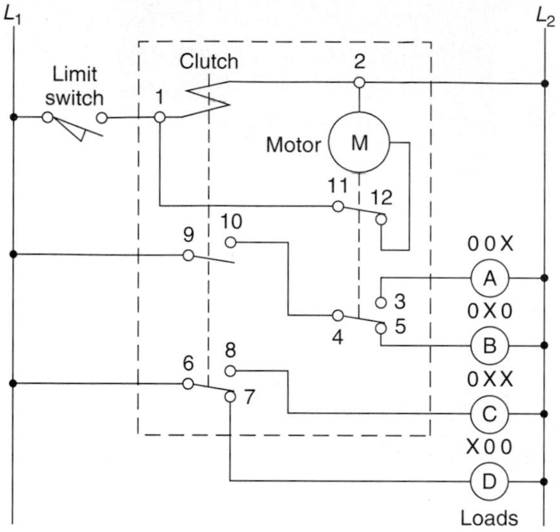

Figure 10-19 shows a typical application of a timer used in a plywood manufacturing operation to vary the amount of time the in-line plywood press is closed. The loading and unload-

Fig. 10-18 **Motor-driven reset timer schematic.**

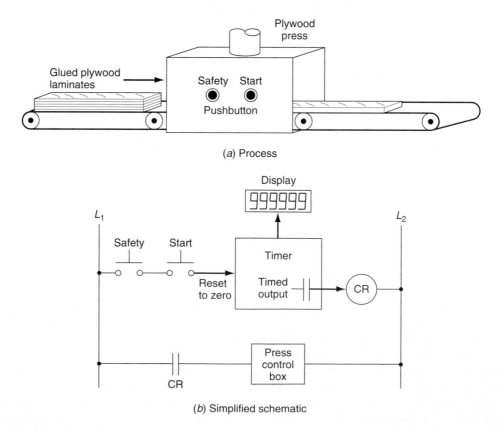

(a) Process

Display

(b) Simplified schematic

Fig. 10-19 **Controlling length of press time.**

ing of the plywood are done manually on a conveyor belt system, and the amount of press time varies because of the width and height of the plywood. Because of this requirement, manual activation of the press cycle is necessary. The time the press must be closed varies according to the number of stacked laminations (thickness). The timer operates in the one-shot mode with manual reset to zero. The preset time value is set to the necessary value for the plywood thickness. The timed output contact is used to activate control coil CR. One of the normally open contacts of the control relay is used to activate the press control box. The safety and start buttons are connected in series with the timer reset input. The operation is as follows:

- The operator loads the stack of glued plywood laminations into the plywood press.
- The operator momentarily presses both safety and start buttons at the same time. This resets the timer to zero.
- At the release of the buttons, the timer timed output contact closes.
- The control relay activates and causes the plywood press to close and stay closed until the timer reaches its preset value.
- When the preset value is reached, the timer timed output contact opens.
- This deactivates the control relay, opening the plywood press.
- The press remains open until the operator presses the safety and start buttons for the next stack of laminations.

Self-Test

Answer the following questions.

26. The _____ contacts are the switch contacts in a timer that switch according to the timing circuits.
27. When voltage is applied to the inhibit terminals of a timer, the timing cycle _____.
28. True or false? The load rating of a timer refers to the power required to operate the unit.
29. True or false? The repeat accuracy of a timer is always 100 percent.
30. Refer to Fig. 10-16. Assume a set of normally open timed contacts are used in place of the normally closed timed contacts. What will happen when the start-up button is pressed?

31. Most motor driven timers use _____ motors for their operation.
32. True or false? Reset timers measure a piece of time, being inactive both before and after timing.
33. Refer to Fig. 10-18. State the condition (ON or OFF) of output A during reset, timing, and timed out conditions.
34. Refer to Fig. 10-19. Assume the operator keeps the safety and start buttons pressed continually closed. What will happen?

10-5 COUNT CONTROL

Counters are devices that will receive a string of count pulses from a machine operation and perform an output function based on a number of counts predetermined by the user. Most counters, like timers, can have interval and delay operation. *Interval operation* means that a load will be actuated when the unit is counting. *Delay operation* means that a load will be actuated at the end of the counting cycle. Solid-state and electromechanical versions are available.

Counters are generally thought of as devices that tabulate or count "things" such as bottles, cans, boxes, castings, and so on. In many industrial control systems, it is necessary to count something that affects a controlled process. When the count reaches a certain number, a control action is initiated. Figure 10-20 on page 292 illustrates a number of typical counter applications and types of counters used.

In a mechanical counter, every time the actuating lever is moved over, the counter adds one number, and the actuating lever returns automatically to its original position. Resetting to zero is done by a pushbutton located on the side of the unit. In an electromechanical counter, the count setpoint can be adjusted by the knob on the front of the unit. A progress pointer, indicating the count progression, advances clockwise, from setpoint to zero. A solid-state counter has high-speed pulse operation with 100 percent accuracy and has many programmable features. Counter output action occurs when the count total indicated by the thumbwheel switches is reached.

The wiring diagram for a typical electromechanical counter is shown in Fig. 10-21 on page 293. The sustained control switch is closed to start the counter and opened to reset the counter. This counter is constructed so that it is similar to the electromechanical reset

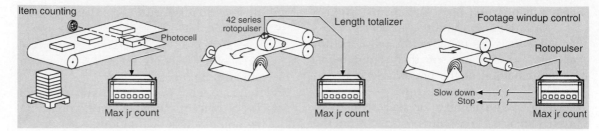

(*Courtesy of Dynapar Corporation, Gurnee, Illinois*)

(*a*) Counter applications

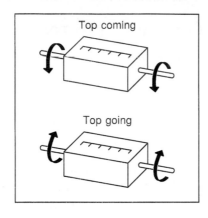

(*b*) Mechanical counter

(*c*) Electromechanical counter

(*d*) Solid-state counter

Fig. 10-20   Counters.

timer, except that the synchronous motor of the timer is replaced by a solenoid-operated pawl feed count motor. The count motor advances one step each time it receives a pulse. The delayed switches transfer at count out. A minimum specified OFF time is required between pulses. The circuit operates as follows:

- The sustained control switch is closed to energize clutch and enable the counter to receive and register counts.

- Instantaneous contacts transfer.
- Each time the count switch is momentarily closed, a pulse is applied to the count motor to register a count by moving the count progress pointer toward the zero point on the dial.
- When the progress pointer reaches zero, the unit is counted out and the delay switch operates to turn output A ON and output B OFF.
- Additional counts will not be registered until the unit is reset.

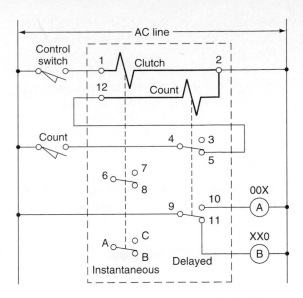

Fig. 10-21 Wiring diagram—electromechanical counter.

- Opening the control switch to remove power from the clutch resets the counter.

Most solid-state counters can count up, count down, or be combined to count up and down. An *up-counter* will count up or increment by 1 each time the counted event occurs. A *down-counter* will count down or decrement by 1 each time the counted event occurs. Normally the down-counter is used in conjunction with the up-counter to form an *up/down-counter* equipped with separate count-up and count-down inputs.

Figure 10-22 illustrates the application of a solid-state up-counter used in a batch counting operation to sort parts automatically for quality control. The counter is programmed to count up from zero to the preset value. When the preset value is reached, the output relay closes for the amount of time programmed. The counter will automatically reset to zero at the beginning of the timed output and continue to accumulate counts. The counter is installed to divert one part out of every 1000 parts for quality control or inspection purposes. The circuit operates as follows:

- A proximity sensor counts the parts as they pass by on the conveyor.
- When 1000 is reached, the counter's output relay contact closes, energizing the CR coil.

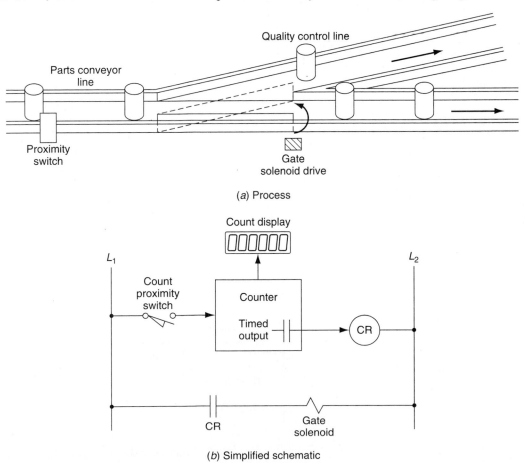

(a) Process

(b) Simplified schematic

Fig. 10-22 Up-counter used to sort parts for quality control.

- This activates the gate solenoid, diverting the part to the inspection line.
- The counter's timed output is set for 0.8-s activation. This allows enough time for the part to pass onto the quality control line.
- The gate returns to its normal position when the timed output ends.
- The counter resets to zero and continues to accumulate counts.

Figure 10-23 illustrates the operation of a solid-state up/down counter used to provide continuous monitoring of items "in-process." An in-feed photoelectric sensor counts raw parts going into the system and an out-feed photoelectric sensor counts finished parts leaving the machine. The number of parts between the in-feed and out-feed is indicated by the display. Counts applied to the up-input are added while counts applied to the down-input are subtracted. The operation of the circuit can be summarized as follows:

- Before start-up, the system is completely empty of parts and the counter is manually reset to zero.
- When the operation begins, raw parts move through the in-feed sensor with each part generating an "up" count.
- After processing, finished parts appearing at the out-feed sensor generate "down" counts so the counter continuously displays the number of "in-process" parts.

In addition to counting "things," counters can be used for *length* and *position* measurement. Figure 10-24 illustrates the operation of a solid-state counter used in length measurement. This system accumulates the total length of random pieces of bar stock being moved on a conveyor. The operation of the circuit can be summarized as follows:

- Count input pulses are generated by the magnetic sensor that detects passing teeth on a conveyor drive sprocket. If 10 teeth per

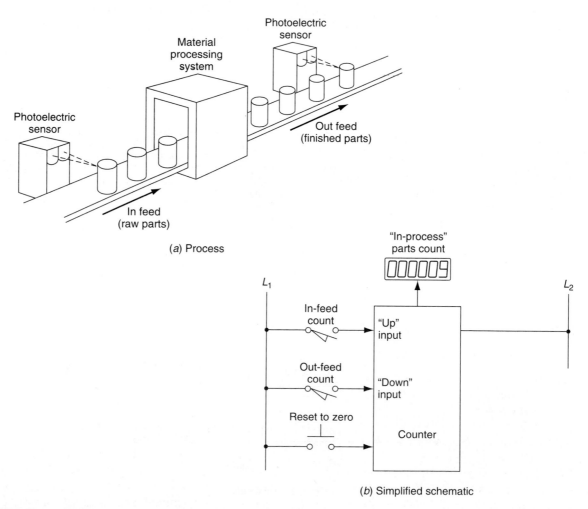

(a) Process

(b) Simplified schematic

Fig. 10-23 Up/down counter used in an in-process monitoring system.

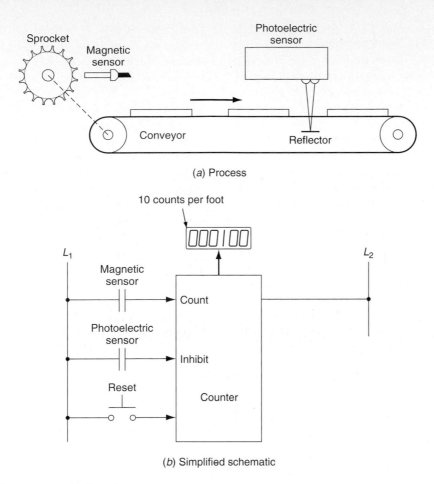

(*a*) Process

10 counts per foot

(*b*) Simplified schematic

Fig. 10-24 **Counter used for length measurement.**

foot of conveyor motion pass the sensor, the counter will display feet in tenths.

- The photoelectric sensor controls the inhibit input and monitors a reference point on the conveyor. When activated, it prevents the unit from counting, thus permitting the counter to accumulate counts only when bar stock is moving by.
- Counter is zeroed by closing the reset button.

Self-Test

Answer the following questions.

35. Delay operation of a counter means that a load will be actuated at the _____ of the counting cycle.

36. True or false? Solid-state counters have more control features than electromechanical types.

37. Refer to Fig. 10-21. Assume that the counter control switch is momentarily opened with a count of 5 registered. What will happen?

38. A(n) _____-counter will increment by 1 each time the counted event occurs.

39. Refer to Fig. 10-22. What counter setting should be changed to divert two parts to the inspection line when 1000 is reached?

40. Refer to Fig. 10-23. Assume that the signal from the out-feed photoelectric sensor is lost with a count of 50 registered. What will happen?

41. In addition to counting, counters can be used for _____ and _____ measurement.

42. Refer to Fig. 10-24. Assume that the signal from the photoelectric sensor is lost. How will this affect the length of measurement?

10-6 SEQUENCE CONTROL

Sequence control involves controlling the *sequence* in which certain events will occur. Many machine tools and process machines require that a predetermined sequence be followed when starting and/or running the equipment.

An example would be a power distribution system that does not have sufficient capacity to start several motors simultaneously. If several motors are to be started from the same push-button station under these conditions, a time delay can be provided between the operation of the motor starters. The sequence control circuit of Fig. 10-25 illustrates such a case. When the START button is pushed, the first starter is energized along with a timing relay. When the timing relay times out, it operates a contact that closes the control circuit of the second starter.

Figure 10-26 shows a sequence control system designed to alternate the use of two pumps so that they both get the same amount of usage over their lifetime. The main control element in the circuit is the *alternating relay* A (also known as an *impulse* or *flip-flop relay*). This relay is bistable, which means that it has two stable states. Contacts of the relay A change state each time the relay coil receives a voltage pulse. The operation of the circuit can be summarized as follows:

- Float switch F1 closes, as water enters a tank, to apply power to the circuit.
- As the water level rises, float switch F2 closes and energizes the alternator relay coil A.
- Assuming that the contacts are initially in the position shown, after the coil A is energized, current can flow through contact A1 to starter coil 1M, which energizes pump number 1. Contact 1M1 closes and forms a holding circuit for pump 1.
- If the water level continues to rise and pump 1 cannot keep up, then float switch F3 will close to energize starter coil 2M and start pump 2. Contact 2M1 closes, forming a holding contact for pump 2.

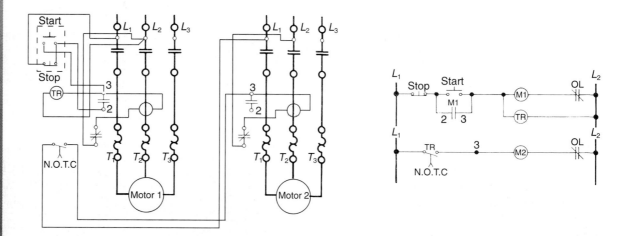

Fig. 10-25   Starting two motors in sequence. *(Courtesy Allen-Bradley Canada)*

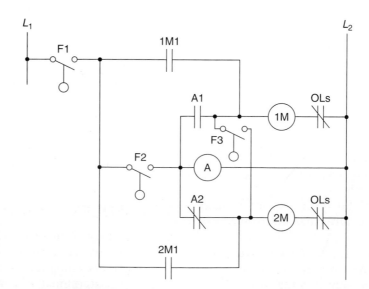

Fig. 10-26   Pump alternating control circuit.

- As the water level decreases, float switch F3 opens but has no effect, because of the holding circuits.
- If the water level continues to fall, float switch F2 will open, but it too has no effect because of the holding circuits.
- Finally, float switch F1 opens and the entire circuit is deenergized—the pumps are turned off.
- This completes one cycle.
- The next time float switch F1 closes, the alternating relay contacts A2 are now closed and contacts A1 remain open.
- Pump 2 will start first and pump 1 will follow.
- If float switch F3 does not close, then only one pump will run.

Sequencers (also referred to as *drum switches, rotary switches, stepper switches,* or *cam switches*) are devices that provide ON/OFF action for a number of output channels in a predefined sequential pattern. The operation can be based on either time or step inputs. *Time sequencers* have a total time base of operation, and the individual load circuits can be programmed to turn on and off anywhere within that time base. *Step sequencers* operate from electrical pulses initiated by such input devices as limit switches, photoelectric sensors, and counters. These input pulses step a camshaft through a fraction of a revolution for each input pulse.

Electromechanical sequencers are basically simple in design. Figure 10-27 illustrates the operation of a cam-operated sequencer switch. An electric motor is used to drive the cams. A series of leaf-spring-mounted contacts interacts with the cam so that in different degrees of rotation of the cam, various contacts are closed and opened to energize and deenergize various electrical devices.

Various control capabilities are possible through the use of sequencers. The operation of a sequencer can be shown as a series of steps, each occurring in a predetermined sequence to complete the operational cycle. Figure 10-28 on page 298 illustrates a typical mechanical drum sequencer switch. The switch consists of a series of contacts that are operated by pegs located on a motor-driven drum. The pegs can be placed at random locations around the circumference of the drum to operate contacts. When the drum is rotated, contacts that align with the pegs will close, and where there are no pegs, the contacts will remain open. Each

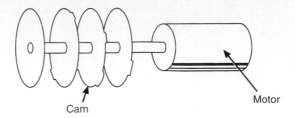

(a) Cam-driving mechanism

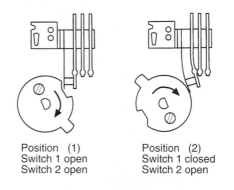

Position (1)
Switch 1 open
Switch 2 open

Position (2)
Switch 1 closed
Switch 2 open

(b) Cam and contact operation

(Courtesy of Allen-Bradley Company, Inc.)

(c) Typical industrial rotating cam limit switch is a pilot circuit device used with machinery having a repetitive cycle of operation

Fig. 10-27 Sequencer switch.

row of pegs represents one step. The number of individual steps are identified, along with the input condition required to terminate a step. This input may be a timer timing out or a contact closing, and so on. While a specific step is being executed, predetermined outputs are energized or deenergized.

In Fig. 10-28(*b*), the input form is shown as an *electrical pulse* to the step mechanism. For each pulse, the program drum advances one

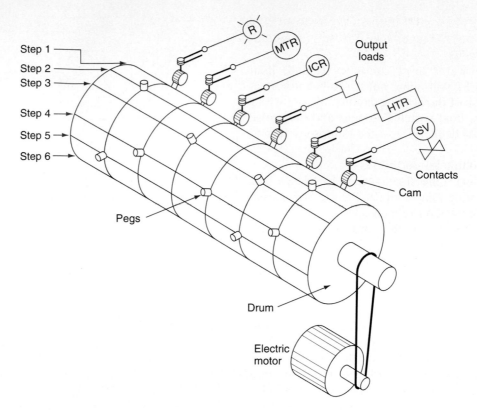

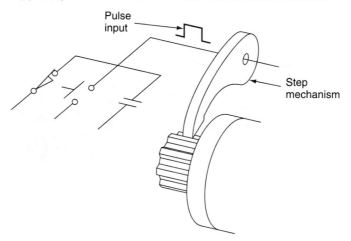

(a) A sequencer with a set of individual load contacts or switches

(b) Input form is an electrical pulse

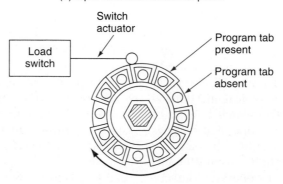

(c) Programmable cam segment

Fig. 10-28 Drum sequencer.

step. Motor circuitry is provided to ensure that the motor advances in increments, representing each step, rather than being allowed to run continuously.

In a typical programmable cam segment [Fig. 10-28(c)], the presence or absence of a program tab in any particular step position will determine the resultant switch state. If the tab is present, the circuit switch is actuated—that is, normally closed contacts are open, and normally open contacts are closed.

Sequencers are used in place of complex relay circuitry to provide sequential control action. An alternative control concept for sequential applications could result in a more complex control scheme involving interwiring of many relays and other control elements.

Figure 10-29 shows a step sequencer that increments by one step for each pulse applied to it. Different isolated inputs are used to advance the stepswitch. A tapswitch, operated by the stepping mechanism, is used to isolate each input, so that other inputs are locked out, except the one which is to advance the stepswitch from one particular interval to the next. The rotary wiper of the tapswitch is advanced so that the wiper is in contact with terminal 1 in step position 1; terminal 2 in step position 2; and so on.

The position shown in Fig. 10-29(a) indicates that the stepswitch is in step with position 3. Only input CR2 will advance the stepswitch from this position. All other inputs are locked out, as they do not complete a circuit to the stepping mechanism. To be recognized as a required step pulse, the stepswitch input must be present *and* absent for a minimum specified time, as indicated in Fig. 10-29(b). Depending upon stepswitch design, the stepswitch advances either on the *application* or on the *removal* of the input signal.

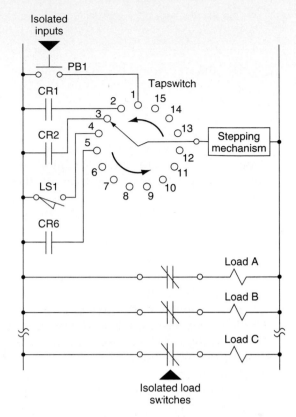

(a) Stepswitch is in step with position 3

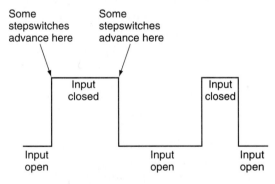

(b) To be recognized as step pulse, stepswitch input must be present *and* absent for a minimum specified time

Fig. 10-29 Step sequencer.

Self-Test

Answer the following questions.

43. Refer to Fig. 10-25. Assume that seal-in contacts M1 fail to close when coil M1 is energized. What will happen when the start button is momentarily pressed?

44. Refer to Fig. 10-26. Assume that wiring continuity to float switch F3 is lost. How will this affect the operation of the circuit?

45. The operation of a sequencer can be based on either _____ or _____ inputs.

46. While a specific step of a sequencer is being executed, predetermined _____ are energized or deenergized.

47. Refer to Fig. 10-28(c). If the camshaft is rotated 360° from the position shown, what ON/OFF sequence would be produced by a normally open load switch contact?

48. To be recognized as a step pulse, the input to a step sequencer must be present and absent for a minimum specified _____.

49. Refer to Fig. 10-29. What input will advance the stepswitch at step position 5?

1. Position information is provided by means of a pilot device which provides an adequate electrical signal.

2. Position information may be used for indication and control.

3. Bridge cranes allow loads to be moved from one point to another, anywhere within the rectangular coverage of the bridge and runway.

4. Monorail crane and hoist systems allow for movement along fixed routes.

5. Jib cranes are used for dedicated workstations.

6. The control system normally applied to the crane hoist drive differs somewhat from that applied to the traverse drive.

7. On lowering, when the load on the crane hoist is heavy enough to overcome the friction of the drive, the load begins to overhaul the motor.

8. Programmable motion controllers are microprocessor- or computer-based systems used to implement precise motion control.

9. A programmable motion controller determines the actual task to be performed by the motor and sets things like speed, distance, direction, and acceleration rate.

10. Common terms and parameters often referred to in conjunction with programmable motion control include acceleration, accuracy, closed loop, damping, duty cycle, encoder, friction, holding torque, home, hysteresis, incremental motion, inertia, input/output, leadscrew drives, limits, microstepping, open loop, pulse rate, ramping, regeneration, repeatability, resolution, servo, and torque.

11. Pressure controls may either control or limit pressure.

12. Pressure is defined as the "force per unit area."

13. Common units of pressure are pounds per square inch (psi), inches of water column in a manometer (wc), inches of mercury in a manometer (Hg), and kilopascal (kPa).

14. Important pressure switch specifications include adjustable operating range, adjustable differential range, set point repeatability, enclosure type, electrical rating, and switch arrangement.

15. The pressure switch operates ON and OFF to transfer information concerning pressure to an electric circuit.

16. The pressure transducer outputs a signal that is proportional to the pressure.

17. Temperature control may be used to maintain a specified temperature within a process or to protect against overtemperature conditions.

18. A temperature controller accepts a temperature sensor as input, compares the actual temperature to the desired control temperature or set point, and provides an output to a control element.

19. There are three modes of temperature control: ON/OFF, proportional, and PID.

20. Control outputs are available with temperature controllers which include relay and solid-state relay (SSR) as well as voltage or current outputs.

21. Temperature alarm outputs include ones for upper limit, lower limit, upper and lower limits, and upper and lower range limits.

22. Dual-output temperature controllers are used in applications requiring two related control actions.

23. A temperature recorder or datalogger makes a permanent record of the process temperature by writing either an analog trace or actual numeric values onto paper.

24. Resistance heating uses current flowing through a conducting material to produce heat.

25. Induction heating uses electromagnetic induction to produce heat. The work coil acts as the primary winding of a transformer, and the workpiece being heated acts as a single-turn secondary winding that is short-circuited.

26. Dielectric heating is based on the principle of capacitance. An insulating material is placed between two conducting plates connected to a high-frequency ac supply.

27. Arc heating uses an electric arc to produce heat. The electric arc is produced by ionizing the air between two electrodes.

28. The main function of a timer is to place information about elapsed time into a control circuit.

29. Time-delay relays are devices having a timing function after the timer coil has been energized or deenergized.

30. A timer opens or closes electrical circuits to selected operations according to a timed program.

31. Terms associated with the selection and operation of timers and time-delay relays include analog timer, automatic reset, delayed contacts, digital setting, electrical service life, external/electrical reset, inhibit function, instantaneous contacts, load rating, maintained start, manual reset, memory protective function during power failure, minimum setting, momentary start, operating voltage range, power reset, repeat accuracy, repeat cycle operation, reset, and timing chart.

32. Timers may be classified according to their method of timing as being motor-driven or solid-state.

33. Timers may be classified according to method of operation as being manually set timers, reset timers, and repeat cycle timers.

34. Counters are generally thought of as devices that tabulate or count "things."

35. Interval counter operation means that a load will be actuated when the unit is counting.

36. Delay counter operation means that a load will be actuated at the end of the counting cycle.

37. Most solid-state counters can count up, count down, or be combined to count up and down.

38. An up-counter will count up or increment by 1 each time the counted event occurs.

39. A down-counter will count down or decrement by 1 each time the counted event occurs.

40. In addition to counting, counters can be used for length and position measurement.

41. Sequencers are devices that provide ON/OFF action for a number of output channels in a predefined sequential pattern.

42. Drum switches, rotary switches, stepper switches, and cam switches can be classified as sequencers.

43. Time sequencers have a total time base of operation and the individual load circuits can be programmed to turn on and off anywhere within that time base.

44. Step sequencers operate from electrical pulses initiated by input devices. These input pulses step a camshaft through a fraction of a revolution for each input pulse.

45. Sequencers are used in place of complex relay circuitry to provide sequential control action.

CHAPTER REVIEW QUESTIONS

Answer the following questions.

10-1. What is most often used to supply position information?

10-2. State two uses made of position information.

10-3. List four types of electrical drive systems used for cranes and hoists.

10-4. Compare the function of bridge, monorail, and jib cranes.

10-5. What does basic crane motion control involve?

10-6. In what way does the hoist control requirement for hoisting differ from that of lowering?

10-7. List and describe the three components of a programmable motion system.

10-8. What are two basic uses for pressure controls?

10-9. List four common units of measurement for pressure.

10-10. Compare the operation of a pressure switch and pressure transducer.

10-11. What are two basic uses for temperature controls?

10-12. List the three basic types of temperature mode control.

10-13. What types of control outputs are available with temperature controllers?

10-14. Explain the function of a temperature recorder or data logger.

10-15. Explain how heat is produced in each of the following processes:
   a. resistance heating   **c.** dielectric heating
   b. induction heating   **d.** arc heating

10-16. What is the main function of a timer?

10-17. What distinction can be made between timers and time-delay relays?

10-18. How are timers classified according to their method of timing?

10-19. How are timers classified according to their method of operation?

10-20. Explain the function of a counter as part of a machine operation.

10-21. Compare timer interval and delay operation.

10-22. In addition to counting, what two other measurements can counters be used for?

10-23. What does sequence control involve?

10-24. List four types of sequencer switches.

10-25. Compare the operation of time and step sequencers.

10-26. What would be involved in implementing an alternative control for a sequential machine operation?

CRITICAL THINKING QUESTIONS

10-1. Refer to Fig. 10-1. Assume during normal operation that limit switch LS2 gets stuck in the actuated position. How would this affect the operation of the conveyor?

10-2. Refer to Fig. 10-2. Assume that, with the correct power and pressure applied, the piston will not move forward from position A when the start button is pressed. List seven electrical faults that might cause this problem.

10-3. Refer to Fig. 10-6. Assume that you are asked to install a safety pushbutton that must be pressed, in addition to the start button, to start the stamping operation. What type of pushbutton would be used, and how would it be connected into the circuit?

10-4. Refer to Fig. 10-7. Assume that the normally closed contact CR1-3 is defective and, as a result, remains permanently open. How would this affect the operation of piston A and piston B?

10-5. Refer to Fig. 10-9. Assume that the temperature takes an excessive time to rise to the set point. An open in one of the heater banks is suspected. What electrical testing could you perform to confirm this?

10-6. Refer to Fig. 10-16. Assume that the start-up button has been pressed and the timer has timed out. State the value of the normal voltage (line or zero) that would be indicated by a voltmeter connected across each of the following:

    a. CR coil   **e.** TD1 contacts

    b. CR1 contacts   **f.** CR3 contacts

    c. CR2 contacts   **g.** horn

    d. TD coil

10-7. Refer to Fig. 10-18. Assume that a pilot light is to be connected to denote the timing period and OFF at all other times. To what terminal number of the timer would this lamp be connected?

10-8. Refer to Fig. 10-21. Assume that a pilot light is to be connected into the circuit so that it is turned on and off for each count pulse received after the unit has counted out. To what terminal numbers of the timer would this lamp be connected?

10-9. Refer to Fig. 10-26. Assume that the circuit must be modified to provide either automatic or manual control by means of an auto/manual selector switch and start-stop pushbutton station. When in the manual mode, motor number 1 is to be operated ON and OFF by means of the pushbutton station (regardless of the state of the float switches). Design a circuit to implement this change.

10-10. Refer to Fig. 10-29. Assume that relay CR1 contacts fail to close during a normal cycle of operation. How would this affect the operation of the stepswitch?

Answers to Self-Tests

1. true
2. By first pressing the start and then the reverse pushbutton.
3. The piston would be spring returned to position A.
4. false
5. microprocessor or computer
6. expected; actual
7. incremental
8. false
9. false
10. false
11. control; limit
12. Seal-in circuit made up of relay coil CR and contacts CR1.
13. false
14. contacts
15. Cylinder A would operate normally but cylinder B would not operate at all.
16. The motor would be switched ON at all times.
17. SCR
18. operator
19. true

20. true
21. temperature; time
22. resistance
23. induction
24. capacitance
25. ionizing
26. delayed
27. stops
28. false
29. false
30. The horn would come ON and remain ON after a 10-s delay period.
31. synchronous
32. true
33. OFF, OFF, and ON
34. The timer will remain reset at zero with the timed output contact open.
35. end
36. true
37. The counter will be reset to zero and additional counts will be registered beginning with zero.
38. up

39. Set the counter's timed output for 1.6 seconds.
40. Additional in-feed counts will increment the counter from 50, but no decrement of the counter will occur.
41. length; position
42. The counter will accumulate counts whether or not bar stock is moving.
43. Motor 1 will operate for as long as the button is pressed. Motor 2 will not operate unless the start button is pressed closed longer than the time delay period.
44. Pumps 1 and 2 will still alternate ON and OFF, but only one pump will operate regardless of the water level.
45. time; step
46. outputs
47. ON–OFF–ON–OFF–ON–OFF–ON
48. time
49. LS1

CHAPTER 11

Process Control Systems

∎

CHAPTER OBJECTIVES

This chapter will help you to:

1. *Discuss* the operation of continuous process, batch production, and individual products production.
2. *Compare* individual, centralized, and distributive control systems.
3. *Explain* the function of the major components of a process control system.
4. *Describe* the difference between ON/OFF, proportional, derivative, and integral types of control.
5. *Outline* the function of the different parts of a data acquisition system.
6. *Discuss* common terms associated with the selection, operation, and connection of a data acquisition system.

This chapter serves as an introduction to the kinds of industrial processes to be controlled, the types of systems, and the methods available to perform the control functions. The acquisition of data is included as a type of process. Open-loop and closed-loop control systems are defined, and the fundamental characteristics of each are discussed.

∎

11-1 TYPES OF PROCESSES

The types of processes carried out in modern manufacturing industries can be grouped into three general areas, in terms of the kind of operation that takes place, as:

- Continuous process
- Batch production
- Individual products production

A *continuous process* is one in which raw materials enter one end of the system and the finished product comes out the other end of the system; the process itself runs continuously. Once the process commences, it is continuous for a relatively long period of time. The time period may be measured in minutes, days, or even months, depending upon the process.

Figure 11-1 shows a continuous process engine assembly line. Engine blocks are fed into one end of the system and completed engines exit at the other end. In the continuous-type process, the product material is subjected to different treatments as it flows through the process (in this case, assembly, adjustment, and inspection). Auto assembly involves the use of automated machines or robots. At each station, parts are supplied as needed.

In *batch processing* there is no flow of product material from one section of the process to another. Instead, a set amount of each of the inputs to the process is received in a batch, and then some operation is performed on the batch to produce a finished product or an intermediate product that needs further processing. The process is carried out, the finished product is stored, and another batch of product is produced. Each batch of product may be different.

Some processes combine the features of the batch and continuous types. In such processes, several product materials are treated and stored in batch operation. Then these stored materials are drawn off as required into a continuous process.

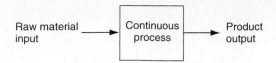

(a) Block diagram: the process runs for a
relatively long time

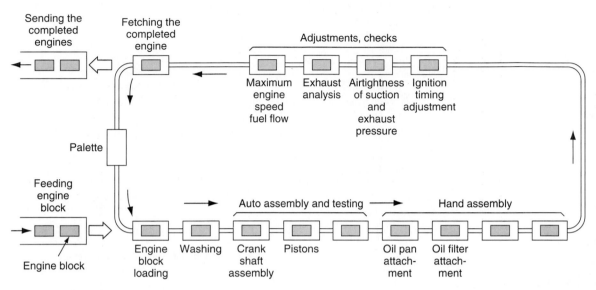

(b) Engine assembly line: involves assembly,
adjustment, and inspection

Fig. 11-1   Continuous process.

From page 304:

**Continuous
process**

Batch processing

On this page:

**Individual product
production process**

Individual control

Many chemically based products are manufactured by using batch processes. Figure 11-2 on page 306 shows a batch process system. Two ingredients are added together, mixed, heated, a third ingredient is added, processed, and then stored. Each batch made may have differing characteristics by design.

The *individual product production process* is the most common of all processing systems. With this manufacturing process, a series of operations produces a useful output product. The item being produced may be required to be bent, drilled, welded, and so on, at different steps in the process. The workpiece is normally a discrete part that must be handled on an individual basis. Figure 11-3 on page 306 shows an individual product production process.

In the modern automated industrial plant, the operator merely sets up the operation and initiates a start, and the operations of the machine are accomplished automatically. These automatic machines and processes were developed to mass-produce products, control very complex operations, or to operate machines accurately for long periods of time. They replaced much human decision, intervention, and observation.

Machines were originally mechanically controlled, then they were electromechanically controlled, and today they are often controlled by purely electrical or electronic means through programmable logic controllers (PLCs) or computers. The control of machines or processes can be divided into the following categories:

- Electromechanical control
- Hardwired electronic control
- Programmable hardwired electronic control
- Programmable logic control (PLC)
- Computer control

Possible control configurations include individual control, centralized control, and distributed control. *Individual control* is used to control a single machine. This type of control does not normally require communication with any other controllers. Figure 11-4 on page 307 shows an individual control application for an industry manufacturing aluminum handrails for indoor and outdoor applications. The operator enters the feed length and batch count via the operator's interface control panel and then presses the start button to initiate the process. Rail lengths vary widely. The operator needs

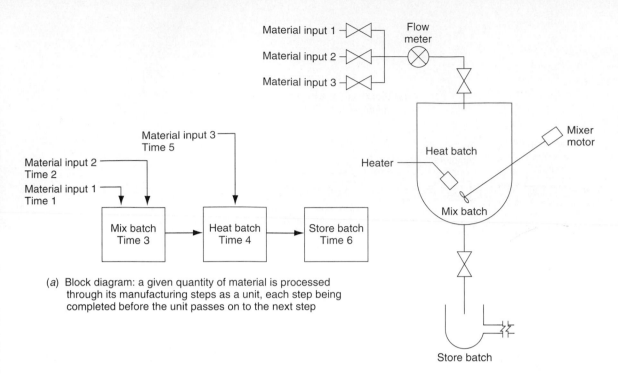

(a) Block diagram: a given quantity of material is processed
through its manufacturing steps as a unit, each step being
completed before the unit passes on to the next step

(b) Multicomponent/multiformula batching system

Fig. 11-2   Batch process.

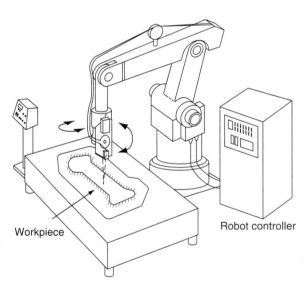

Industrial robot used in an individual item process

Fig. 11-3   Individual product production.

to select the rail length and the number of rails
to cut.

Centralized control is used when several machines or processes are controlled by one central controller (Fig. 11-5). This control layout
utilizes a *single* large control system to control
many diverse manufacturing processes and operations. Each individual step in the manufac-

turing process is handled by a central control
system controller. No exchange of controller
status or data is sent to other controllers. Some
processes require central control because of
the complexity of decentralizing the control
tasks into smaller ones. One disadvantage of
centralized control is that if the main controller fails, the whole process is stopped.

A central control system is especially useful
in a large interdependent process plant where
many different processes must be controlled
for efficient use of the facilities and the raw
materials. It also provides a central point
where alarms can be monitored and processes
can be changed without the need to travel constantly throughout the plant.

The *distributive control system* (DCS) differs
from the centralized system in that each
machine is handled by a *dedicated control system*. Each dedicated control is totally independent and could be removed from the overall
control scheme if it were not for the manufacturing functions it performs.

Distributive control involves two or more
computers communicating with each other to
accomplish the complete control task. This
type of control typically employs local area networks (LANs) in which several computers control different stages or processes locally and are

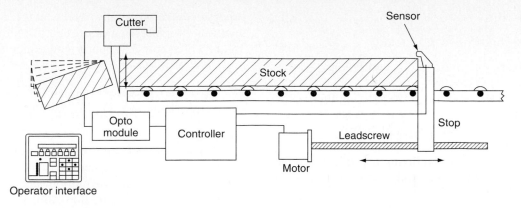

Fig. 11-4   Individual control.

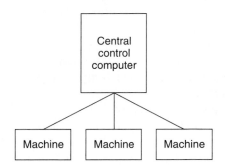

Fig. 11-5   Centralized control.

constantly interchanging information and reporting status on the process. Communication among the computers is done through single coaxial cables or fiber optics at very high speeds.

Because of their flexibility, distributive control systems have emerged as the system of choice for a wide range of batch and continu-ous process automation requirements. Distrib-utive control permits the distribution of the processing task among several control ele-ments. Instead of just one computer located at a central control point doing all the processing, each local loop controller, placed very close to the point being controlled, has processing capability.

Figure 11-6 shows a DCS that is supervised by a host computer. The host computer could be a personal computer and would handle such things as program up and down load, alarm reporting, data storage, and operator interac-tion facilities. Remote computer operators overseeing the system can be located some dis-tance from the industrial environment in dust-free and temperature-controlled conditions. Each PLC controls its associated machine or process. Depending on the process, in many instances, one PLC failure would not stop the complete system.

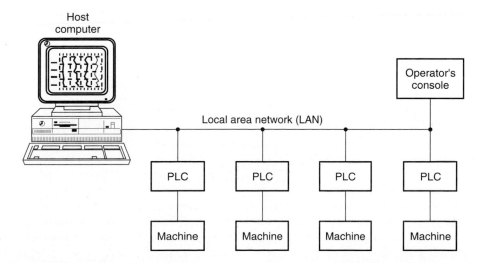

Fig. 11-6   Distributive control system (DCS).

Self-Test

Answer the following questions.

1. True or false? A continuous process involves the flow of product material from one section of the process to another.
2. In a(n) _____ process, a set amount of product is received and then some operation is performed on the product.
3. True or false? Individual product production is not a commonly used processing system.
4. True or false? With most automated machines, the operator merely sets up operation and initiates a start.
5. _____ control is used to control a single machine.
6. _____ control is used when several machines are controlled by one controller.
7. _____ control involves two or more computers communicating with each other to accomplish the complete control task.

11-2 STRUCTURE OF CONTROL SYSTEMS

A *process control system* can be defined as the functions and operations necessary to change a material either physically or chemically.

Process control normally refers to the manufacturing or processing of products in industry. Figure 11-7 shows the major components of a process control system.

▐▌▶ **YOU MAY RECALL** much of the following information from previous chapters.

Sensors:
- Provide inputs from the process and from the external environment
- Convert physical information such as pressure temperature, flow rate, and position into electrical signals
- Are related to a physical variable in a known way so that their electrical signal can be used to monitor and control a process

Operator-Machine Interface:
- Allows inputs from a human to set up the starting conditions or alter the control of a process
- Allows human inputs through various types of switches, controls, and keypads
- Operates using supplied input information that may include emergency shutdown or changing the speed, the type of process to be run, the number of pieces to be made, or the recipe for a batch mixer

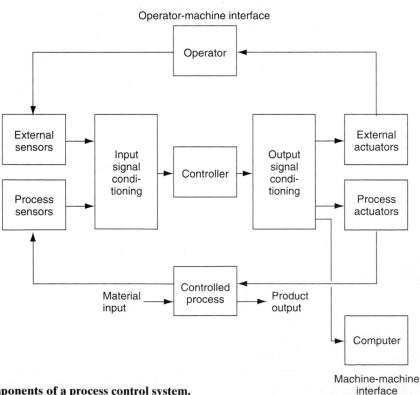

Fig. 11-7 Components of a process control system.

Signal Conditioning:
- Involves converting input and output signals to a usable form
- Includes signal conditioning techniques such as amplification, attenuation, filtering, A/D converters, and D/A converters

Actuators:
- Convert system output electrical signals into physical action
- Have process actuators that include flow control valves, pumps, positioning drives, variable speed drives, clutches, brakes, solenoids, stepping motors, and power relays
- Through external actuators such as meters, cathode-ray tube (CRT) monitors, printer, alarms, and pilot lights, indicate the state of the process or the value of certain process variables (machine-operator interface)
- Can send outputs directly from the controller to a computer for storage of data and analysis of results (machine-machine interface)

Controller:
- Makes the system's decisions based on the input signals
- Generates output signals which operate actuators to carry out the decisions

Control systems can be classified as open-loop or closed-loop. Figure 11-8 depicts a typical *open-loop control system.* The process is controlled by inputting to the controller the desired *set point* (also known as *command* and *reference*) believed necessary to achieve the ideal operating point for the process and accepting whatever output results. Since the only input to the controller is the set point, it is apparent that an open-loop system controls the process blindly. That is, the controller receives no information concerning the present status of the process or the need for any corrective action. Open-loop control systems are considerably cheaper and less complex than their closed-loop counterparts. The inevitable result of their use, however, is poor process control.

Open-loop speed control of motors is still found in many industrial applications. Figure 11-9 on page 310 shows open-loop armature voltage control of a permanent magnet dc motor. The armature voltage is varied to control the speed of the motor. However, with a set armature voltage, the motor speed changes when the motor load changes. Variation of motor speed as the load on the motor is changed is essentially the inherent regulation characteristic of the particular motor as illustrated by its torque-speed curves. The motor will operate along the load line by jumping from one torque-speed curve to another. Open-loop, or nonfeedback, control is only as stable as the load and the individual components of the system.

A *closed-loop control system* is one in which the output of a process affects the input. The system measures the actual output of the process and compares it to the desired output. Adjustments are made continuously by the control system until the difference between the desired and actual output is as small as practical.

Figure 11-10 on page 310 depicts a typical closed-loop control system. The actual output is sensed and fed back (hence the name *feedback control*) to be subtracted from the set point input that indicates what output is desired. If a difference occurs, a signal to the controller causes it to take action to change the actual output until the difference is zero.

The system illustrated in Fig. 11-10 has the following components:

Set Point
- Is the input that determines the desired operating point for the process
- Is normally provided by a human operator, although it may also be supplied by another electronic circuit

Process Variable
- Is the signal that contains information about the current process status
- Refers to the feedback signal
- Ideally, matches the set point (indicating that the process is operating exactly as desired)

Error Amplifier
- Determines whether the process operation matches the set point

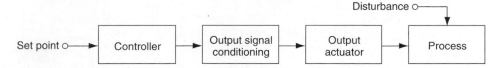

Fig. 11-8 Open-loop control system.

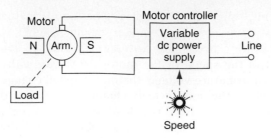

(a) Motor speed is set by amount of armature voltage, and for any given setting, motor speed changes when motor load changes

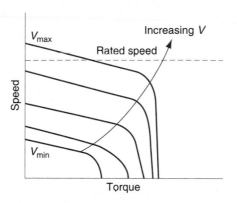

(b) Motor torque-speed curves

Fig. 11-9   Open-loop motor speed control.

- Quite often a differential amplifier circuit provides output referred to as the *error signal* or the *system deviation signal*
- The magnitude and polarity of the error signal will determine how the process will be brought back under control

Controller

- Produces the appropriate corrective output signal based on the error signal input

Output Actuator

- Is the component that directly affects a process change
- Has motors, heaters, fans, and solenoids that are all examples of output actuators

The container-filling process illustrated in Fig. 11-11 is an example of a closed-loop process. An empty box is moved into position and filling begins. The weight of the box and contents is monitored. When the actual weight equals the desired weight, filling is halted.

In this operation:

- A sensor attached to the scale weighing the container generates the voltage signal or digital code that represents the weight of the container and contents
- The sensor signal is subtracted from the voltage signal or digital code that has been input to represent the desired weight
- As long as the difference between the input signal and feedback signal is greater than zero, the controller keeps the solenoid gate open
- When the difference becomes zero, the controller outputs a signal that closes the gate

Regulators are used in many industrial applications to provide precise regulation of voltage, speed, current, tension, position, temperature, and the like. Regulators may be defined as a closed-loop system that maintains a steady level or value for some quantity.

Figure 11-12 on page 312 shows a loop regulator used for tension control on a textile machine. This process requires a slack loop in the continuously moving strip of cloth. When slack loops are used in continuous processing, the regulating problem involved is to synchronize the speed of one side of the loop with that of the other side, to maintain the length of the loop constant.

During this operation:

- A change in the feeding speed of materials is detectable from the tension roller position by the three photoelectric sensors
- The set point reference, in this case, selects the desired loop position in the light system
- Any change in loop length changes the

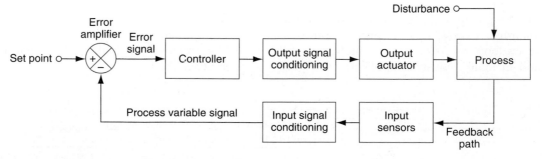

Fig. 11-10   Closed-loop control system.

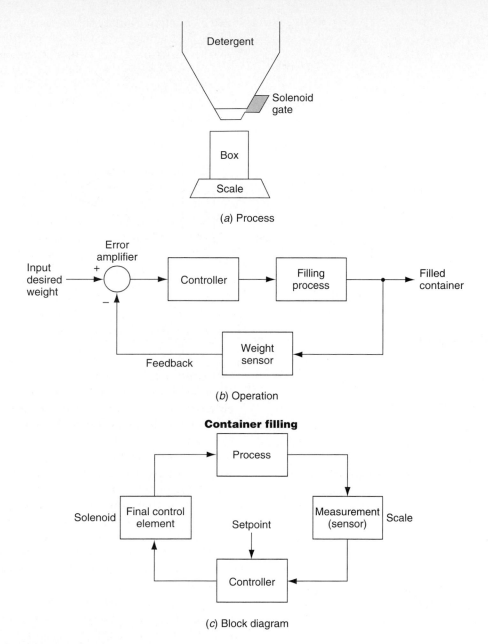

(a) Process

(b) Operation

Container filling

(c) Block diagram

Fig. 11-11   Container-filling closed-loop process.

amount of light intercepted, resulting in a signal change from the photoelectric sensors
- Sensor signal changes are received by the controller which changes the speed of the motor, in a direction tending to restore the preselected loop length

Figure 11-13 on page 313 shows an example of a closed-loop motor speed controller used in conjunction with an eddy-current drive. The reference signal from the speed potentiometer in the operator's station tells the drive at what speed to run. An internal generator in the drive supplies a feedback signal that lets the controller know the actual speed of the drive

output shaft. Automatic corrective action takes place continuously during operation.

During this operation:

- The eddy-current controller varies the excitation to the clutch field coil, changing magnetic field strength, which in turn proportionally affects torque developed by the clutch.
- To maintain constant speed, drive torque must be constantly modulated to match the torque demanded by the load. Too little torque will allow the speed to decrease, and too much torque will increase speed. The eddy-current controller performs this modulation function.

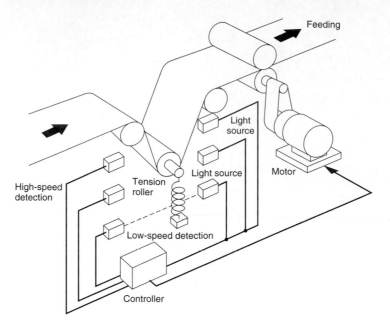

Fig. 11-12   Loop regulator.

Self-Test

Answer the following questions.

8. Process control can be defined as the functions and operations necessary to change a material either _____ or _____.
9. True or false? Actuators convert physical information into electrical signals.
10. True or false? A pushbutton switch could be classified as a type of operator-machine interface.
11. Signal _____ involves converting input and output signals to a usable form.
12. True or false? A sensor could be classified as a type of controller.
13. Control systems can be classified as _____-loop or _____-loop.
14. A(n)_____-loop system is one in which the output of a process affects the input.
15. True or false? A regulator would be classified as an open-loop system.

11-3 CONTROLLER RESPONSES

The four most popular types of control response used in the process industry are:

- ON/OFF
- Proportional (P)
- Integral (I)
- Derivative (D)

The names of these actions describe the response of the final control element. With *ON/ OFF* control (also known as *two-position* and *bang-bang control*), the final control element is either ON or OFF—one for the occasion when the value of the measured variable is above the set point, and the other for the occasion when the value is below the set point. The controller will never keep the final control element in an intermediate position.

Figure 11-14 shows a system using ON/OFF control, in which a liquid is heated by steam. If the liquid temperature goes below the set point, the steam valve opens and the steam is turned ON. When the liquid temperature goes above the set point, the steam valve closes and the steam is shut OFF. The ON/OFF cycle will continue as long as the system is operating.

Figure 11-15 on page 314 illustrates the control response for the ON/OFF controller. When the measured variable goes above the set point, the final control element is turned OFF. It will stay OFF until the measured variable goes below the set point. Then, the final control element will turn ON. The measured variable will oscillate around the set point at an amplitude and frequency that depend on the process's capacity and time response. This oscillation is typical of the ON/OFF controller. Oscillations may be reduced in amplitude by increasing the sensitivity of the controller. This will cause the controller to turn ON and OFF more often, a possibly undesirable result.

This type of control is inexpensive but not accurate enough in most process and machine control applications. ON/OFF control almost

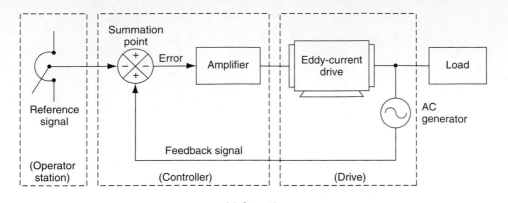

(a) Operation

(b) Typical physical layout

Fig. 11-13   Closed-loop motor speed controller.

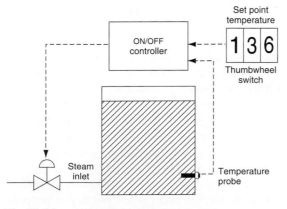

Fig. 11-14   ON/OFF liquid heating system.

always means overshoot and resultant system cycling. A *deadband* is usually required around the set point to prevent relay chatter at set point. ON/OFF control does not adjust to the time constants of a particular system.

Proportional controls are designed to eliminate the hunting or cycling associated with ON/OFF control. It allows the final control element to take intermediate positions between ON and OFF. This permits *analog control* of the final control element to vary the amount of energy to the process, depending upon how much the value of the measured variable has shifted from the desired value.

CHAPTER 11   PROCESS CONTROL SYSTEMS   **313**

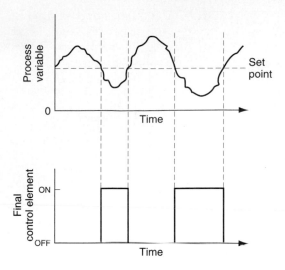

Fig. 11-15 ON/OFF control.

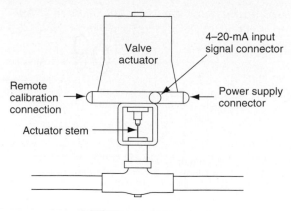

(a) Proportional control valve

Actuator current (mA)	Valve response (% open)
4	0
6	12.5
8	25
10	37.5
12	50
14	62.5
16	75
18	87.5
20	100

(b) Changes in valve response in reaction to increases in actuator current

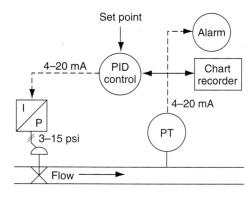

(c) Pressure reduction control application

Fig. 11-16 Motor-driven proportional control valve.

Proportioning action can also be accomplished by turning the final control element ON and OFF for short intervals (see Fig. 11-16). This *time proportioning* (also known as *proportional pulsewidth modulation*) varies the ratio of ON time to OFF.

A proportional controller allows tighter control of the process variable because its output can take on any value between fully ON and fully OFF, depending on the magnitude of the error signal. Consider the application of the liquid heating system using steam and a proportional control. A motor-driven proportional control valve is used as the final control element (Fig. 11-16). Typically, it receives an input current of between 4 mA and 20 mA from the controller; in response, it provides linear movement of the valve. As the temperature reduces or increases, the steam valve will close or open accordingly, in an attempt to maintain the liquid temperature. This has the effect of slowing down the heat so that it will not overshoot the set point but will approach the set point and maintain a stable temperature.

In Fig. 11-16(c), the pressure transmitter (PT) senses the downstream pressure from the diaphragm control valve. This signal is compared to the set point pressure (SP), and the valve is throttled to maintain the desired pressure. It is frequently applied to steam control. Accessories, including recorders and alarms, are generally used to provide information to the operator and warn of malfunctions.

The proportioning action occurs within a "proportional band" around the set point temperature. Outside this band, the controller functions as an ON/OFF unit, with the output either fully ON (below the band) or fully OFF (above the band). However, within the band, the output is turned ON and OFF in the ratio of the measurement difference from the set point. At the set point (the midpoint of the proportional band), the output ON:OFF ratio is 1:1; that is, the ON time and OFF time are equal. If the temperature is further from the set point, the ON and OFF times vary in proportion to the temperature difference. If the temperature is below the set point, the output will be ON longer; if the temperature is too high, the output will be OFF longer.

The proportional band is usually expressed as a percentage of full scale, or degrees, as shown in Fig. 11-17. It may also be referred to as gain, which is the reciprocal of the bandwidth. Note that in time-proportioning control, full power is applied to the heater but cycled ON and OFF, so that the average time is varied. In most units, the cycle time and/or proportional band are adjustable, so that the controller may be better matched to a particular process.

| Time proportional | | | | 4–20 mA proportional | |
Percent on	On time seconds	Off time seconds	Temp. (°F)	Output level	Percent output
0.0	0.0	20.0	over 540	4 mA	0.0
0.0	0.0	20.0	540.0	4 mA	0.0
12.5	2.5	17.5	530.0	6 mA	12.5
25.0	5.0	15.0	520.0	8 mA	25.0
37.5	7.5	12.5	510.0	10 mA	37.5
50.0	10.0	10.0	500.0	12 mA	50.0
62.5	12.5	7.5	490.0	14 mA	62.5
75.0	15.0	5.0	480.0	16 mA	75.0
87.5	17.5	2.5	470.0	18 mA	87.5
100.0	20.0	0.0	460.0	20 mA	100.0
100.0	20.0	0.0	under 460	20 mA	100.0

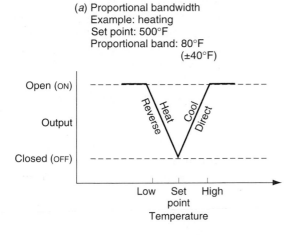

(a) Proportional bandwidth
Example: heating
Set point: 500°F
Proportional band: 80°F
(±40°F)

(b) Control action can be adjusted

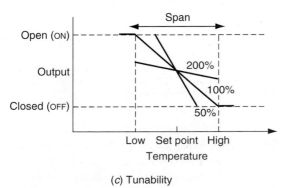

(c) Tunability

Fig. 11-17   Proportional control.

A proportional controller adjusts its control action according to the need of the process system. Proportional control can be heat-acting (reverse) or cool-acting (direct) in temperature systems. For other applications, either reverse- or direct-acting control may be appropriate. Proportional controllers are tunable. Their response to process changes can be adjusted to suit the time constants of the specific process system.

In theory, a proportional controller should be all that is needed for process control. Any change in system output is corrected for by an appropriate change in controller output. Unfortunately, the operation of a proportional controller leads to process deviation known as *offset* or *droop*. This steady-state error is the difference between the attained value of the controller and the required value, as shown in Fig. 11-18. It may require an operator to make a small adjustment (manual reset) to bring the temperature to set point on initial start-up, or whenever the process conditions change significantly.

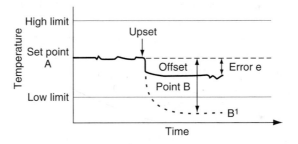

Fig. 11-18   Proportional control action (P).

For example, in Fig. 11-18, if you assume process equilibrium at set point A, any change in the process such as a change in flow rate will result in a new equilibrium at point B or event point B^1.

Proportional control is usually used in conjunction with derivative control, integral control, or a combination of both. Processes with long time lags and a large maximum rate of rise (for example, a heat exchanger) require wide proportional bands to eliminate oscillation. The wide band can result in large offsets with changes in the load. To eliminate these offsets, automatic reset (integral) can be used. Derivative (rate) action can be used on processes with long time delays to speed recovery after a process disturbance.

The *integral action*, sometimes termed *reset action*, responds to the size and time duration of the error signal; therefore, the output signal from an integral controller is the mathematical integral of the error. An error signal exists when there is a difference between the process variable and the set point, so the integral action will cause the output to change and continue to change until the error no longer exists. Integral action eliminates steady-state error. The amount of integral action is measured as minutes per repeat or repeats per minute, which is the relationship between changes and time.

The *derivative mode controller* responds to the speed at which the error signal is changing—that is, the greater the error change, the greater the correcting output. The derivative action is measured in terms of time.

Proportional plus integral (PI) control combines the characteristics of both types of control (Fig. 11-19). A step change in the measurement causes the controller to respond in a proportional manner followed by the integral response, which is added to the proportion response. Because the integral mode determines the output changes as a function of time, the more integral action in the control, the faster the output changes. The PI control mode is used in situations in which changes in the process load do not happen very often, but when they do happen, changes are small.

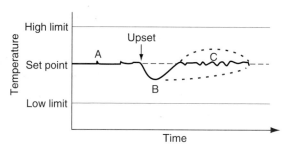

Fig. 11-19   Proportional plus integral control action (PI).

In Fig. 11-19:

- To eliminate the offset error, the controller needs to change its output until the process variable error is zero.
- Reset (integral) control action changes the controller output by the amount needed to drive the process variable back to the set point value.
- The new equilibrium point after reset action is at point C.
- Since the proportional controller is constrained to always operate on its propor-

tional band, the proportional band must be shifted to include the new point C. A controller with reset does this automatically.

Rate action acts on error just like reset does, but rate action is a function of the rate of change rather than the magnitude of error. Rate action is applied as a change in output for a selectable time interval, usually stated in minutes. Rate-induced change in controller output is calculated from the derivative of the change in input. Input change rather than proportional control error change is used to improve response. Rate action quickly positions the output where proportional action alone would eventually position the output. In effect, rate action puts the brakes on any offset or error by quickly shifting the proportional band.

Proportional plus derivative (PD) control is used in process-control systems that have errors that change very rapidly. By adding derivative control to proportional control, we get a controller output that responds to the measurement's rate of change as well as to its size.

Proportional-integral-derivative (PID) controllers produce outputs that depend on the magnitude, duration, and rate of change of the system error signal (Fig. 11-20). Sudden system disturbances are met with an aggressive attempt to correct the condition. A PID controller can reduce the system error to zero faster than any other controller. Because it has an integrator and a differentiator, however, this controller must be *custom-tuned* to each process being controlled. The PID controller is the most sophisticated and widely used type of process controller.

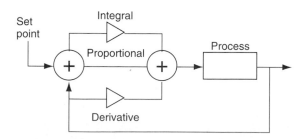

Fig. 11-20   PID control loop.

In Fig. 11-20:

- During setup the set point, proportional band, reset (integral), and rate (derivative), as well as the output limits are specified.
- All these can be changed during operation, to tune the process.

- The integral term improves accuracy, and the derivative reduces overshoot for transient upsets.
- The output can be used to control valve positions, temperature, flow metering equipment, and so on.

The long-term operation of any system, large or small, requires that there be a mass-energy balance between input and output. If a process were operated at equilibrium at all times, control would be simple. Since change does occur, the critical parameter in process control is time—that is, how long it takes for a change in any input to appear in the output. System time constants can vary from fractions of a second to many hours. The PID controller has the ability to tune its control action to specific process time constants and therefore to deal with process changes over time. PID control changes the amount of output signal in a mathematically specified way that accounts for amount of error and rate of signal change. A common PID equation for obtaining PID control is:

$$Co = K\,(\overbrace{E}^{\text{Proportional}} + \overbrace{1/Ti\int_0^t Edt}^{\text{Integral}}$$

$$+ \underbrace{KD\,[E - E\,(n-1)]/dt)}_{\text{Derivative}} + bias$$

where Co = control output
K = controller gain (no units)
$1/Ti$ = reset gain constant (repeats per minute)
KD = rate gain constant (minutes)
dt = time between samples (minutes)
bias = output bias
E = error; equal to measured minutes set point or set point minus measured minutes
$E(n-1)$ = error from last sample

Basically, PID controller tuning consists of determining the appropriate values for the gain (proportional band), rate (derivative), and reset (integral) time tuning parameters (control constants) that will give the control required. Depending on the characteristics of the deviation of the process variable from the set point, the tuning parameters interact to alter the controller's output and produce changes in the value of the process variable. In general, three methods of controller tuning are used:

Manual:
- The operator estimates the tuning parameters required to give the desired controller response.

- The proportional, integral, and derivative terms must be individually adjusted or "tuned" to a particular system, using a trial and error method.

Semiautomatic or Autotune:
- The controller takes care of calculating and setting PID parameters.

  Measure sensor
  Calculate error, sum of error, rate of change of error
  Calculate desired power with PID equation
  Update control output

Fully Automatic or Intelligent:
- It is also known in the industry as *fuzzy logic control.*
- It uses artificial intelligence to readjust PID tuning parameters continually as necessary.
- It evaluates rules based on inputs that can have intermediate values between true and false. Fuzzy logic is used to convert variables, such as temperature, from traditional units, such as degrees, into fuzzy variables. Fuzzy variables are then used to evaluate fuzzy rules. The resulting fuzzy output is converted back into traditional units, such as valve position, in a process known as *defuzzification.* Fuzzy logic control is capable of providing superior performance over traditional PID control strategies in terms of integrated absolute error and overshoot for both set point changes and external load disturbances.

Self-Test

Answer the following questions.

16. With ON/OFF control the measured variable will _____ around the set point.
17. True or false? Proportional control alone produces the type of cycling associated with ON/OFF control.
18. _____ proportioning varies the ratio of ON time to OFF.
19. Proportioning action occurs within a proportional _____ around the set point temperature.
20. The operation of a proportional controller leads to process deviation known as _____ or _____.
21. True or false? Integral action eliminates steady-state error.

22. Derivative action responds to the _____ at which the error signal is changing.
23. The outputs produced by proportional-integral-derivative (PID) controllers depend on the _____, _____, and _____ of change of the system error signal.
24. True or false? A PID controller must be custom tuned to each process being controlled.
25. True or false? Fuzzy logic uses artificial intelligence to continuously readjust PID tuning parameters.

11-4 DATA ACQUISITION SYSTEMS

Data acquisition is the collection, analysis, and storage of information by a computer-based system. Initially these systems were only used with large mainframe computers and installations with thousands of input data channels. Today, however, powerful personal computers have opened up the advantages of data acquisition to all manufacturers at very modest cost.

A data acquisition or computer interface system (Fig. 11-21) is a device that allows you to feed data from the real world to your computer. It takes the signals produced by temperature sensors, pressure transducers, flow meters, and so on, and converts them into a form that your computer can understand. With an acquisition system, you can use your computer to gather, monitor, display, and analyze your data. If the acquisition system has control output capabilities, you can also use your computer to accurately control your processes for maximum efficiency.

The great advantage of data acquisition is that data is stored automatically in a form that can be retrieved without error or additional work for later analysis. It is easy to examine the effect of factors other than those originally anticipated because the cost of measuring some additional points is low. Measurements are made under computer control and then displayed on screen and stored to disk. Accurate measurements are easy to obtain, and there are no mechanical limitations to measurement speed.

Data acquisition software works with the hardware interface equipment to take full

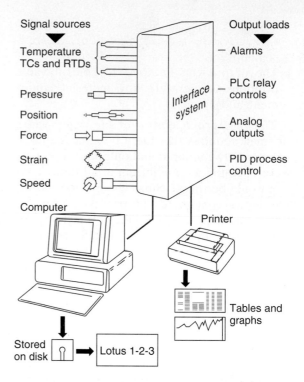

Fig. 11-21 Data acquisition and control system.

advantage of your computer system. In addition to controlling the data collection by the interface, the software is designed to display the data in tables like a spreadsheet or in histogram charts, pie charts, or line plots, and to analyze the data. Most data acquisition systems designed to work with a personal computer share a basic overall similarity. Major system components (Fig. 11-22) include:

Sensors and Transducers. Device used to convert a mechanical or electrical signal into a specific signal such as a 4- to 20-mA loop, or a 0- to 5-V signal.

Analog Multiplexer. A device that accepts several signals at once and allows the user to pick the signal to be examined. Much like a TV that has 99 channels but only displays one at a time. Used to reduce system cost by time-sharing expensive elements.

Amplifier. Device used to increase the magnitude of a signal. Amplifiers can increase the voltage or current a sensor produces.

Signal Conditioner. Changes a signal in a desired manner to make it easier to measure, or more stable (i.e., current to voltage converter, V/F converter, or a filter).

A/D Converter. A special type of signal conditioner that converts analog signals into digital signals.

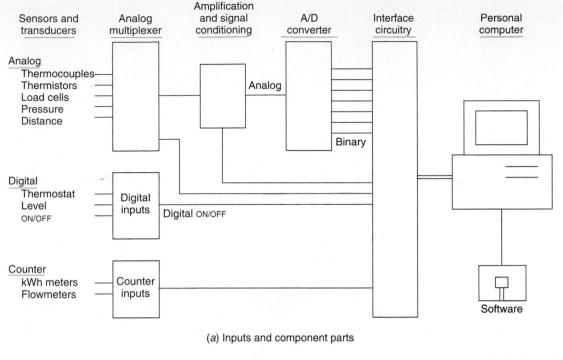

(a) Inputs and component parts

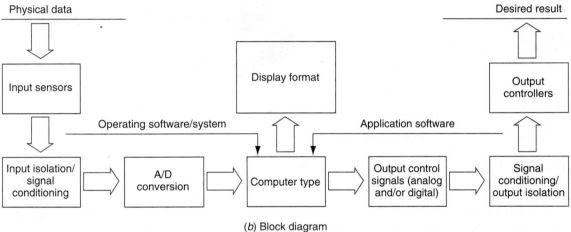

(b) Block diagram

Fig. 11-22   Major components of a data acquisition system.

Digital and Counter Inputs. Signals that have a specific number of possible values (1 bit = 2 values; 4 bits = 16 values).

Basically, there are two kinds of data acquisition interface systems: *plug-in* and *stand-alone*. These two extremes in providing signal conditioning to a data acquisition system involve putting the signal conditioning circuitry directly on a data acquisition board, or using a stand-alone data acquisition instrument with built-in signal conditioning.

Figure 11-23 on page 320 shows a modular plug-in data acquisition system that uses separate modules for signal conditioning and data acquisition. The advantage of a modular system is its ability to utilize the latest in plug-in data acquisition hardware, software, and PC technology, while maintaining the proper signal conditioning performance of stand-alone data acquisition units.

Special signal conditioning is needed for many sensors and applications. For example, thermocouples require cold junction compensation. RTDs, thermistors, and strain gauges need current or voltage excitation. Strain gauges often need bridge completion resistors. Most sensors need amplification in order to match the input range of the A/D converter. Isolation and filtering are recommended to protect your equipment and make accurate measurements in industrial environments. Antialiasing filters

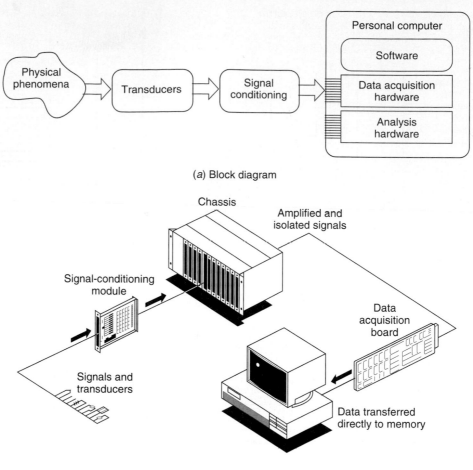

(a) Block diagram

(b) External instrumentation front-end chassis for signal conditioning
boards sends conditioned signals directly to a plug-in data
acquisition board for high-speed input into computer memory

Fig. 11-23 Modular plug-in data acquisition system.

are useful for filtering out high-frequency
noise from a measurement. Input multiplexers
supply an economical means of connecting to a
large number of signals. External amplification
allows low-level signals to be amplified for better transmission to the data acquisition board.
Discrete inputs from switches or high-current
relays need isolation and conditioning before
reaching the computer.

Data acquisition systems can get caught in
ground loops (Fig. 11-24). A *ground loop*
results when there are two or more connections to ground on the same electrical path,
causing different amounts of current to flow
through some ground wires. The value
received at the data acquisition board represents the voltage from the desired signal, as
well as the voltage value from the additional
ground currents in the system. Isolation circuitry on signal-conditioning accessories will
break ground loops and eliminate the effects
of ground currents on measurements.

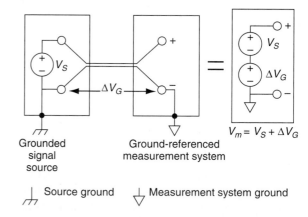

Fig. 11-24 Ground loops.

Important terms associated with the selection, operation, and connection of data acquisition systems include:

Absolute Signals. Signals that do not require a
reference quantity.

Accuracy. Is *not* equal to resolution (common perception is that these are the same). Accuracy is the total of the errors caused by:

Resolution + gain + offset + noise

Alias. A false lower-frequency component that appears in sampled data acquired at too low a sampling rate.

Analog. A signal that is continuous, having an infinite number of possible values.

Analog Ground. Provides a reference for signals. No shared circuit currents should flow in an analog ground.

Baud Rate. Serial communications data transmission rate expressed in bits per second (b/s).

Calibration. Ensuring that the output signal from a sensor is proportional (and the proportion is known) to the quantity the sensor is measuring. The manufacturer will usually give the calibration of the transducer on a data sheet or on a nameplate. For example:

Distance Sensor	
Full scale	2.0 mm
Output at full scale	5.0 mV
Excitation voltage	10.0 V

- In this case, a 5.0-mV output will correspond to a full-scale travel of 2.0 mm.
- The offset factor, often called the *zero,* is the output when the sensor is set at 0 mm travel.

Control Update Interval. Refers to how often the computer checks the measured value and adjusts the control output. If it's *too* slow, the control may not work. If it's *much faster* than the rate the system changes, computer time is wasted. The update rate is generally set to 5 to 10 times faster than the

time it takes the controlled system to droop by the desired accuracy.

Data Storage Values. Data acquisition systems can usually be programmed to save various values, which include:

- *Raw samples:* as measured each sample time
- *Averages:* the average value of all measurements made in a certain interval
- *Maximums:* the maximum value of all measurements made in a certain interval
- *Minimums:* the minimum value of all measurements made in a certain interval
- *Totals:* the total of all measurements made in a certain interval

Floating Signals and Common Mode Voltages. A *floating signal* is one which is not connected to any other systems or voltages. An example of truly floating voltage is a standard 12-V car battery with no cables attached:

- The voltage between the – and + terminals is 12 V.
- The voltage between either terminal and any other system is 0 (or undefined).

Consider a string of batteries in series with one end connected to ground (as shown in Fig. 11-25):

- The voltage at the beginning is 0 with respect to ground.
- The differential voltage across any battery is only 5 V.
- The voltage at the end of the chain is 40 V.
- The *common mode* of a differential measurement across the third battery will be 10 V.
- In measuring the voltage across the last battery, the 5-V signal has 35 V of common voltage (Fig. 11-25).

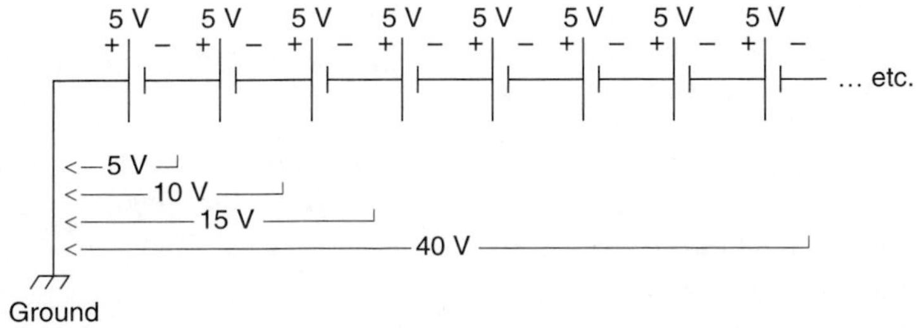

Fig. 11-25

Gain. The factor by which a signal is amplified, sometimes expressed in decibels (dB).

Graphic User Interface (GUI). An intuitive, easy-to-use means of communicating information to and from a computer program by means of graphical screen displays. GUIs can resemble the front panels of instruments or other objects associated with a computer program.

Grounds. There are two types of grounds in data acquisition. *System* or *power ground* is used to provide a path for current flow. *Analog* or *signal ground* is used to provide a voltage reference for signal conditioners and converters. In general, no power current should flow in analog ground connections. Analog grounds should never be connected to each other or to power grounds, as this would form a loop for current to flow through.

Input/Output (I/O). The transfer of data to or from a computer system involving communications channels, operator interface devices, and/or data acquisition and control interfaces.

Isolation. Refers to the process of measuring a signal without *direct* electrical connection. Solid-state relays isolate using an optical technique (Fig. 11-26).

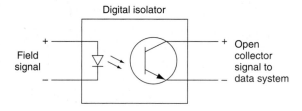

Digital isolator

Field signal

+

−

Open collector signal to data system

+

−

Fig. 11-26

Multitasking. A property of an operating system in which several processes can be run simultaneously.

Noise. Refers to undesirable interference with an otherwise normal signal. Common sources of noise include power lines (60 Hz), fluorescent lights, and large motors.

Real Time. A property of an event or system in which data is processed as it is acquired instead of being accumulated and processed at a later time.

Resolution. The smallest signal increment that can be detected by a measurement system. Resolution can be expressed in bits, in proportions, or in percentage of full scale.

For example, a system can have 12-bit resolution, one part in 4096 resolution, and 0.0244 percent of full scale.

Sample and Hold. Most fast (kHz) A/D converters use a sample and hold amplifier in the system. This device takes a very fast "snapshot" of the input voltage and freezes it before the A/D converts. This prevents the A/D from measuring a signal that is changing, which can be inaccurate. With *simultaneous* sample and hold:

- More than one channel is captured at the same time.
- Each channel needs its own sample and hold amplifier.
- A global "hold" signal freezes all channels.
- The A/D converter steps through each channel one at a time.
- The user is thus able to compare data samples from different sensors taken at the same instant.

Save Interval or Save Period. Specifies the time between data saves (Fig. 11-27). Note the following about the save interval:

- It can be a constant, always the same.
- It can vary with time, or as a function of certain measurements.
- In many data logging applications, a fixed save rate can result in too much data—a variable save rate could be set up to save data only when something changes.

Track-and-Hold (T/H). A circuit that tracks an analog voltage and holds the value on command.

Transfer Rate. The rate, measured in bytes, at which data is moved from source to destination after software initialization and set-up operations; the maximum rate at which the hardware can operate.

Triggering. Is used to start and/or stop data acquisition. The system waits until it gets a start signal (i.e., a trigger), and then it measures and saves data, and waits for a stop signal.

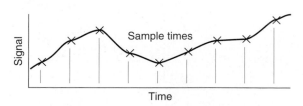

Signal

Sample times

Time

Fig. 11-27

26. Data acquisition is the _____, _____, and _____ of information by a computer-based system.
27. True or false? Data acquisition systems are only practical for installations with thousands of input data channels.
28. The two kinds of data acquisition interface systems are _____ and _____.
29. True or false? Input multiplexers supply an economical means of connecting to a large number of signals.
30. True or false? Ground loops can produce inaccurate data.
31. True or false? The accuracy and resolution of a measurement system are equal.
32. An _____ signal is one which is not connected to any other systems or voltages.
33. _____ refers to the process of measuring a signal without direct electrical connection.
34. True or false? A variable save rate could be set up to save data only when something changes.
35. _____ is used to start and/or stop data acquisition.

SUMMARY

1. A continuous process is one in which raw materials enter one end of the system and the finished product comes out the other end of the system, while the process itself runs continuously.
2. In batch processing, a set amount of each of the inputs to the process is received in a batch, and then some operation is performed on the batch to produce a finished product or an intermediate product that needs further processing.
3. With individual product production, a series of operations produces a useful output product.
4. Automatic machines and processes were developed to mass-produce products, control very complex operations, or operate machines accurately for long periods of time.
5. Individual control is used to control a single machine. This type of control does not normally require communication with any other controllers.
6. Centralized control is used when several machines or processes are controlled with one central controller. No exchange of controller status or data is sent to other controllers.
7. The distributive control system (DCS) differs from the centralized system in that each machine is handled by a dedicated control system. Each dedicated control is independent and involves two or more computers communicating with each other to accomplish the complete control task.
8. A process control system can be defined as the functions and operations needed to change a material physically or chemically.
9. Sensors provide inputs from the process and from the external environment.
10. Operator-machine interfaces are inputs from a human to set up the starting conditions or alter the control of a process.
11. Signal conditioning involves converting input and output signals to a usable form.
12. Acutators convert system output electrical signals into physical action.
13. The controller makes the system's decisions based on the input signals and generates output signals that operate actuators to carry out the decisions.
14. Control systems can be classified as open loop or closed loop.
15. With an open-loop system, the controller receives no information concerning the present status of the process or the need for any corrective action.
16. A closed-loop control system is one in which the system measures the actual output of the process and adjustments are made continually until the difference between the desired and actual output is as small as practical.
17. Regulators may be defined as closed-loop systems that maintain a steady level or value for some quantity.
18. The four most popular types of control response used in the process industry are:
   - ON/OFF
   - Proportional (P)
   - Integral (I)
   - Derivative (D)

19. With ON/OFF control the final control element is either ON or OFF—one for the occasion when the value of the measured variable is above the set point, and the other for the occasion when the value is below the set point.

20. With ON/OFF control, the measured variable will oscillate around the set point at an amplitude and frequency that depends on the process's capacity and time response.

21. Proportional controls are designed to eliminate the cycling associated with ON/OFF control.

22. Proportioning action can also be accomplished by analog control or time proportioning control of the final control element.

23. A proportional controller allows tighter control of the process variables because its output can take on any value between fully ON and fully OFF, depending on the magnitude of the error signal.

24. The proportioning action occurs within a "proportional band" around the set point, and outside this band, the controller functions as an ON/OFF unit.

25. The operation of a proportional controller leads to process deviation known as *offset* or *droop,* which is the difference between the attained value of the controller and the required value.

26. The integral action, sometimes termed reset action, responds to the size and time duration of the error signal.

27. The derivative mode controller responds to the speed at which the error signal is changing.

28. Proportional-integral-derivative (PID) controllers produce outputs that depend on the magnitude, duration, and rate of change of the system error signal.

29. The PID controller is the most sophisticated and widely used type of process controller. It can reduce the system error to zero faster than any other controller.

30. Because a PID controller has an integrator and a differentiator, however, this controller must be custom tuned to each process being controlled.

31. PID control changes the amount of output signal in a mathematically specified way that accounts for amount of error and rate of signal change.

32. Basically, PID controller tuning consists of determining the appropriate values for the gain (proportional band), rate (derivative), and reset (integral) time tuning parameters.

33. With manual PID tuning, the operator estimates the tuning parameters required to give the desired controller response.

34. With semiautomatic, or autotune, PID tuning, the controller takes care of calculating and setting PID parameters.

35. With fully automatic, or intelligent, PID tuning, artificial intelligence continually readjusts PID tuning parameters as needed.

36. Data acquisition is the collection, analysis, and storage of information by a computer-based system.

37. A data acquisition interface system takes the signals produced by sensors and converts them into a form that your computer can understand.

38. The great advantage of data acquisition is that data is stored automatically in a form that can be retrieved without error or additional work for later analysis.

39. Basically, there are two kinds of data acquisition interface systems: plug-in and stand-alone. These two extremes involve putting the signal conditioning circuitry directly on a data acquisition board, or using a stand-alone data acquisition instrument with built-in signal conditioning.

CHAPTER REVIEW QUESTIONS

Answer the following questions.

11-1. List the three types of processes carried out in industries.
11-2. State the type of process used for each of the following applications:
    a. Mixing ingredients to manufacture chemically based products
    b. Assembly of TV sets
    c. Manufacturing of electronic chassis
11-3. List three reasons why automatic machines and processes were developed.

11-4. Compare individual, centralized, and distributive control systems.

11-5. Define what is meant by the term *process control system.*

11-6. State the basic function of each of the following as part of a process control system:

    a. Sensors

    b. Operator-machine interface

    c. Signal conditioning

    d. Actuators

    e. Controller

11-7. Compare open-loop and closed-loop control systems.

11-8. What are regulators, and for what purpose are they used?

11-9. List the four most popular types of control response used in the process industry.

11-10. Outline the operating sequence of an ON/OFF controller.

11-11. Explain how a proportional controller eliminates the cycling associated with ON/OFF control.

11-12. What process error or deviation is produced by a proportional controller?

11-13. How does integral control work to eliminate offset?

11-14. What does the derivative action of a controller respond to?

11-15. List three factors of a system error signal that influence the output response of a proportional-integral-derivative (PID) controller.

11-16. Why must a PID controller be custom-tuned to each process being controlled?

11-17. Compare manual, autotune, and intelligent tuning of a PID controller.

11-18. Define data acquisition.

11-19. Outline how a typical data acquisition and control system operates.

11-20. Explain the function of each of the following devices with reference to a data acquisition system:

    a. Transducer

    b. Analog multiplexer

    c. Amplifier

    d. Signal conditioner

    e. A/D converter

11-21. Explain each of the following terms as it applies to the operation of a data acquisition system:

    a. Sensor calibration

    b. Floating signal

    c. Gain

    d. Isolation

    e. Real time

    f. Resolution

    g. Save interval

    h. Transfer rate

CRITICAL THINKING QUESTIONS

11-1. Give an example of an open-loop control system. Explain why the system is open-loop and identify the controller and final control element.

11-2. Give an example of a closed-loop control system. Explain why the system is closed-loop and identify the sensor, the controller, and the final control element.

11-3. How would an ON/OFF controller respond if the deadband were too narrow?

11-4. In a home heating system with ON/OFF control, what will be the effect of widening the deadband?

11-5. a. Calculate the proportional band of a temperature controller with a 5 percent bandwidth and a set point of 500° F.

    b. Calculate the upper and lower limits beyond which the controller functions as an ON/OFF unit.

    c. Calculate the proportion gain:

$$\text{Gain} = \frac{100}{\text{Percentage bandwidth}}$$

11-6. What effect, if any, would a higher process gain have on offset errors?

11-7. Describe the advantages of a 4- to 20-mA current loop as an input signal compared to a 0- to 5-V input signal.

11-8. How might a data acquisition system be applied to a petrochemical metering station whose purpose is measurement and flow control?

Answers to Self-Tests

1. true
2. batch
3. false
4. true
5. individual
6. centralized
7. distributed
8. physically; chemically
9. false
10. true
11. conditioning
12. false
13. open; closed
14. closed
15. false
16. oscillate
17. false
18. time
19. band
20. offset; droop
21. true
22. speed
23. magnitude; duration; rate
24. true
25. true
26. collection; analysis; storage
27. false
28. plug-in; standalone
29. true
30. true
31. false
32. floating
33. isolation
34. true
35. triggering

CHAPTER 12

Programmable Logic Controllers (PLCs)

■

CHAPTER OBJECTIVES

This chapter will help you to:

1. *Identify* the main parts of a PLC and *describe* their functions.
2. *Define* the different binary numbering systems and be able to *convert* from one numbering or coding system to another.
3. *Understand* how ladder diagram language is used to communicate information to the PLC.
4. *Analyze* and *interpret* typical PLC timer and counter instructions.
5. *Interpret* data manipulation instructions as they apply to a PLC program.
6. *Interpret* and *explain* information associated with a PLC shift register and sequencer instruction.

In the world of automation, the programmable logic controller (PLC) has become a standard for control. It now not only replaces the earlier relay controls but has taken over many additional control functions. This chapter is devoted to the principles upon which all PLCs operate. All basic instructions are discussed, and conversion from relay ladder diagrams to logic ladder diagrams is emphasized.

■

12-1 OVERVIEW

A *programmable controller* is a computer designed for use in machines. Unlike a computer, it has been designed to operate in the industrial environment and is equipped with special inputs/outputs and a control programming language. The common abbreviation used in industry for these devices, *PC*, can be confusing because it is also the abbreviation for *personal computer.* Therefore, some manufacturers refer to their programmable controller as a *PLC*, which is an abbreviation for *programmable logic controller.*

Initially the PLC was used to replace relay logic, but its ever-increasing range of functions means that it is found in many and more complex applications. As the structure of a PLC is based on the same principles as those employed in computer architecture, it is capable of performing not only relay switching tasks, but also other applications such as counting,

calculating, comparing, and the processing of analog signals (Fig. 12-1 on page 328).

Programmable controllers offer several advantages over a conventional relay type of control. Relays have to be *hard-wired* to perform a specific function (Fig. 12-2 on page 328). When the system requirements change, the relay wiring has to be changed or modified, which requires time. In extreme cases, such as in the auto industry, complete control panels had to be replaced since it was not economically feasible to rewire the old panels with each model changeover. The programmable controller has eliminated much of the hand wiring associated with conventional relay control circuits. It is small and inexpensive compared to equivalent relay-based process control systems. Programmable controllers also offer solid-state reliability, lower power consumption, and ease of expandability. If an application has more than a half-dozen relays, it probably will be less expensive to install a

Fig. 12-1 Programmable logic controller. *(Courtesy Allen-Bradley Company, Inc.)*

PLC. Simulating a hundred relays, timers, and counters is not a problem even on small PLCs.

A personal computer can be made into a programmable controller if you provide some way for the computer to receive information from devices such as pushbuttons or switches. You also need a program to process the inputs and decide the means of turning OFF and ON load devices. A typical PLC can be divided into three parts, as illustrated in the block dia-

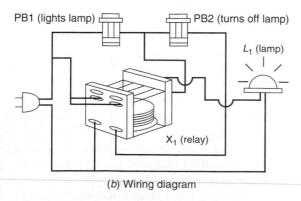

(a) Schematic diagram

(b) Wiring diagram

Fig. 12-2 Hard-wired relay type of control.

gram of Fig. 12-3. These three components are the *central processing unit* (*CPU*), the *input/output* (*I/O*) *section*, and the *programming device*. The programmable controller is an event-driven device, which means that an event taking place in the field will result in an operation or output taking place.

The central processing unit (CPU) is the heart of the PLC system. A typical central processing unit or *processor* is shown in Fig. 12-4. The CPU is a microprocessor-based system that replaces control relays, counters, timers, and sequencers. A processor appears only once in a PLC, and it can be either a one-bit or a word processor. One-bit processors are adequate for dealing with logic operations. PLCs with word processors are used when processing text and numerical data, calculations, gauging, controlling, and recording, as well as the simple processing of signals in binary code, are required. The principle of operation of a CPU can be briefly described as follows:

- The CPU accepts (reads) input data from various sensing devices, executes the stored user program from memory, and sends appropriate output commands to control devices.
- A direct current (dc) power source is required to produce the low-level voltage used by the processor and the I/O modules. This power supply can be housed in the CPU unit or may be a separately mounted unit, depending on the PLC system manufacturer.

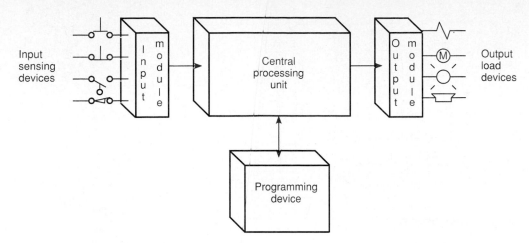

Fig. 12-3 PLC block diagram.

Memory

Volatile memory

Nonvolatile memory

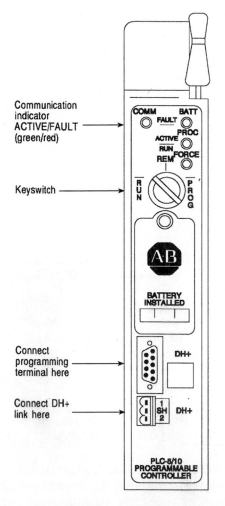

Fig. 12-4 Typical processor unit. *(Courtesy Allen-Bradley Company, Inc.)*

- Most CPUs contain backup batteries that keep the operating program in storage in the event of a plant power failure. Typical retentive backup time is one month to one year.

The CPU contains various electrical parts and receptacles for connecting the cables that go to the other units as well as to operational key switches.

Typical operation key switch positions are:

- OFF: System cannot be run or programmed.
- Run: Allows the system to run, but no program alterations can be made.
- Program: Disables outputs and allows creating, modifying, and deleting of programs.

The processor memory module is a major part of the CPU housing (Fig. 12-5 on page 330). *Memory* is where the control plan or program is held or stored in the controller. The information stored in the memory relates to the way the input and output data should be processed. The complexity of the program determines the amount of memory required. Memory elements store individual pieces of information called *bits* (for *binary digits*).

The actual control program is held within electronic memory storage components, such as the RAMs and EEPROMs. The processing unit scans data from the input and output modules and stores their conditions in the memory. The processor unit then scans the user program stored in the memory and makes decisions that cause outputs to change.

Memory can be placed into two categories: *volatile* and *nonvolatile*. Volatile memory will lose its stored information if all operating power is lost or removed. Volatile memory is easily altered and quite suitable for most applications when supported by battery backup. Nonvolatile memory can retain stored information when power is removed accidentally or intentionally. PLCs make use of many different types of volatile and nonvolatile memory

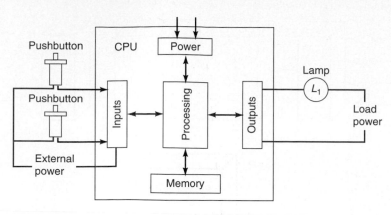

Fig. 12-5   Processor memory.

devices. Following is a generalized description of a few of the more common types:

RAM. Random access memory (RAM) is designed so that information can be written into or read from the memory. Today's controllers, for the most part, use the CMOS-RAM with battery support for user program memory. RAM provides an excellent means for easily creating and altering a program.

ROM. Read-only memory (ROM) is designed so that information stored in memory can only be read and, under ordinary circumstances, cannot be changed. Information found in the ROM is placed there by the manufacturer for the internal use and operation of the PLC.

EPROM. Erasable programmable read-only memory (EPROM) is designed so that it can be reprogrammed after being entirely erased with the use of an ultraviolet light source.

EEPROM. Electrically erasable programmable read-only memory (EEPROM) is a nonvolatile memory that offers the same programming flexibility as does RAM. It provides permanent storage of the program but can be easily changed using standard programming devices.

The I/O section of a PLC consists of input modules and output modules. The I/O system forms the interface by which field devices are connected to the controller. The purpose of this interface is to condition the various signals received from or sent to external field devices. Input devices such as pushbuttons, limit switches, sensors, selector switches, and thumbwheel switches are hard-wired to terminals on the input modules. Output devices such as small motors, motor starters, solenoid valves, and indica-

tor lights are hard-wired to the terminals on the output modules. These devices are also referred to as "field" or "real-world" inputs and outputs. The terms *field* or *real-world* are used to distinguish actual external devices that exist and must be physically wired from the internal user program that duplicates the function of relays, timers, and counters. Some programmable controllers have separate modules for inputs and outputs; others have the inputs and outputs connected as an integral part of the controller (Fig. 12-6). When the module is slid into the rack, it makes an electrical connection with a series of contacts called the *backplane,* located at the rear of the rack. The PLC processor is also connected to the backplane and can communicate with all the modules in the rack [see Fig. 12-6(*b*)].

Input interface modules accept signals from the machine or process devices (e.g., 120 V ac) and convert them into signals (e.g., 5 V dc) that can be used by the controller. Output interface modules convert controller signals (e.g., 5 V dc) into external signals (e.g., 120 V ac) used to control the machine or process. There are many types of inputs and outputs that can be connected to a programmable controller, and they can all be divided into two groups: digital (also known as discrete) and analog.

▶ YOU MAY RECALL   that digital inputs and outputs are those that operate because of changes in a discrete state or level. Analog inputs and outputs change continuously over a variable range.

The most common type of I/O interface module is the discrete type. This type of interface connects field input devices of the ON/OFF nature such as selector switches, pushbuttons, and limit switches. Likewise, output control is

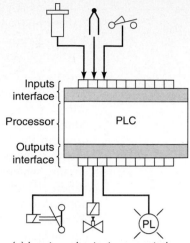

Inputs interface

Processor

Outputs interface

PLC

(a) Inputs and outputs connected as
integral part of the controller

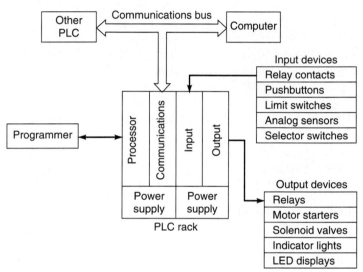

(b) I/O modules connected to PLC processor
through rack system

Fig. 12-6   Input/output (I/O) systems.

limited to devices such as lights, small motors, solenoids, and motor starters that require simple ON/OFF switching. Analog inputs and outputs are used in more complex control applications such as furnace temperature control (Fig. 12-7).

Each I/O module is powered by some field-supplied voltage source (Fig. 12-8 on page 332). Since these voltages can be of different magnitudes or types, I/O modules are available at various ac and dc voltage and current ratings. Both voltage and current must match the electrical requirements of the system to which it is connected. There are typically 4, 8, 12, 16, or 32 terminals per module. PLC manufacturers have a wide variety of input and out-

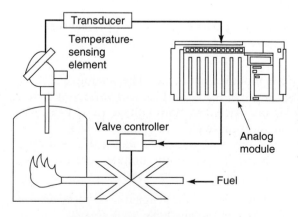

Furnace temperature control

Fig. 12-7   Analog I/O module.

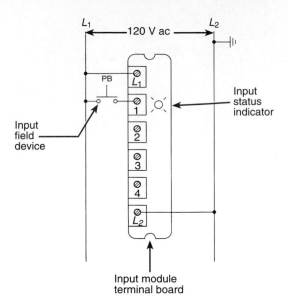

(a) Typical discrete four-point, 120-V ac, input module

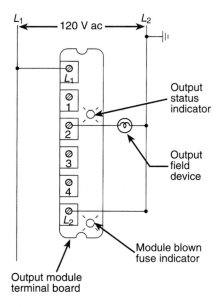

(b) Typical discrete four-point, 120-V ac, 4-A output module

Fig. 12-8   I/O module powered by a field-supplied voltage source.

put modules available. The analog I/O modules provide an interface to a variety of analog signals, including both voltage (e.g., 1- to 5-V) and current (e.g., 4- to 20-mA) ranges.

Signals are connected to the PLC through input modules. Input modules perform four tasks in the PLC control system:

- They sense when a signal is received from a sensor on the machine.
- They convert the input signal to the correct voltage level for the particular PLC.

- They isolate the PLC from fluctuations in the input signal's voltage or current.
- They send a signal to the PLC indicating which sensor originated the signal.

Figure 12-9 shows a simplified block and wiring diagram for one input of a typical ac interface input module. The input circuit is composed of two basic sections: the *power section* and the *logic section*. The power and logic sections are normally coupled together with a circuit, which electrically separates the two. When the pushbutton is closed, 120 V ac is applied to the bridge rectifier through resistors R_1 and R_2. This produces a low-level dc voltage, which is applied across the LED of the *optical isolator*. The zener diode (Z_D) voltage rating sets the minimum level of voltage that can be detected. When light from the LED strikes the phototransistor, it switches into conduction, and the status of the pushbutton is communicated in logic or low-level dc voltage to the processor. The optical isolator not only separates the higher ac input voltage from the logic circuits but also prevents damage to the processor due to line voltage transients. Optical isolation also helps reduce the effects of electrical noise, common in the industrial environment, which can cause erratic operation of the processor. Coupling and isolation can also be accomplished by use of a pulse transformer or reed relay.

The output interface module of a programmable controller acts as a switch to supply power from the user power supply to operate the output. The output under the control of the program is fed from the processor to a logic circuit that will receive and store the processor command that is required to make an output become active. The output switching devices most often used to switch power to the load in programmable controllers are:

- Relay for ac or dc loads
- Triac for ac loads only
- Transistors for dc loads only

The output module has a function similar to that of the input module except in reverse order. Figure 12-10 shows a simplified block and wiring diagram of a typical ac interface output module. As part of its normal operation, the processor sets the output status according to the logic program. When the processor calls for an output, a voltage is applied across the LED of the isolator. The LED then emits light, which switches the phototransistor into conduction. This in turn switches the triac into

332   CHAPTER 12   PROGRAMMABLE LOGIC CONTROLLERS (PLCs)

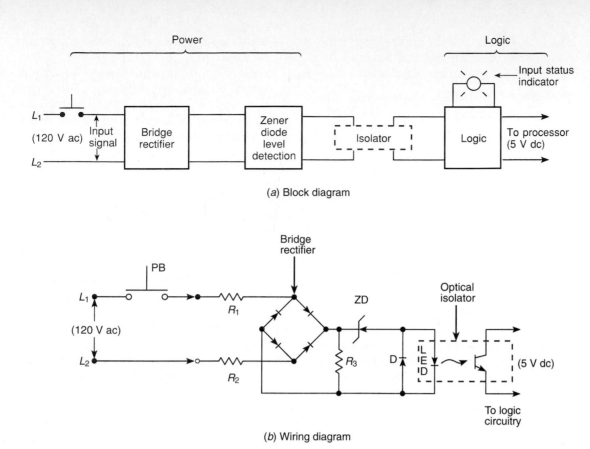

(a) Block diagram

(b) Wiring diagram

Fig. 12-9   Alternating current interface input module.

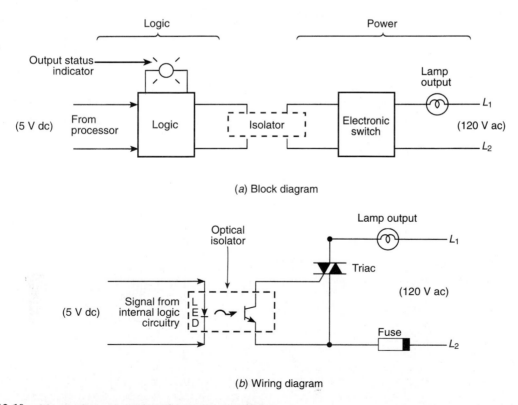

(a) Block diagram

(b) Wiring diagram

Fig. 12-10   Alternating current interface output module.

conduction, which, in turn, turns on the lamp. Since the triac conducts in either direction, the output to the lamp is alternating current.

▐▐▶ YOU MAY RECALL that the triac, rather than having ON and OFF status, actually has LOW and HIGH resistance levels, respectively. In its OFF state (HIGH resistance), a small current leakage of a few milliamperes still flows through the triac.

As with input circuits, the output interface is usually provided with LEDs that indicate the status of each output. In addition, if the module contains a fuse, a fuse status indicator may also be used.

Output interface modules are normally designed to handle currents in the 2- to 3-A range. To protect the output module circuits, specified current ratings should not be exceeded. For controlling larger loads, such as large motors, a standard control relay is connected to the output module. The contacts of the relay can then be used to control a larger load or motor starter as shown in Fig. 12-11. When a control relay is used in this manner, it is called an *interposing relay*.

Analog input interface modules contain the circuitry necessary to accept analog voltage or current signals from analog field devices. These inputs are converted from an analog to a digital value by an *analog-to-digital (A/D) converter circuit*. The conversion value, which is proportional to the analog signal, is expressed as a 12-bit binary or as a three-digit binary-coded decimal (BCD) for use by the processor. Analog input sensing devices include temperature, light, speed, pressure, and position transducers.

The analog output interface module receives digital data from the processor that is converted into a proportional voltage or current to control an analog field device. The digital data is passed through a *digital-to-analog (D/A) converter circuit* to produce the necessary analog form. Analog output devices include small motors, valves, analog meters, and seven-segment displays.

Each port or terminal on input and output modules is assigned a unique *address* number (Fig. 12-12). This address is used by the processor to identify the location of the device in order to monitor or control it. The type of the module and the actual physical location of the terminal determines the programming address. The addressing format of inputs and outputs depends on the particular PLC used and can normally be found in the particular PLC's user's manual. These addresses can be represented in decimal, octal, or hexadecimal terms depending upon the number system used by the PLC.

The micro PLC format illustrated in Fig. 12-12(*a*) uses a limited number of points of control. Each input and output device must have a specific address. With the large PLC rack installation shown in Fig. 12-12(*b*), the location of a module within a rack and the terminal number of a module to which an input or output device is connected will determine the address of the device.

The programmable controller system requires two power supplies. One provides the power necessary for field devices and output loads to operate and is provided by the programmable controller user. The second power supply is provided internally or as a module

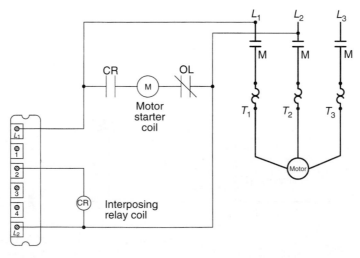

Fig. 12-11   Interposing relay connection.

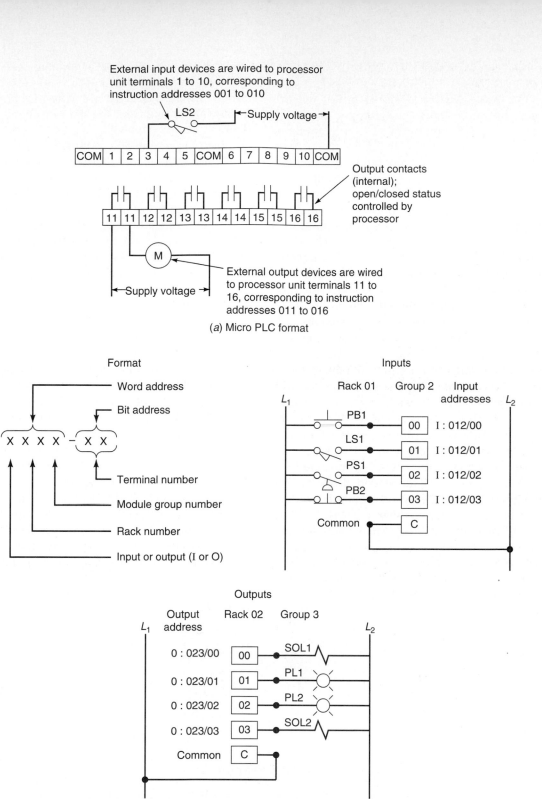

External input devices are wired to processor unit terminals 1 to 10, corresponding to instruction addresses 001 to 010

LS2 — Supply voltage

| COM | 1 | 2 | 3 | 4 | 5 | COM | 6 | 7 | 8 | 9 | 10 | COM |

Output contacts (internal); open/closed status controlled by processor

| 11 | 11 | 12 | 12 | 13 | 13 | 14 | 14 | 15 | 15 | 16 | 16 |

M

External output devices are wired to processor unit terminals 11 to 16, corresponding to instruction addresses 011 to 016

Supply voltage

(*a*) Micro PLC format

Format

Word address
Bit address

X X X X – X X

Terminal number
Module group number
Rack number
Input or output (I or O)

Inputs

Rack 01   Group 2   Input addresses

		Input address	
PB1	00	I : 012/00
LS1	01	I : 012/01
PS1	02	I : 012/02
PB2	03	I : 012/03
Common	C	

Outputs

Output address   Rack 02   Group 3

Output address		
0 : 023/00	00	SOL1
0 : 023/01	01	PL1
0 : 023/02	02	PL2
0 : 023/03	03	SOL2
Common	C	

(*b*) Format used with large PLC rack installations

Fig. 12-12 Typical addressing formats.

that is part of the programmable controller system. This power supply provides internal direct current to operate the processor logic circuitry and I/O assemblies. The voltage that it provides will depend on the type of integrated circuits (ICs) within the system. If the system is made up of transistor-transistor logic (TTL) ICs, the internal power supply will be 5 V, but if the integrated circuit is a complementary metal oxide semiconductor (CMOS) type, the power supply voltage will be in the range of 3 V to 18 V.

The *programming device* provides the primary means by which the user can communicate with the circuits of the programmable controller (Fig. 12-13). It allows the user to enter, edit, and monitor programs by connecting into the processor unit and allowing access to the user memory. The programming unit can be a liquid crystal display (LCD) handheld terminal, a single-line LED display unit, or a keyboard and a video display unit. The video display offers the advantage of displaying large amounts of logic on the screen, simplifying the interpretation of the program. The programming unit communicates with the processor via a serial or parallel data communications link. If the programming unit is not in use, it may be unplugged and removed. Removing the programming unit will not affect the operation of the user program. A personal computer with appropriate software can also act as a program terminal, making it possible to carry out the programming away from the physical location of the programmable controller. When the program is complete, it is saved to some form of mass storage and downloaded to the programmable controller when required.

A handheld device is shown in Fig. 12-13(*a*). The display is usually an LED or liquid-crystal display (LCD), and the keyboard consists of numeric keys, programming instruction keys, and special function keys. To change the program in the PLC, it is necessary either to enter a new program directly from the keyboard or to download one from the hard disk while online.

When programmable controllers were first introduced, the *ladder diagram* was selected to be the fundamental programming format because it was well known in the electrical and electronics industry.

To get an idea of how a PLC operates, consider the simple process control application illustrated in Fig. 12-14. Here a mixer motor is to be used to automatically stir the liquid in a vat when the temperature and pressure reach preset values. In addition, direct manual operation of the motor is provided by means of a separate pushbutton station. The process is monitored with temperature and pressure sensor switches that close their respective contacts when conditions reach their preset values.

This control problem can be solved using the relay method for motor control shown in the relay ladder diagram. The motor starter coil (M) is energized when both the pressure and temperature switches are closed or when the manual pushbutton is pressed.

Now let's look at the way a PLC might be used for this application. The same input field devices (pressure switch, temperature switch, and pushbutton) are used. These devices are hard-wired to an appropriate input module ad-

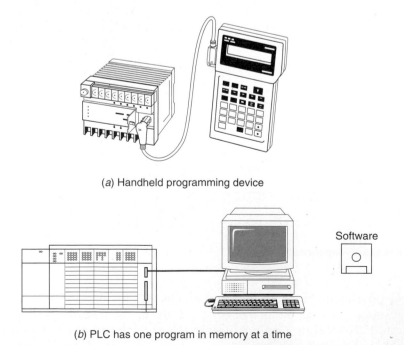

(*a*) Handheld programming device

Software

(*b*) PLC has one program in memory at a time

Fig. 12-13 **Programming devices.**

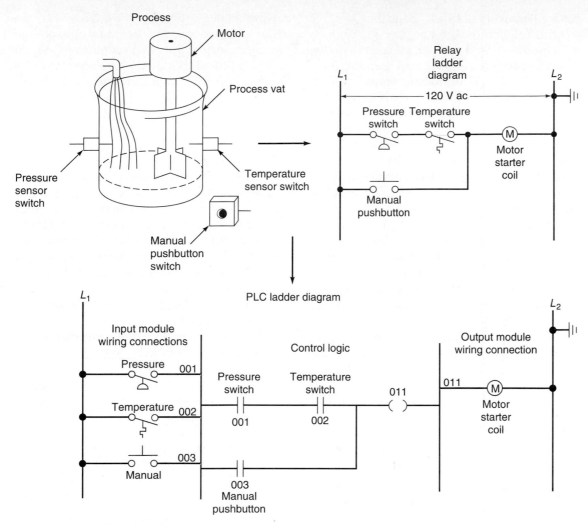

Fig. 12-14   Process control application.

dress according to the manufacturer's address-ing format. The same output field device (mo-tor starter coil) is also used. This device is hard-wired to an appropriate output module address according to the manufacturer's ad-dressing format.

⚹ Next, the PLC ladder control logic diagram is constructed and programmed into the mem-ory of the CPU. A typical ladder logic diagram for this process is shown. The format used is similar to the layout of the hard-wired relay ladder circuit. The individual symbols repre-sent *instructions* and the numbers represent the *instruction addresses.* When programming the controller, these instructions are entered one by one into the processor memory from the operator terminal keyboard. Instructions are stored in the user program portion of the processor memory.

To operate the program, the controller is placed in the RUN mode, or operating cycle. During each operating cycle, the controller

examines the status of input devices, executes the user program, and changes outputs accord-ingly. Each ⊣⊢ can be thought of as a set of nor-mally open (NO) contacts. The ⊣( )⊢ can be considered to represent a coil that, when ener-gized, will close a set of contacts. In the ladder logic diagram shown, the coil 011 is energized when contacts 001 and 002 are closed or when contact 003 is closed. Either of these condi-tions provides a continuous path from left to right across the rung that includes the coil.

The RUN operation of the controller can be described by the following sequence of events. First, the inputs are examined and their status is recorded in the controller's memory (a closed contact is recorded as a signal that is called a logic 1 and an open contact by a signal that is called a logic 0). Then the ladder dia-gram is evaluated, with each internal contact given OPEN or CLOSED status according to the record. If these contacts provide current path from left to right in the diagram, the output coil memory location is given a logic 1 value and the output module interface contacts will close. If there is no conducting path on the program rung, the output coil memory location is set to logic 0 and the output module interface contacts will be open. The completion of one cycle of this sequence by the controller is called a *scan*. The *scan time*, the time required for one full cycle, provides a measure of the speed of response of the PLC.

As mentioned, one of the important features of a PLC is the ease with which the program can be changed. For example, assume that our original process control circuit for the mixing operation must be modified as shown in the *relay ladder diagram* of Fig. 12-15. The change requires that the manual pushbutton control be permitted to operate at any pressure *but not unless* the specified temperature setting has been reached.

If a relay system were used, it would require some rewiring of the system to achieve the desired change [Fig. 12-15(*a*)]. However, if a PLC system were used, *no* rewiring would be necessary. The inputs and outputs are still the same. All that is required is to change the PLC ladder logic diagram as shown [Fig. 12-15(*b*)].

Self-Test

Answer the following questions.

1. Initially, programmable logic controllers were designed to replace _____ logic.
2. True or false? The design of a PLC is similar to that of an electromechanical relay.
3. The PLC _____ accepts input data, executes the program, and sends appropriate output commands to control devices.
4. True or false? Memory is where the control plan or program is held or stored in the controller.
5. Memory elements store individual pieces of information called _____.

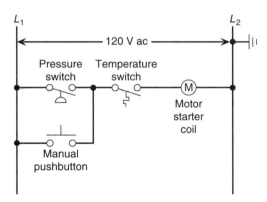

(*a*) Hard-wired relay circuit requires rewiring for changed application

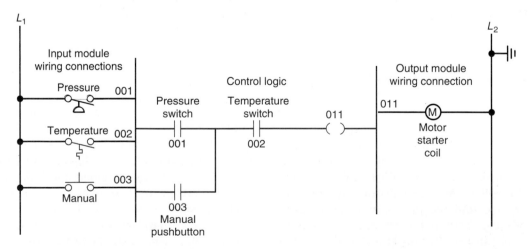

(*b*) PLC system requires only that control logic be changed; all input and output module connections remain the same

Fig. 12-15   Modified process control application.

Decimal number
system

Base of a number
system

Binary numbering
system

6. True or false? Volatile memory is easily altered.

7. The programmable controller I/O system forms the _____ by which field devices are connected to the controller.

8. Discrete I/O modules allow only _____-type devices to be connected.

9. True or false? Each I/O module is powered by some field-supplied voltage source.

10. The I/O module circuitry can be divided into two basic sections: the _____ section and the _____ section.

11. Output switching devices most often used to switch power to the load in a PLC are: _____ for ac or dc loads; _____ for ac loads only; _____ for dc loads only.

12. True or false? The maximum current rating for the individual outputs of an output interface module is usually in the range of 20 to 30 A.

13. True or false? A thermocouple would be classified as an analog input sensing device.

14. The I/O address is used by the processor to identify where the device is _____.

15. True or false? Removing the PLC programming unit will affect the operation of the user program.

16. The _____ diagram is the fundamental programming format used with PLCs.

17. With reference to a PLC ladder control logic diagram, the individual symbols represent _____ and the numbers represent _____.

18. Scan time is a measure of the _____ of response of the PLC.

19. True or false? The control logic of a PLC system can be changed without any rewiring of the system.

12-2 NUMBER SYSTEMS AND CODES

Knowledge of number systems other than the decimal numbering system is quite useful when working with PLCs or with almost any type of digital equipment. This is true because a basic requirement of these devices is to represent, store, and operate on numbers. In general, PLCs work on binary numbers in one form or another; these are used to represent various codes or quantities. Often the programmer needs to be able to perform conversions between the systems, and to perform math functions within each system.

The *decimal system,* which is most common to us, has a base of 10. The *base* of a number system determines the total number of different symbols or digits used by that system. For instance, in the decimal system, 10 unique numbers or digits—the digits 0 through 9—are used. Figure 12-16 shows a comparison between four common numbering systems: decimal (base 10), octal (base 8), hexadecimal (base 16), and binary (base 2). Note that all numbering systems start at zero.

Decimal	Octal	Hexadecimal	Binary
0	0	0	0
1	1	1	1
2	2	2	10
3	3	3	11
4	4	4	100
5	5	5	101
6	6	6	110
7	7	7	111
8	10	8	1000
9	11	9	1001
10	12	A	1010
11	13	B	1011
12	14	C	1100
13	15	D	1101
14	16	E	1110
15	17	F	1111
16	20	10	10000
17	21	11	10001
18	22	12	10010
19	23	13	10011
20	24	14	10100

Fig. 12-16 Number system comparisons.

The value of a decimal number depends on the digits that make up the number and the place value of each digit. A place (weight) value is assigned to each position that a digit would hold from right to left. In the decimal system the first position, starting from the furthest right position, is 0; the second is 1; and so on up to the last position. The weighted value of each position can be expressed as the base (10 in this case) raised to the power of the position. For the decimal system, then, the position weights are 1, 10, 100, 1000, and so on. Figure 12-17 on page 340 illustrates the way the value of a decimal number can be calculated by multiplying each digit by the weight of its position and summing the results.

The *binary numbering system* (base 2) is the basis of all digital systems. Two states exist in digital equipment, an ON state, which is representative of one (1), and an OFF condition, which is representative of zero (0). The ON condition in a circuit is approximately equal to

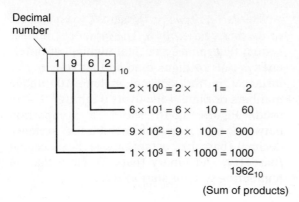

Fig. 12-17 Weighted value in the decimal system.

supply voltage, and the OFF to zero volts or ground. A third state may exist in some logic circuits to produce tri-state logic. This condition is a high-impedance or no-voltage state and is not considered in the binary system.

The only allowable digits in a binary numbering system are zero (0) and one (1). Since the binary system uses only two digits, each position of a binary number can go through only two changes, and then a 1 is carried to the immediate left position.

The decimal equivalent of a binary number is calculated in a manner similar to that used for a decimal number. This time the weighted values of the positions are 1, 2, 4, 8, 16, 32, 64, and so on. Instead of being 10 raised to the power of the position, the weighted value is 2 raised to the power of the position. Figure 12-18 shows how the binary number 10101101 is converted to its decimal equivalent: 173.

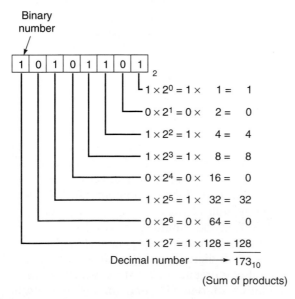

Fig. 12-18 Converting a binary number to a decimal number.

Each digit of a binary number is known as a *bit*. In a PLC the processor-memory element consists of hundreds or thousands of locations. These locations, or *registers*, are referred to as *words*. Each word is capable of storing data in the form of *binary digits*, or *bits*. The number of bits that a word can store depends on the type of PLC system used. Eight-bit and sixteen-bit words are the most common. As the technology continues to develop, 32-bit or larger words will be possible. Bits can also be grouped within a word into *bytes*. Usually a group of 8 bits is a byte, and a group of 1 or more bytes is a word. Figure 12-19 illustrates a 16-bit word made up of 2 bytes. The *least significant bit (LSB)* is the digit that represents the smallest value and the *most significant bit (MSB)* is the digit that represents the largest value. A bit within the word can exist only in two states: a logical 1 or ON condition or a logical 0 or OFF condition.

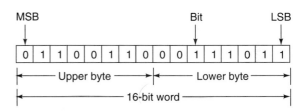

Fig. 12-19 A 16-bit word.

The size of the programmable controller *memory* relates to the amount of user program that can be stored. If a memory size is 884 words, then it can actually store 14,144 (884 × 16) bits of information using 16-bit words or 7072 (884 × 8) using 8-bit words. Therefore, when comparing different PLC systems, you must know the number of bits per word of memory in order to determine the relative capacity of the systems' memories. Normally programmable controllers do not require storage space above 128 K and in many instances need a memory size of only 1 K to 2 K.

To convert a decimal number to its binary equivalent, we must perform a series of divisions by 2. Figure 12-20 illustrates the conversion of the decimal number 47 to binary. We start by dividing the decimal number by 2. If there is a remainder, it is placed in the LSB of the binary number. If there is no remainder, a 0 is placed in the LSB. The result of the division is brought down, and the process repeated until the result of successive divisions has been reduced to 0.

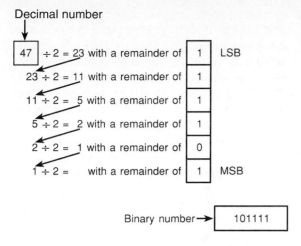

Fig. 12-20 Converting a decimal number to a binary number.

To express a number in the binary system requires many more digits than in the decimal system. Too many binary digits can become cumbersome to read or write. To solve this problem, other related numbering systems are brought into use.

The *octal numbering system*, a base 8 system, is often used in microprocessor, computer, and programmable controller systems because 8 data bits make up a byte of information which can be addressed by the PLC user or programmer. In some instances, programmable controller manufacturers use the octal system to number wiring terminals, programmable controller racks, and other PLC hardware. The octal number system makes use of 8 digits: 0 through 7. As in all other number systems, each digit in an octal number has a weighted decimal value according to its position. Figure 12-21 illustrates how the octal number 462 is converted to its decimal equivalent: 306.

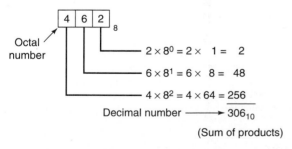

Fig. 12-21 Converting an octal number to a decimal number.

As mentioned, octal is used as a convenient means of handling large binary numbers. For example, the octal number 462 can be con-

verted to its binary equivalent by assembling the 3-bit groups as illustrated in Fig. 12-22. Thus, octal 462 is binary 100110010 and decimal 306. Notice the simplicity of the notation. The octal 462 is much easier to read and write than its binary equivalent.

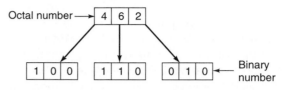

Fig. 12-22 Converting an octal number to a binary number.

The *hexadecimal (hex) number system* provides even shorter notation than octal. Hexadecimal uses a base of 16. It employs 16 digits: numbers 0 through 9, and letters A through F, with A through F being substituted for numbers 10 through 15, respectively. The techniques for converting hexadecimal to decimal and decimal to hexadecimal are the same as those used for binary and octal (Fig. 12-23).

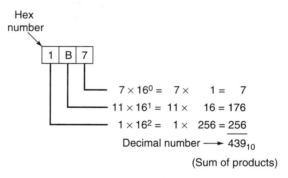

(a) Converting a hexadecimal number to a decimal number

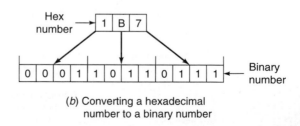

(b) Converting a hexadecimal number to a binary number

Fig. 12-23 Hexadecimal numbering system.

The *binary coded decimal (BCD) system* provides a convenient means to handle large numbers that need to be input to or output from a PLC. The BCD system represents decimal numbers as patterns of 1s and 0s. This system provides a means of converting a code readily handled by humans (decimal) to a code readily handled by the equipment (binary).

The BCD code employs 4 binary bits, with the weights of 1, 2, 4, and 8, to represent each numeral in the decimal system. This is referred to as the *8421 code,* since 8421 is the natural binary progression. The BCD representation of a decimal number is obtained by replacing each decimal digit by its BCD equivalent. To distinguish the BCD numbering system from a binary system, a BCD designation is placed to the right of the units digit. The BCD representation of the decimal 7863 is illustrated in Fig. 12-24.

Scientific calculators are available to convert numbers back and forth between decimal, binary, octal, and hexadecimal. They are inexpensive and easy to use, for example, in converting a number displayed in decimal to one in binary. This simply involves one key stroke to change the display mode from decimal to binary. In addition, many PLCs contain num-

ber conversion functions for converting numbers back and forth as illustrated in Fig. 12-25. As shown in Fig. 12-25(a), BCD-to-binary conversion is required for the input. Binary-to-BCD conversion is required for the output. In Fig. 12-25(b) the convert-to-decimal instruction will convert the binary bit pattern at the source address, N7:23, into a BCD bit pattern of the same decimal value as the destination address, 0:20. The instruction executes every time it is scanned and the instruction is true.

Self-Test

Answer the following questions.

20. PLCs work on _____ numbers in one form or another to represent various codes or quantities.
21. True or false? The decimal system has a base of 9.
22. The only allowable digits in the binary system are _____ and _____.

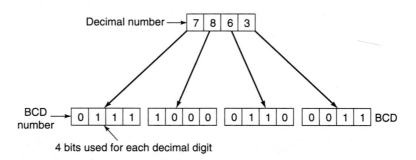

Fig. 12-24   The BCD representation of a decimal number.

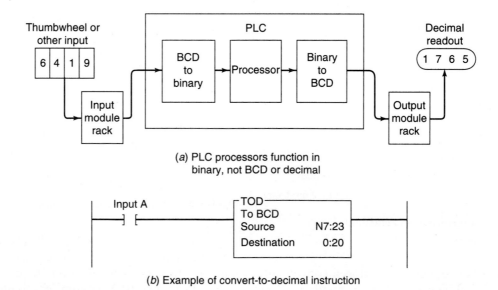

(a) PLC processors function in binary, not BCD or decimal

(b) Example of convert-to-decimal instruction

Fig. 12-25   PLC number conversion.

23. Each digit of a binary number is known as a(n) _____.
24. The _____ _____ bit in a word is the digit that represents the smallest value.
25. True or false? To express a number in binary requires fewer digits than in the decimal system.
26. True or false? The octal number system consists of digits 0, 1, 2, 3, 4, 5, 6, and 7. There are no 8s or 9s.
27. The octal number 153 expressed as a binary number would be _____.
28. The hexadecimal number system uses a base of _____.
29. Hexadecimal 2F equals _____ in decimal.
30. The decimal number 29 equals _____ in binary and _____ in BCD.

12-3 BASICS OF PLC PROGRAMMING

To program a programmable controller, it is necessary to have some knowledge of how its memory is organized. Figure 12-26 shows a typical PLC memory organization known as a *memory map*. The individual sections, their order, and the sections' length will vary and may be fixed or variable depending on the manufacturer and model.

The *user program* is where the programmed logic ladder diagram is entered and stored (Fig. 12-27). The user program will account for most

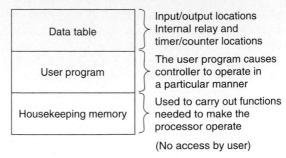

(a) General organization

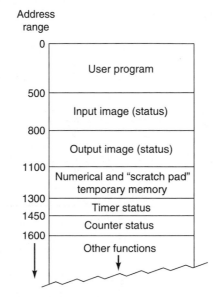

(b) Memory map shows how memory is organized

Fig. 12-26   Memory organization.

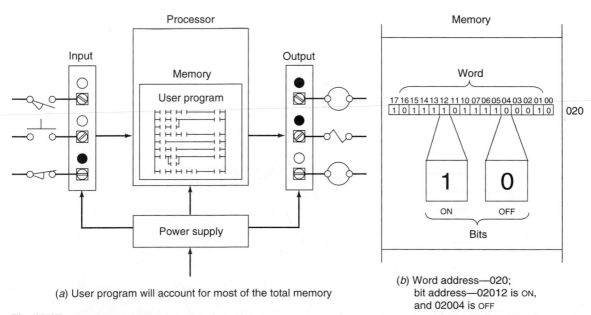

(a) User program will account for most of the total memory

(b) Word address—020; bit address—02012 is ON, and 02004 is OFF

Fig. 12-27   User program.

of the total memory of a given PLC system. It contains the logic that controls the machine operation. This logic consists of *instructions* that are programmed in a ladder logic format. Most instructions require one *word* of memory.

The *data table* stores the information needed to carry out the user program. This includes such information as the status of input and output devices (Fig. 12-28), timer and counter values, data storage, and so on. Contents of the data table can be divided into two categories: *status data* and *numbers* or *codes*. Status is ON/OFF type of information represented by 1s and 0s, stored in unique bit locations.

Number or code information is represented by groups of bits stored in unique *register* or *word* locations. The *address number* assigned to an instruction associates it with a particular status bit. This bit will be either ON (logic 1) or OFF (logic 0), indicating whether the instruction is TRUE or FALSE.

The data table can be divided into the following three sections according to the type of information to be remembered: input image table, output image table, and timer and counter storage. The *input image table* stores the status of digital inputs, which are connected to input interface circuits. Figure 12-29

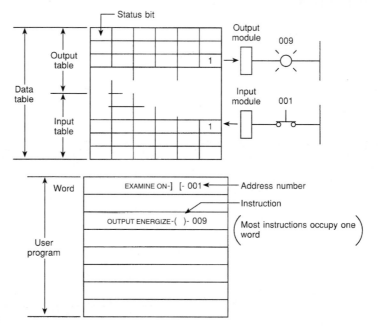

Fig. 12-28 Data table.

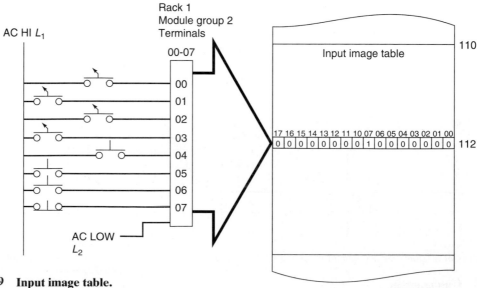

Fig. 12-29 Input image table.

shows typical connections to the input image table through the input module. When the switch is closed, the processor detects a voltage at the input terminal and records that information by storing a binary 1 in the proper bit location. Each connected input has a bit in the input image table that corresponds exactly to the terminal to which the input is connected. The input image table is constantly being changed to reflect the current status of the switch. If the input is ON (switch closed), its corresponding bit in the table is set to 1. If the input is OFF (switch open), the corresponding bit is "cleared," or reset to 0.

The *output image table* is an array of bits that controls the status of digital output devices, which are connected to output interface circuits. Figure 12-30 shows a typical connection of lights to the output image table through the output module. The status of the lights (ON/OFF) is controlled by the user program and indicated by the presence of 1s (ON) and 0s (OFF). Each connected output has a bit in the output image table that corresponds exactly to the terminal to which the output is connected. If the program calls for a specific output to be ON, its corresponding bit in the table is set to 1. If the program calls for the output to be OFF, its corresponding bit in the table is set to 0.

During each operating cycle, the processor reads all the inputs, takes these values, and according to the user program energizes or de-energizes the outputs. This process is known as a *scan*.

The scan is normally a continuous and sequential process of reading the status of inputs, evaluating the control logic, and updating the outputs. Figure 12-31 on page 346 illustrates this process. A scan cycle is the time required for a PLC to scan its inputs and generate appropriate control responses at its outputs. Scan time varies with program content and length. A scan can take from about 1 to 20 ms. If a controller has to react to an input signal that changes states twice during the scan time, it is possible that the PLC will never be able to detect this change. The scan time of a PLC should be known to ensure that scanning is faster than any field device operation.

The term *PLC programming language* refers to the method by which the user communicates information to the PLC. *Relay ladder logic* was the first and most popular language available on the PLC, and it is still popular. Most PLCs on the market today can be programmed either exclusively or partially in relay ladder logic.

Relay ladder logic is a graphical programming language designed to closely represent the appearance of a wired relay system. It offers considerable advantages for PLC control. Not only is it reasonably intuitive, especially for technicians with relay experience, it is particularly effective in an on-line mode when the PLC is actually performing control. Operation of the logic is apparent from the highlighting of the various relay contacts and coils on screen, identifying the logic state in real time (Fig. 12-32 on page 346).

The ladder diagram language is basically a *symbolic* set of instructions used to create the controller program. These ladder instruction symbols are arranged to obtain the desired

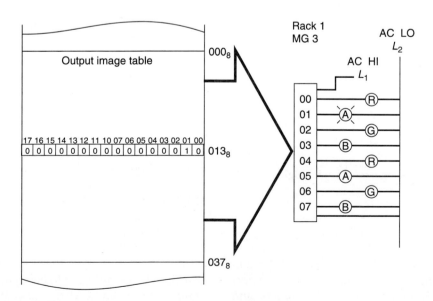

Fig. 12-30   Output image table.

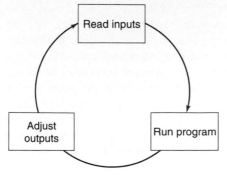

(a) Typical scan cycle

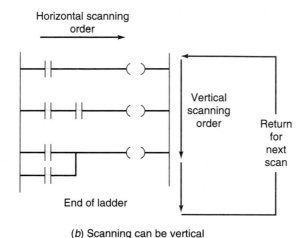

(b) Scanning can be vertical or horizontal

Fig. 12-31   Scan sequence.

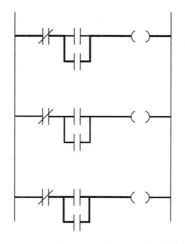

Fig. 12-32   Monitoring a relay ladder logic diagram.

control logic that is to be entered into the memory of the PLC. Because the instruction set is composed of contact symbols, ladder diagram language is also referred to as *contact symbology*. Representations of contacts and coils are the basic symbols of the logic ladder

diagram instruction set (Fig. 12-33). The following three are the basic symbols used to translate relay control logic to contact symbolic logic:

Instruction	Symbol	Mnemonic
EXAMINE ON	┤ ├	XIC
EXAMINE OFF	┤/├	XIO
OUTPUT ENERGIZE	-( )-	OTE

The main function of the logic ladder diagram program is to control outputs based on input conditions. This control is accomplished through the use of what is referred to as a *ladder rung*. In general, a rung consists of a set of input conditions, represented by contact instructions, and an output instruction at the end of the rung represented by the coil symbol (Fig. 12-34 on page 349). Each contact or coil symbol is referenced with an address number that identifies what is being evaluated and what is being controlled. The same contact instruction can be used throughout the program whenever that condition needs to be evaluated. For an output to be activated or energized, at least *one* left-to-right path of contacts must be closed. A complete closed path is referred to as having *logic continuity*. When logic continuity exists in at least one path, the rung condition is said to be TRUE. The rung condition is FALSE if no path has continuity.

To complete the entry of a relay-type instruction, you must assign an *address number* to it. This number will indicate what PLC input is connected to what input device and what PLC output will drive what output device. The assignment of I/O address is sometimes included in the I/O connection diagram as shown in Fig. 12-35 on page 349. Inputs and outputs are typically represented by squares and diamonds, respectively.

Branch instructions are used to create parallel paths of input condition instructions. This allows more than one combination of input conditions (OR logic) to establish logic continuity in a rung. Figure 12-36 on page 349 illustrates a simple branching condition. The rung will be TRUE if either instruction 110/00 or 110/01 is TRUE. A branch *START instruction* is used to begin each parallel logic branch. A single branch *CLOSE instruction* is used to close the parallel branch.

In *some* PLC models the programming of a branch circuit within a branch circuit or a *nested branch* cannot be done directly. It is possible, however, to program a logically equivalent

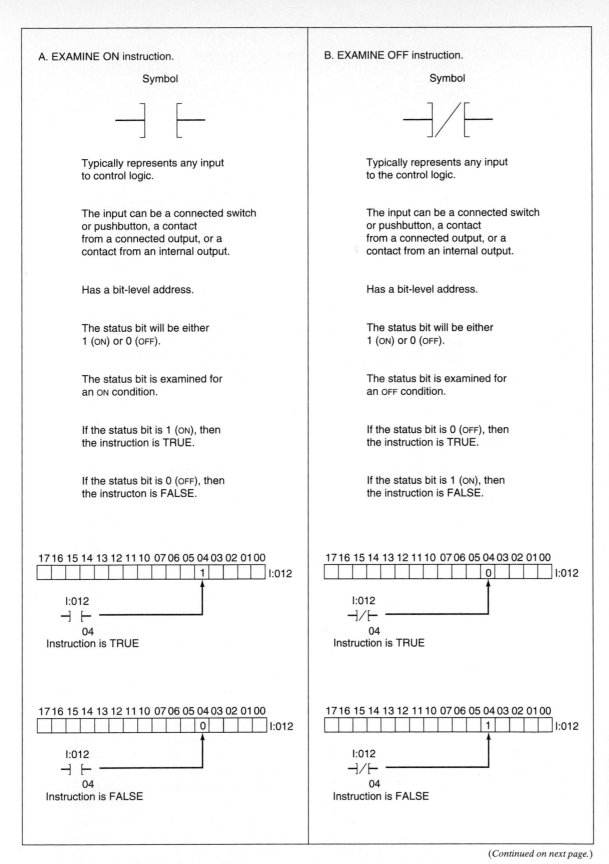

A. EXAMINE ON instruction.

Symbol

Typically represents any input to control logic.

The input can be a connected switch or pushbutton, a contact from a connected output, or a contact from an internal output.

Has a bit-level address.

The status bit will be either 1 (ON) or 0 (OFF).

The status bit is examined for an ON condition.

If the status bit is 1 (ON), then the instruction is TRUE.

If the status bit is 0 (OFF), then the instructon is FALSE.

17 16 15 14 13 12 11 10 07 06 05 04 03 02 01 00
| | | | | | | | | | | |1| | | | | I:012

I:012
—] [—
04
Instruction is TRUE

17 16 15 14 13 12 11 10 07 06 05 04 03 02 01 00
| | | | | | | | | | | |0| | | | | I:012

I:012
—] [—
04
Instruction is FALSE

B. EXAMINE OFF instruction.

Symbol

Typically represents any input to the control logic.

The input can be a connected switch or pushbutton, a contact from a connected output, or a contact from an internal output.

Has a bit-level address.

The status bit will be either 1 (ON) or 0 (OFF).

The status bit is examined for an OFF condition.

If the status bit is 0 (OFF), then the instruction is TRUE.

If the status bit is 1 (ON), then the instruction is FALSE.

17 16 15 14 13 12 11 10 07 06 05 04 03 02 01 00
| | | | | | | | | | | |0| | | | | I:012

I:012
—]/[—
04
Instruction is TRUE

17 16 15 14 13 12 11 10 07 06 05 04 03 02 01 00
| | | | | | | | | | | |1| | | | | I:012

I:012
—]/[—
04
Instruction is FALSE

(*Continued on next page.*)

Fig. 12-33   Basic set of instructions that perform functions similar to relay functions.

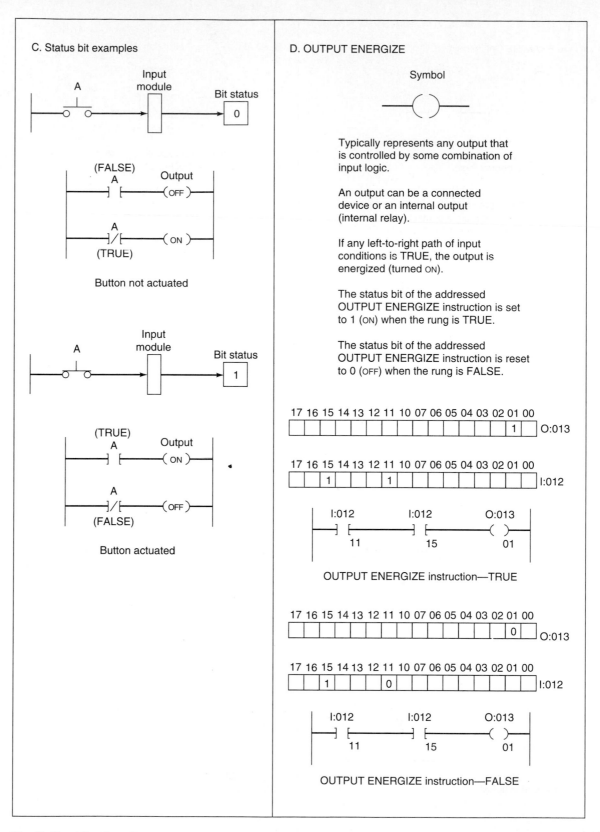

C. Status bit examples

A — Input module — Bit status 0

(FALSE)
A — Output
─┤ ├─ ─(OFF)─

A
─┤/├─ ─(ON)─
(TRUE)

Button not actuated

A — Input module — Bit status 1

(TRUE)
A — Output
─┤ ├─ ─(ON)─

A
─┤/├─ ─(OFF)─
(FALSE)

Button actuated

D. OUTPUT ENERGIZE

Symbol

─( )─

Typically represents any output that is controlled by some combination of input logic.

An output can be a connected device or an internal output (internal relay).

If any left-to-right path of input conditions is TRUE, the output is energized (turned ON).

The status bit of the addressed OUTPUT ENERGIZE instruction is set to 1 (ON) when the rung is TRUE.

The status bit of the addressed OUTPUT ENERGIZE instruction is reset to 0 (OFF) when the rung is FALSE.

17 16 15 14 13 12 11 10 07 06 05 04 03 02 01 00
[1] O:013

17 16 15 14 13 12 11 10 07 06 05 04 03 02 01 00
[1   1] I:012

I:012   I:012   O:013
─┤ ├─   ─┤ ├─   ─( )─
  11   15   01

OUTPUT ENERGIZE instruction—TRUE

17 16 15 14 13 12 11 10 07 06 05 04 03 02 01 00
[0] O:013

17 16 15 14 13 12 11 10 07 06 05 04 03 02 01 00
[1   0] I:012

I:012   I:012   O:013
─┤ ├─   ─┤ ├─   ─( )─
  11   15   01

OUTPUT ENERGIZE instruction—FALSE

Fig. 12-33 (Continued).

branching condition. Figure 12-37(*a*) on page 350 shows an example of a circuit that contains a nested contact D. To obtain the required logic the circuit would be programmed as shown in Fig. 12-37(*b*). The duplication of contact C eliminates the nested contact D.

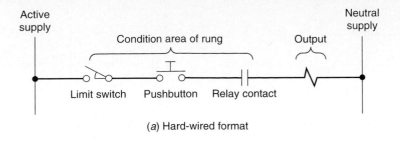

(a) Hard-wired format

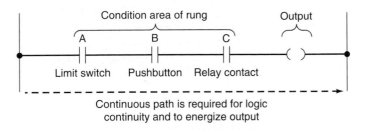

Continuous path is required for logic
continuity and to energize output

(b) PLC ladder diagram format

Fig. 12-34   Ladder rung.

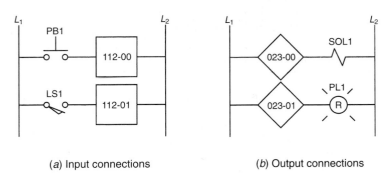

(a) Input connections   (b) Output connections

Fig. 12-35   I/O connection diagram.

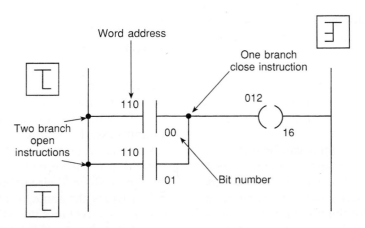

Fig. 12-36   Parallel path (branch) instructions.

For some PLC models there is a limit to the number of series contact instructions that can be included in one rung of a ladder diagram, as well as a limit to the number of parallel branches. Figure 12-38 on page 350 shows a typical matrix limitation diagram for a PLC.

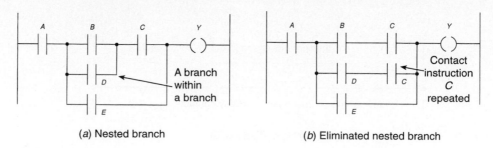

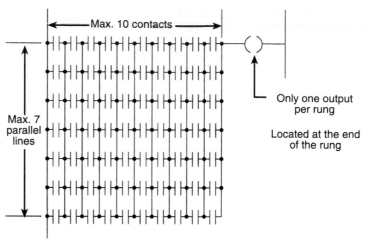

(a) Nested branch

(b) Eliminated nested branch

Fig. 12-37 Nested branch.

Fig. 12-38 Typical PLC matrix limitation diagram.

Programming more than the allowable series elements or parallel branches will result in an error message. Also, there is a further limitation with some PLCs: there can be only one output per rung, and the output must be located on the right-hand end of a rung.

Most PLCs have an area of the memory allocated for what are known as *internal storage bits*. These storage bits are also called *internal outputs, internal coils, internal control relays,* or just *internals*. The internal output operates just as any output that is controlled by programmed logic; however, the output is used strictly for internal purposes. In other words, the internal output does not directly control an output device.

An internal control relay can be used when a circuit requires more series contacts than the rung allows. Figure 12-39 shows a circuit that allows for only 7 series contacts when 12 are actually required for the programmed logic. To solve this problem, the contacts are split into two rungs as shown. The first rung contains 7 of the required contacts and is programmed to an internal relay. The address of the internal relay would also be the address of the first EXAMINE ON contact on the second rung.

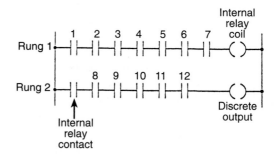

Fig. 12-39 Internal control relay.

The remaining 5 contacts are programmed, followed by the discrete output. The advantage of an internal storage bit is that it does not unnecessarily occupy a physical output.

A simple program using the EXAMINE ON instruction is shown in Fig. 12-40. This figure shows a hard-wired circuit and a user program that provides the same results. You will note that *both the NO and the NC* pushbuttons are represented by the EXAMINE ON symbol. This is because the normal state of an input (NO or NC) does not matter to the controller. What does matter is that if contacts need to *close* to energize the output, then the EXAMINE ON instruction is used. Since both PB1

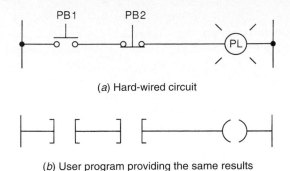

(a) Hard-wired circuit

┤ ├ ┤ ├ ┤ ├ ─( )─

(b) User program providing the same results

Fig. 12-40 Simple program using the EXAMINE ON instruction.

and PB2 must be closed to energize the pilot light, the EXAMINE ON instruction is used for both.

A simple program using the EXAMINE OFF instruction is shown in Fig. 12-41. Again, both the hard-wired circuit and user program are shown. In the hard-wired circuit, when the pushbutton is *open,* relay coil CR is de-energized and contacts CR1 close to switch the pilot light ON. When the pushbutton is *closed,* relay coil CR is energized, and contacts CR1 open to switch the pilot light OFF. The pushbutton is represented in the user program by an EXAMINE OFF instruction. This is because the rung must be TRUE when the external pushbutton is open, and FALSE when the pushbutton is closed. Using an EXAMINE OFF instruction to represent the pushbutton satisfies these requirements. The NO or NC mechanical action of the pushbutton is not a consideration. It is important to remember that the user program is not an electrical circuit but a *logic* circuit. In effect we are interested in logic continuity when establishing an output.

The START/STOP program shown in Fig. 12-42 can be used to start or stop a motor, or process. Note that the stop pushbutton input is programmed as an EXAMINE ON instruction. This is because the stop pushbutton is

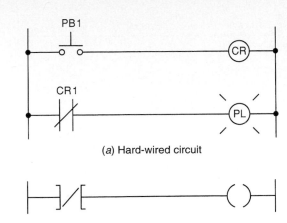

(a) Hard-wired circuit

┤ ├ ┤/├ ─( )─

(b) User program providing the same results

Fig. 12-41 Simple program using the EXAMINE OFF instruction.

wired normally closed. Since the start pushbutton (normally open) is a momentary device (allows continuity only when closed), a contact from the output coil is used to seal in the circuit. Often, the seal-in contact is an input from the motor starter auxiliary contacts.

▐▶ **YOU MAY RECALL** that electromagnetic latching relays are used where it is necessary for contacts to stay open and/or closed even though the coil is energized only momentarily.

An electromagnetic latching relay function can be programmed on a PLC to work like its real-world counterpart. The use of the output LATCH and output UNLATCH coil instructions is illustrated in the ladder program of Fig. 12-43 on page 352. Both the latch (L) and the unlatch (U) coil have the *same* address (0:013/10). When the ON button is momentarily actuated, the latch rung becomes TRUE and the latch status bit (10) is set to 1, and so the output is switched ON. This status bit *will remain set to 1* when logical continuity of the latch rung is lost. When the unlatch rung

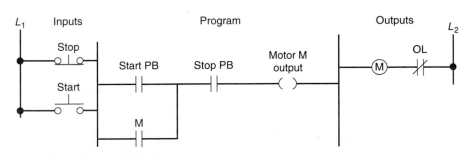

Fig. 12-42 START/STOP program.

LATCH/UNLATCH
instruction

CLEAR MEMORY
mode

PROGRAM mode

TEST mode

RUN mode

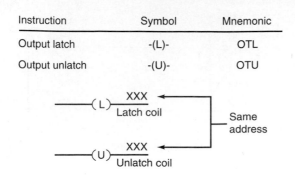

Instruction	Symbol	Mnemonic
Output latch	-(L)-	OTL
Output unlatch	-(U)-	OTU

(a) Latch and unlatch coils have same address

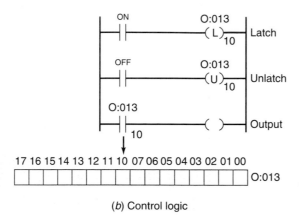

(b) Control logic

Fig. 12-43 Output latch and output unlatch instructions.

becomes TRUE (OFF button actuated), the status bit (10) is reset to 0 and so the output is switched OFF. The *LATCH/UNLATCH instruction* is *retentive* on power lost. That is to say, if the relay is latched, it will remain latched if power is lost and then restored.

The method of entering a program is through the operator terminal keyboard or handheld programming device. The ladder diagram is entered by pushing keys on the keyboard in a prescribed sequence. The results are displayed on either the CRT of a desktop pro-

grammer or with an LED or LCD for a handheld programmer. Because hardware and programming techniques vary with each manufacturer, it is necessary to refer to the programming manual for a specific PLC to determine how the instructions are entered.

The programming device can also be used to select the various processor modes of operation. Again, the number of different operating modes and the method of accessing them varies with the manufacturer. Regardless of PLC model, some common operating modes are *CLEAR MEMORY, PROGRAM, TEST,* and *RUN.*

Mode	Description
CLEAR MEMORY	Used to erase the contents of the on-board RAM memory
PROGRAM	Used to enter a new program or update an existing one in the internal RAM memory
TEST	Used to operate or monitor the user program without energizing any outputs
RUN	Used to execute the user program
	Input devices are monitored and output devices are energized accordingly

After you enter the rungs in your ladder program, you may want to document what the instruction or rung does for someone else who may be maintaining the program. The use of personal computers and programmable controller manufacturer software allows quality documentation to be produced (Fig. 12-44). Documentation may include rung, instruction, and address comments as well as rung cross references where repeatedly used addresses are cross-referenced to rung numbers.

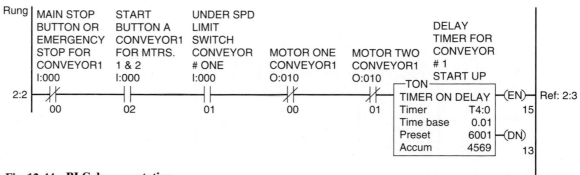

Fig. 12-44 PLC documentation.

A *programmable graphic interface* can be connected to communicate with a PLC to replace pushbuttons, pilot lights, thumbwheels, and other operator control panel devices (Fig. 12-45). These luminescent touchscreens provide an operator interface that operates like traditional hard-wired control panels. Features include:

- Pushbuttons and pilot lights are replaced by realistic-looking icons. The machine operator needs only to touch the display panel to activate the pushbuttons.
- Sequential operations can be shown in graphic format for easier viewing of the operation.
- The operator can change timer and counter presets by touching the numeric keypad graphic on the touchscreen.
- The screen can show alarms, complete with time of occurrence and location.
- Variables can be displayed as they change over time.

Self-Test

Answer the following questions.

31. The memory organization of a PLC is often called a memory _____.
32. Most of the total PLC memory is used for the _____ _____.
33. Contents of the data table can be divided into two categories: _____ and _____.
34. If a switch connected to an input module is closed, a binary _____ is stored in the proper bit location.

35. True or false? If the program calls for a specific output to be OFF, its corresponding bit in the data table is set to 1.
36. The _____ is normally a continuous and sequential process of reading the status of inputs, evaluating the control logic, and updating the outputs.
37. True or false? The most common PLC programming language is BASIC.
38. The ladder diagram language is basically a(n) _____ set of instructions used to create the controller program.
39. In general, a logic ladder rung consists of input conditions represented by _____ instructions and an output instruction represented by the _____ symbol.
40. True or false? The addressing format or inputs and outputs is standard for all PLC models.
41. True or false? For some PLC models, only one output is allowed per rung.
42. True or false? An internal relay cannot directly control an output device.
43. True or false? Both normally open and normally closed pushbuttons can be represented by the EXAMINE ON instruction.
44. True or false? In determining continuity, the rules that apply to an electrical circuit also apply to a logic circuit.
45. True or false? The latching relay instruction is retentive in that, if the relay is latched, it will unlatch if power is lost and then restored.
46. To execute the user program, the PLC must be placed in the _____ mode.
47. _____ _____ are used for documentation of PLC programs.

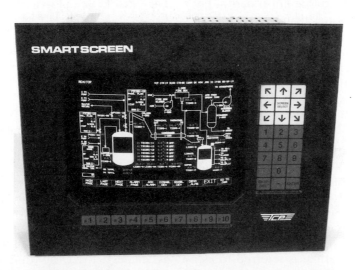

Fig. 12-45 Programmable graphic interface. *(Courtesy Total Control Products, Inc., Melrose Park, Illinois)*

48. True or false? Programmable graphic interfaces operate like traditional hard-wired control panels.

12-4 TIMER AND COUNTER INSTRUCTIONS

The advantage of PLC timers and counters is that their settings can be easily altered, or the number of them used in a circuit can be increased or decreased, by programming changes without wiring changes. Counter and timer addresses are usually specified by the programmable controller manufacturer and are located in a specific area of the data organization table. The number of timers and counters that can be programmed depends upon the model of PLC you are using. However, the availability usually far exceeds the requirement.

Timers and counters are output instructions. In general, there are three different timers: the *on-delay timer (TON)*, the *off-delay timer (TOF)*, and the *retentive timer on (RTO)*. There are two different counter instructions: the *count-up counter (CTU)* and the *count-down counter (CTD)*. A *reset instruction* is required to reset both retentive timer instructions and the counter instructions.

PLC *timers* provide the same functions as mechanical and electronic timing relays.

⫸ **YOU MAY RECALL** that timers are used to activate or de-activate a device after a preset interval of time.

There are two methods used to represent a timer instruction within a PLC's logic ladder program. The first depicts the timer instruction as a *relay coil* similar to that illustrated in Fig. 12-46. The timer is assigned an address as well as being identified as a timer. Also included as part of the timer instruction is the time base of the timer, the timer's preset value or time-delay period, and the accumulated value or current time-delay period for the timer. When the timer rung has logic continuity, the timer begins counting time-based intervals and times until the accumulated value equals the preset value. When the accumulated time equals the preset time, the output is energized and the timed output contact associated with the output is closed. The timed contact can be used throughout the program as a NO or NC contact as many times as you wish.

The second timer format is referred to as a *block format.* Figure 12-47 illustrates a generic

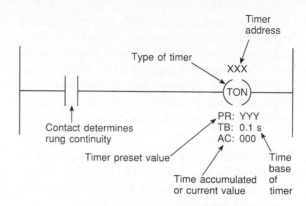

Fig. 12-46 Coil-formatted timer instruction.

block format. The timer block has two input conditions associated with it: the *control* and *reset*. The control line controls the actual timing operation of the timer. The reset line resets the timer's accumulated value to zero. All block-formatted timers provide at least one output signal from the timer. When a single output is provided, it is used to signal the completion of the timing cycle. For dual-output timer instructions, the second output signal operates in the reverse mode. Whenever the timer has *not* reached its timed-out state, the second output is ON and the first output remains OFF. As soon as the timer reaches its timed-out state, the second output is turned OFF and the first output is turned ON.

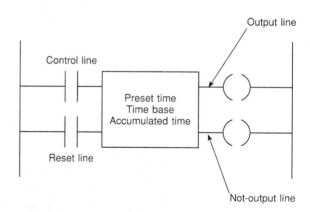

Fig. 12-47 Block-formatted timer instruction.

The on-delay timer (TON) is the most commonly used timer. The on-delay timer operates so that when the rung containing the timer is true, the timer time-out period commences. At the end of the timer time-out period, an output is made active. The timed output becomes active sometime after the timer becomes active; hence the timer is said to have an ON delay. Figure 12-48 shows a ladder diagram containing a typical on-delay timer. The timer instruction consists of three data table words: the control word, the preset word, and the accumulated word.

- *Control word*
  - The control word uses three bits: the enable bit (EN), the timer-timing bit (TT), and the done bit (DN).
  - The *enable bit* is true (has status of 1) whenever the timer instruction is true. When the timer instruction is false, the enable bit is false (has status of 0).
  - The *timer-timing bit* is true whenever the accumulated value of the timer is changing, which means the timer is timing. When the timer is not timing, the accumulated value is not changing, so the timer-timing bit is false.

- The *done bit* changes state whenever the accumulated value reaches the preset value.
- *Preset word*
  - The preset value is the set point of the timer, that is, the value up to which the timer will time.
- *Accumulated word*
  - The accumulated value is the value that increments as the timer is timing. The accumulated value will stop incrementing when its value reaches the preset value.

This timer instruction requires you to enter a *time base,* which is either 1.0 or 0.01 for long or short time delays. The actual preset time interval is the time base multiplied by the value stored in the timer's preset word. The actual accumulated time interval is the time base multiplied by the value stored in the timer's accumulated word.

Control word

Enable bit

Timer-timing bit

Done bit

Preset word

Accumulated word

Time base

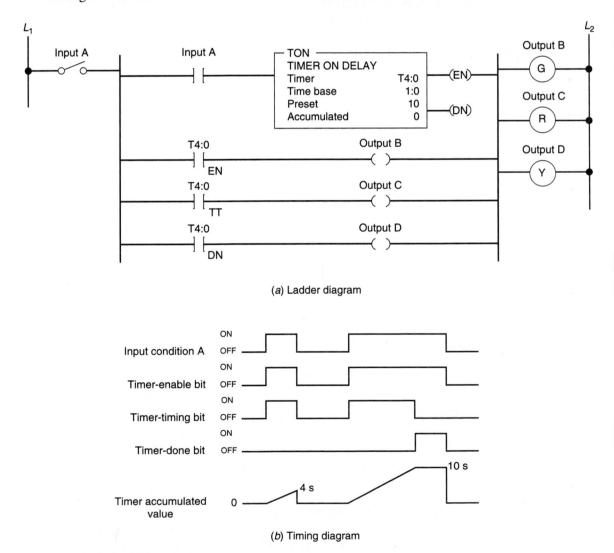

(a) Ladder diagram

(b) Timing diagram

Fig. 12-48 On-delay timer.

The timer is activated by closing the switch. The preset time for this timer is 10 s, at which time output D will be energized. When the switch is closed, the timer begins counting, and counts until the accumulated time equals the preset value; the output is then energized. If the switch is opened before the timer is timed out, the accumulated time is automatically reset to zero. This timer configuration is termed *nonretentive* since loss of power flow to the timer causes the timer instruction to reset. This timing operation is that of an ON-DELAY timer, since output D is switched on 10 s after the switch has been actuated from the OFF to the ON position.

In Fig. 12-48(*b*), the timing diagram first shows the timer timing to 4 s and then going FALSE. The timer resets, and both the timer-timing bit and the enable bit go FALSE. The accumulated value also resets to 0. Input A then goes TRUE again and remains TRUE in excess of 10 s. When the accumulated value reaches 10 s, the done bit (DN) goes from FALSE to TRUE and the timer-timing bit (TT) goes from TRUE to FALSE. When input A goes FALSE, the timer instruction goes FALSE and also resets, at which time the control bits are all reset and the accumulated value resets to 0.

The off-delay timer (TOF) operation will keep the output energized for a timed period after the rung containing the timer has gone false. Figure 12-49 illustrates the programming of an off-delay timer using the coil-formatted timer instruction. If logic continuity is *lost*, the timer begins counting time-based intervals until the accumulated time equals the programmed preset value. When the switch connected to input 001 is first closed, timed output 009 is set to 1 immediately and the lamp is switched ON. If this switch is now opened, logic continuity is lost and the timer begins counting. After 15 s, when the accumulated time equals the preset time, the output is reset to 0 and the lamp switches OFF. If logic continuity is gained before the timer is timed out, the accumulated time is reset to zero. That is why this timer is classified as nonretentive.

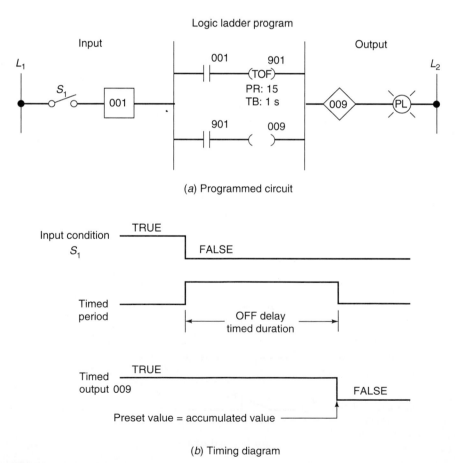

(*a*) Programmed circuit

(*b*) Timing diagram

Fig. 12-49 Off-delay programmed timer.

The PLC-programmed retentive on-delay timer (RTO) operates in the same way as the nonretentive on-delay timer (TON) with one major exception. This is a *retentive timer reset* (*RTR*) instruction. Unlike the TON, the RTO will hold its accumulated value when the timer rung goes FALSE and will continue timing where it left off when the timer rung goes TRUE again. This timer must be accompanied by a timer RESET instruction to reset the accumulated value of the timer to zero. The RTR instruction is the *only* automatic means of resetting the accumulated value of a retentive timer. The RTR instruction must be addressed to the same word as the RTO instruction. Normally, if any RTR rung path has logic continuity, then the accumulated value of the referenced timer is reset to zero.

The program in Fig. 12-50 shows a practical application for an RTO. The RTO timer detects whenever a piping system has sustained a *cumulative* overpressure condition of 60 s. At that point, a horn is automatically sounded to call attention to the malfunction. When alerted, maintenance personnel can silence the alarm by switching the key switch S_1 to the RESET (contact closed) position. After the problem has been corrected, the alarm system can be reactivated by switching the key switch to the ON (contact open) position.

The *counter instructions* allow for the counting of events and the controlling of other events based on the accumulated count. Counters operate on transition rather than length of time. Programmed counters can count up, count down, or be combined to count up and down and require a reset instruction.

PLC counters have programming formats similar to timer formats. A coil programming format, similar to that shown in Fig. 12-51 on page 358, is used by some manufacturers. The counter is assigned an address as well as being identified as a counter. Also included as part of the counter instruction are the counter's *preset value* and the current *accumulated count* for the counter. The up-counter increments its accumulated value by 1 each time the counter run makes a FALSE-to-TRUE transition. When the accumulated count equals the preset count, the output is energized, and the counter output is closed. The counter contact can be used throughout the program as an NO or NC contact as many times as you wish.

A *counter reset instruction,* which permits the counter to be reset, is also used in conjunction with the counter instruction. Up-counters are always reset to zero. Down-counters may be reset to zero or to some preset value. Some manufacturers include the reset function as a part of the general counter instruction, while others dedicate a separate instruction for resetting of the counter. Figure 12-52 on page 358 shows a generic coil-formatted counter instruction with a separate instruction for resetting the counter. When programmed, the counter reset coil (CTR) is given the *same* reference address as the counter (CTU) that it is to reset. The reset instruction is activated whenever the CTR rung condition is TRUE.

Figure 12-53 on page 358 shows a generic block-formatted counter instruction. The instruction block indicates the type of counter (up or down) along with the counter's preset value and accumulated or current value. The counter has two input conditions associated with it: COUNT and RESET. All PLC counters operate, or count, on the leading edge of the input signal. The counter will either increment or decrement whenever the count input

Retentive timer reset (RTR) instruction

Counter instructions

Preset value

Accumulated count

Counter reset instruction

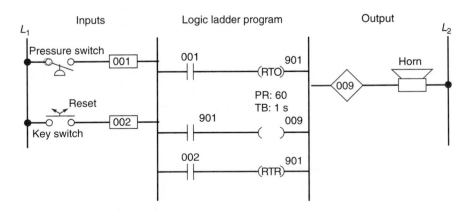

Fig. 12-50 Retentive on-delay alarm program.

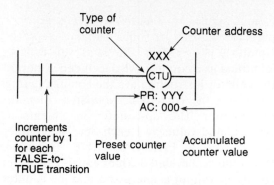

Fig. 12-51   Coil-formatted counter instruction.

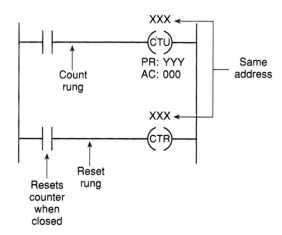

Fig. 12-52   Coil-formatted counter and RESET instructions.

transfers from an OFF state to an ON state. The counter *will not* operate on the trailing edge, or ON-to-OFF transition, of the input condition.

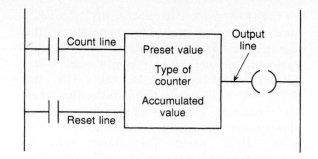

Fig. 12-53   Block-formatted counter instruction.

Most PLC counters are normally retentive. That is to say, whatever count was contained in the counter at the time of a processor shutdown will be restored to the counter on power-up. The counter may be reset, however, if the reset condition is activated at the time of power restoration.

The *up-counter output instruction* will increment by 1 each time the counted event occurs. Figure 12-54 shows the program and timing diagram for a simple up-counter. This control application is designed to turn the green pilot light ON and the red pilot light OFF after an accumulated count of 7. Operating pushbutton PB1 provides the OFF-to-ON transition pulses that are counted by the counter. The preset value of the counter is set for 007. Each FALSE-to-TRUE transition of rung 1 increases the counter's accumulated value by 1. After 7 pulses, or counts, when the preset counter value equals the accumulated counter value, output 901 is energized. As a result,

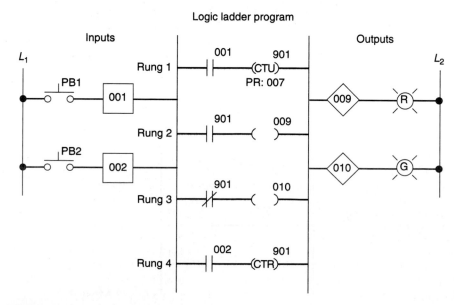

Fig. 12-54   Simple up-counter program.

rung 2 becomes TRUE, energizing output 009 to switch the red pilot light ON. At the same time, rung 3 becomes FALSE, de-energizing output 010 to switch the green pilot light OFF. The counter is reset by closing pushbutton PB2, which makes rung 4 TRUE and resets the accumulated count to zero. Counting can resume when rung 4 goes FALSE again.

Figure 12-55 shows the program for a *one-shot,* or *transitional, contact circuit* often used to automatically clear or reset a counter. The program generates an output pulse that, when triggered, goes OFF. The one-shot can be triggered from a momentary signal or from one that comes ON and stays ON for some time. Whichever is used, the one-shot is triggered by the leading edge (OFF-to-ON) transition of the input signal. It stays ON for one scan and then goes OFF. It stays OFF until the trigger goes OFF, and then comes ON again. The one-shot is perfect for resetting both counters and timers since it stays ON for one scan only.

The *down-counter output instruction* will count down or decrement by 1 each time the counted event occurs. Each time the down-count event occurs, the accumulated value is decremented. Normally a down-counter is used in conjunction with an up-counter to form an up/down-counter. Figure 12-56 on page 360 shows the program for a block-formatted up/down-counter program. Note that the same address is given the *up-counter* instruction, the *down-counter* instruction, and the *reset* instruction. All three instructions will be looking at the *same address* in the counter file. When input A goes from FALSE to TRUE, one count is added to the accumulated value. When input B goes from FALSE to TRUE, one count is subtracted from the accumulated value. To summarize:

- When the CTU instruction is TRUE, C5:2/CU will be TRUE, causing output A to be TRUE.
- When the CTD instruction is TRUE, C5:2/CD will be TRUE, causing output B to be TRUE.
- When the accumulated value is greater than or equal to the preset value, C5:2/DN will be TRUE, causing output C to be TRUE.
- Input C's going TRUE will cause both counters to reset. When reset, by the *reset* instruction, the accumulated value will be reset to 0 and the done bit will be reset.

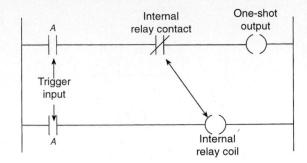

Fig. 12-55 One-shot, or transitional, contact program.

Self-Test

Answer the following questions.

49. Timer and counter instructions may be _____ formatted or _____ formatted.

50. True or false? Only one timer and one counter instruction can be programmed on a PLC.

51. The timer output instruction is energized when the _____ time equals the _____ time.

52. True or false? Loss of power or logic continuity to a nonretentive on-delay will automatically reset the accumulated time to zero.

53. If the preset time of a timer is 100 and the timer base is 0.1 second, the time-delay period is _____ seconds.

54. An off-delay timer will keep the output energized for a timed period after the rung containing the timer has gone _____.

55. A(n) _____ timer must be intentionally reset with a separate signal.

56. True or false? The retentive timer reset instruction is given the same address as the timer it resets.

57. True or false? Normally, the reset input to a timer will override the control input of the timer.

58. Counters operate on _____ rather than length of time.

59. True or false? The output of the counter is energized whenever the accumulated count is less than or equal to the preset count.

60. The up-counter increments its accumulated value by one each time the counter rung makes a(n) _____ to _____ transition.

61. True or false? Most PLC counters are normally nonretentive.

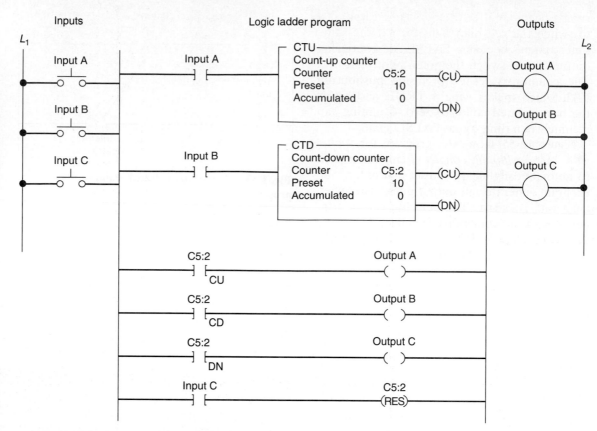

Fig. 12-56   Up-down counter program.

62. True or false? A transitional OFF-to-ON contact will allow logic continuity for one scan and then open, even though the triggering signal may stay ON.

63. True or false? A down-counter output instruction will decrement by one each time the counted event occurs.

64. True or false? In normal use, the down-counter is used in conjunction with the up-counter to form an up/down-counter.

12-5   DATA MANIPULATION INSTRUCTIONS

Data manipulation instructions enable the programmable controller to take on some of the qualities of a computer system. Most PLCs now have this ability to manipulate data that is stored in memory. It is this extra computer characteristic that gives the PLC capabilities that go far beyond the conventional relay equivalent instructions.

Data manipulation involves the transfer of data, operating on data with math functions, data conversions, data comparison, and logical

operations on data. There are two basic classes of instructions to accomplish this: instructions that operate on word data, and those that operate on file data, which is multiple words. Figure 12-57 illustrates the difference between word data and file data.

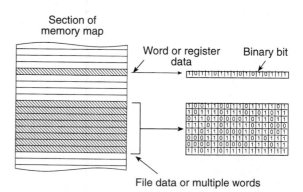

Fig. 12-57   Word data and file data.

Data transfer instructions simply involve the transfer of the contents from one word or register to another. Figure 12-58 illustrates word-level move instructions. Data transfer instructions can address almost any location in the

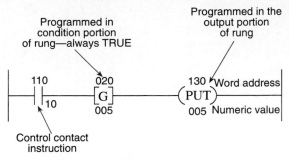

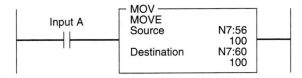

(b) Allen-Bradley PLC-5 protocol

Fig. 12-58 **Word-level move instruction.**

memory. Prestored values can be automatically retrieved and placed in any new location. That location may be the preset register for a timer or counter or even an output register that controls a 7-segment display.

In Fig. 12-58(a), the Allen-Bradley PLC-2 protocol, the *PUT instruction* is used with the *GET instruction* to form a data transfer rung. When input 110/10 is TRUE, the GET/PUT

instructions tell the processor to get the numeric value 005 stored in word 020 and put it into word 130. In every case the PUT instruction must be preceded by a GET instruction. In Fig. 12-58(b), the Allen-Bradley PLC-5 protocol, the block-formatted move (MOV) instruction is used to copy data from a source word to a destination word. In this example the value stored at the address indicated in the source, N7:56, is being copied into the address indicated in the destination, N7:60. This value will be copied every time the instruction is scanned and the instruction is TRUE. When the rung goes FALSE, the destination address will retain the value, unless it is changed elsewhere in the program. The instruction may be programmed with input conditions preceding it, or it may be programmed unconditionally.

Data-compare instructions compare the data stored in two or more words (or registers) and make decisions based on the program instructions (Figs. 12-59 to 12-63). Numeric values in two words of memory can be compared for each of the following conditions depending on the PLC:

Name	Mnemonic	Symbol
LESS THAN	LES	$(<)$
EQUAL TO	EQU	$(=)$
GREATER THAN	GRT	$(>)$
LESS THAN OR EQUAL TO	LEQ	$(\leq)$
GREATER THAN OR EQUAL TO	GEQ	$(\geq)$

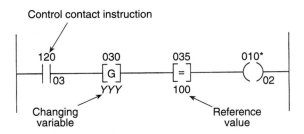

(a) Allen-Bradley PLC-2 protocol: Output 010-02 will be energized when input 120-03 is TRUE and the value in word 030 is equal to the reference value 100 in word 035.

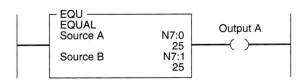

(b) Allen-Bradley PLC-5 protocol: When the value in source A equals the value in source B, the instruction is TRUE, causing output A to go TRUE. Source A or source B could also be a constant stored in the instruction.

Fig. 12-59 **EQUAL instruction.**

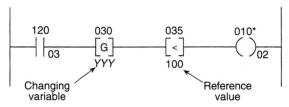

(a) Allen-Bradley PLC-2 protocol: Output 010-02 will be energized when input 120-03 is TRUE and the value in word 030 is less than the reference value 100 in word 035.

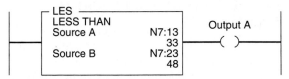

(b) Allen-Bradley PLC-5 protocol: When the value in source A is less than the value stored in source B, the instruction is TRUE, and output A will be TRUE; otherwise, output A will be FALSE.

Fig. 12-60 **LESS THAN instruction.**

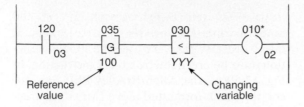

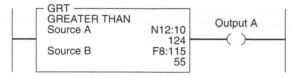

(a) Allen-Bradley PLC-2 protocol: Output 010-02 will be energized when input 120-03 is TRUE and the value in word 030 is greater than the reference value 100 in word 035.

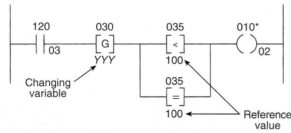

(b) Allen-Bradley PLC-5 protocol: When the value at source A is greater than the value at source B, the instruction is TRUE and output A will be TRUE; otherwise, it is FALSE.

Fig. 12-61 GREATER THAN instruction.

(a) Allen-Bradley PLC-2 protocol: Output 010-02 will be energized when input 120-03 is TRUE and the value in word 030 is less than or equal to the reference value 100 in word 035.

(b) Allen-Bradley PLC-5 protocol: When the value at source A is less than or equal to the value at source B, the instruction is TRUE and output A will be TRUE; otherwise, it is FALSE.

Fig. 12-62 LESS THAN OR EQUAL TO instruction.

Data comparison concepts have already been used with the timer and counter instructions. In both of these instructions an output was turned ON or OFF when the accumulated value of the timer or counter equaled its preset value (AC = PR). What actually occurred was that the accumulated numeric data in one

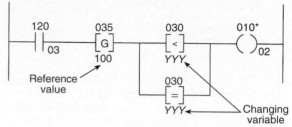

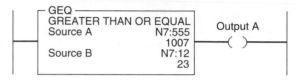

(a) Allen-Bradley PLC-2 protocol: Output 010-02 will be energized when input 120-03 is TRUE and the value in word 30 is greater than or equal to the reference value 100 in word 035.

(b) Allen-Bradley PLC-5 protocol: When the value at source A is greater than or equal to the value at source B, the instruction is TRUE and output A will be TRUE; otherwise, output A will be FALSE.

Fig. 12-63 GREATER THAN OR EQUAL TO instruction.

memory word was *compared* to the reset value of another memory word on each scan of the processor. When the processor saw that the accumulated value was equal to (=) the preset value, it switched the output ON or OFF.

The *limit test instruction* (Fig. 12-64) compares a test value to values in the low limit and in the high limit. It can function in either of two ways:

- If the high limit has a greater value than the low limit, then the instruction is true if the value of the test is between or equal to the values of the high limit and the low limit.
- If the value of the low limit is greater than the value of the high limit, the instruction is true if the value of the test is equal to or less than the low limit or equal to or greater than the high limit.

In Fig. 12-64(a), the high limit has a value of 50, and the low limit a value of 25. The instruction is TRUE, then, for values of the test from 25 through 50. The instruction as shown is TRUE because the value of the test is 48. In Fig. 12-64(b), the high limit has a value of 50, and the low limit a value of 100. The instruction is TRUE, then, for test values of 50 and less than 50, and 100 and greater than 100. The instruction as shown is TRUE because the test value is 125.

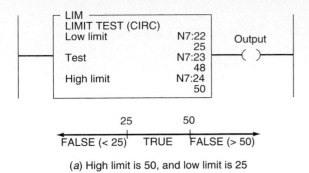

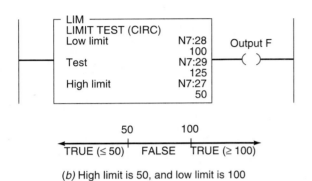

(a) High limit is 50, and low limit is 25

(b) High limit is 50, and low limit is 100

Fig. 12-64 Limit test instruction.

The use of comparison instructions is generally straightforward. However, one common programming error involves the use of these instructions in a PLC program to control the flow of a raw material into a vessel. The receiving vessel has its weight continuously monitored by the PLC program as it fills. When the weight reaches a preset value, the flow is cut off. While the vessel fills, the PLC performs a comparison between the vessel's current weight and a predetermined constant programmed in the processor. If the programmer uses only the *EQUAL TO instruction,* problems may result. As the vessel fills, the comparison for equality will be FALSE. At the instant the vessel weight reaches the desired preset value of the *EQUAL TO* instruction, the instruction becomes TRUE and the flow is stopped. However, should the supply system leak additional material into the vessel, the total weight of the material could rise *above* the preset value, causing the instruction to go FALSE and the vessel to overfill. The simplest solution to this problem is to program the comparison instruction as *GREATER THAN OR EQUAL TO.* This way any excess material entering the vessel will not affect the filling operation. It may be necessary, in some cases, to include additional programming to indicate a serious overfill condition.

Set point control in its simplest form compares an input value, such as an analog or thumbwheel inputs, to a set point value. A discrete output signal is provided if the input value is less than, equal to, or greater than the set point value.

The temperature control program of Fig. 12-65 on page 364 is one example of set point control. Here, a PLC is to provide for simple OFF/ON control of the electric heating elements of an oven. The oven is to maintain an average set point temperature of 600°F, with a variation of about 1 percent between the OFF and ON cycles. Therefore, the electric heaters will be turned ON when the temperature of the oven is 597°F or less and stay ON until the temperature rises to 603°F or more. The electric heaters stay OFF until the temperature drops down to 597°F, at which time the cycle repeats itself. Rung 2 contains a *GET/LESS OR EQUAL TO logic instruction* and rung 3 contains the equivalent of a *GET/GREATER THAN OR EQUAL TO logic instruction.* Rung 4 contains the logic for switching the heaters ON and OFF according to the high and low set points. Rung 1 contains the logic that allows the thermocouple temperature to be monitored by the LED display board.

In complex industrial machine or process control, it is necessary to carry out mathematical functions ranging from basic arithmetic, such as add, subtract, divide, and multiply, to complex mathematics used in flow measurement and PID control. *Math instructions* allow the PLC to perform arithmetic functions on values stored in memory words (Figs. 12-66 to 12-69; see pages 364 and 365).

Figure 12-70 on page 365 shows a program used to convert Celsius temperature to Fahrenheit. In this application, the thumbwheel switch connected to the input module indicates Celsius temperature. The program converts the Celsius temperature in the data table to Fahrenheit values for display. The formula

$$°F = (\frac{9}{5} \cdot °C) + 32$$

forms the basis for the program. In this example, a temperature reading of 60°C is assumed. In rung 1, the GET instruction at address 030 multiplies the temperature (60°C) by 9 and stores the product (540) in address 032. In rung 2, the GET instruction at address 033 divides 5 into 540 and stores the quotient 108 in address 034. In rung 3, the GET instruction at address 035 adds 32 to the value of 108 and

EQUAL TO
instruction

GREATER THAN OR
EQUAL TO
instruction

Set point control

GET/LESS OR
EQUAL TO logic
instruction

GET/GREATER THAN
OR EQUAL TO logic
instruction

Math instructions

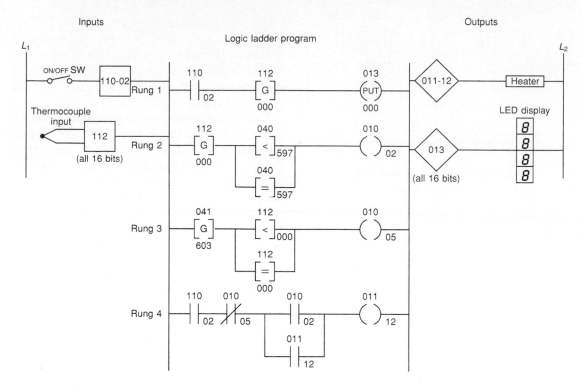

Fig. 12-65 Set point temperature control program.

(a) Allen-Bradley PLC-2 protocol: When input device 111/05 is TRUE, the value of word 030 (105) is added to the value of word 031 (080) and the sum (185) is stored in word 032.

(a) Allen-Bradley PLC-2 protocol: SUBTRACT instruction rung. When input device 111/05 is TRUE, the value of word 031 (080) is subtracted from the value of word 030 (105), and the difference (025) is stored in word 032. Only positive values can be used with some PLCs.

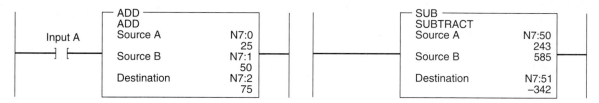

(b) Allen-Bradley PLC-5 protocol: When input device A is TRUE, the value stored at address N7:0 (25) is added to the value stored at address N7:1 (50) and the sum is stored at address location N7:2 (75).

(b) Allen-Bradley PLC-5 protocol: Allows negative values to be used. Here the instruction is programmed unconditionally, i.e., the subtract operation will take place every time the instruction is scanned. The value stored at the address indicated in source B is subtracted from the value stored at the address indicated in source A, and the answer is stored at the address location indicated in the destination. Also, this example shows the instruction with a constant entered in source B. A constant may be entered in either source A or source B, and it is stored in the instruction.

Fig. 12-66 ADD instruction.

sum 140 in address 036. Thus 60°C = 140°F. In rung 4, the GET/PUT instruction pair transfers the converted temperature reading 140°F, to the LED display.

Word-level logical instructions include AND, OR, EXclusive OR (XOR), and NOT instructions. Allen-Bradley PLC-5 protocol for these instructions is illustrated in Figs. 12-71 to 12-74 on page 366.

Fig. 12-67 SUBTRACT instruction.

A *file* is a group of consecutive words in the data table that have a defined start and end

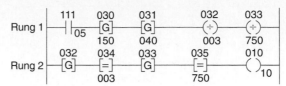

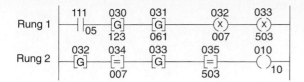

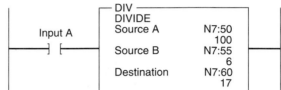

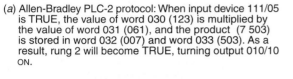

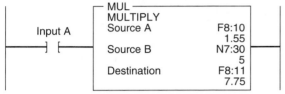

(a) Allen-Bradley PLC-2 protocol: When input device 111/05 is TRUE, the value of word 030 (123) is multiplied by the value of word 031 (061), and the product (7 503) is stored in word 032 (007) and word 033 (503). As a result, rung 2 will become TRUE, turning output 010/10 ON.

(b) Allen-Bradley PLC-5 protocol: Value stored at the address indicated in source A, F8:10, being multiplied by the value stored at the address in source B, N7:30, the result being stored at the address in the destination, F8:11.

Fig. 12-68   MULTIPLY instruction.

(a) Allen-Bradley PLC-2 protocol: When input device 111/05 is TRUE, the value of word 030 (150) is divided by the value of word 031 (040), and the quotient is stored in words 032 and 033 (003.750 or 3.75). As a result, rung 2 will become TRUE, turning output 010/10 ON.

(b) Allen-Bradley PLC-5 protocol: The example divides the value stored at source A's address (N7:50) by the value stored at source B's address (N7:55) and stores the result at the destination's address (N7: 60). Note that the result stored at the destination address is rounded off. If the remainder is 0.5 or above, the result is rounded up. If the remainder is less than 0.5, the answer is rounded down.

Fig. 12-69   DIVIDE instruction.

and are used to store information. For example, a batch process program may contain several separate recipes in different files that could be selected by an operator.

In some instances it may be necessary to shift complete files from one location to another within the programmable controller memory. Such data shifts are termed *file-to-file shifts*. File-to-file shifts are used when the data in one file represents a set of conditions that must interact with the programmable controller program a number of times and, therefore, must remain intact after each operation.

Because the data within this file must also be changed by the program action, a second file is used to handle the data changes, and the information within that file is allowed to be altered by the program. The data in the first file, however, remains constant and can, therefore, be called upon to be used a number of times. Figure 12-75 on page 367 illustrates the file-to-file copy instruction protocol used with the Allen-Bradley PLC-5 family of controllers. Data

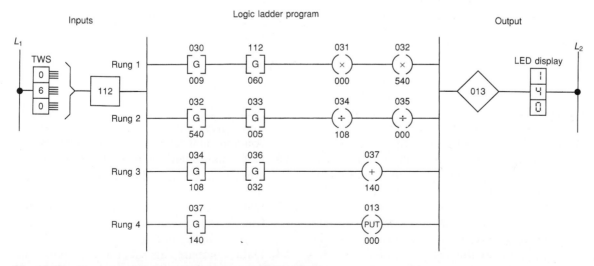

Fig. 12-70   Converting Celsius temperature to Fahrenheit.

Source A	Source B	Destination
0	0	0
1	0	0
0	1	0
1	1	1

(a) Truth table (1 appears in destination only
when both A and B have a 1)

```
┌─ AND ─────────────────────────┐
│  BITWISE AND                  │
│  Source A           B3:5      │
│  1100110011001100             │
│  Source B           B3:7      │
│  1111111100000000             │
│  Destination        B3:10     │
│  1100110000000000             │
└───────────────────────────────┘
```

(b) AND instruction (destination bits are
result of logical AND operation)

Fig. 12-71   ADD instruction.

Source A	Source B	Destination
0	0	0
1	0	1
0	1	1
1	1	1

(a) Truth table (1 does not appear at destination
only when both source A and source B are 0)

```
┌─ OR ──────────────────────────┐
│  BITWISE INCLUS OR            │
│  Source A           B3:1      │
│  1100110011001100             │
│  Source B           B3:2      │
│  1111111100000000             │
│  Destination        B3:20     │
│  1111111111001100             │
└───────────────────────────────┘
```

(b) OR instruction (destination bits are
result of logical OR operation)

Fig. 12-72   OR instruction.

from the expression file #N7:20 will be copied
into the destination file #N7:50. The length of
the two files is set by the value entered in the
control element word R6:1.LEN. In this instruction we have also used the ALL mode,
which means all of the data will be transferred
in the first scan in which the FAL instruction
sees a FALSE-to-TRUE transition. The DN
bit will also come ON in that scan, unless an
error occurs in the transfer of data, in which
case the ER bit will be set, the instruction will
stop operation at that position, and then the
scan will continue at the next instruction.

Source A	Source B	Destination
0	0	0
1	0	1
0	1	1
1	1	0

(a) Truth table (1 appears at destination when source A
or source B have a 1 but not when both have a 1)

```
┌─ XOR ─────────────────────────┐
│  BITWISE EXCLUS OR            │
│  Source A           B3:15     │
│  1100110011001100             │
│  Source B           B3:17     │
│  1111111100000000             │
│  Destination        B3:25     │
│  0011001111001100             │
└───────────────────────────────┘
```

(b) XOR instruction (destination bits are
result of logical XOR operation)

Fig. 12-73   EXclusive OR (XOR) instruction.

Source A	Destination
0	1
1	0

(a) Truth table (bits are inverted between source A
and the destination, with the 0s changed to 1s and
the 1s changed to 0s)

```
┌─ NOT ─────────────────────────┐
│  NOT                          │
│  Source A           B3:25     │
│  1100110011001100             │
│  Destination        B3:27     │
│  0011001100110011             │
└───────────────────────────────┘
```

(b) NOT instruction (destination bits are
result of logical NOT operation)

Fig. 12-74   NOT instruction.

Self-Test

Answer the following questions.

65. True or false? Data manipulation instructions give the PLC capabilities beyond
conventional relay equivalent instructions.

66. Data manipulation instructions can operate on _____ data or on _____
data.

67. Data _____ instructions involve the
transfer of the contents from one word or
register to another.

68. Data compare instructions compare the
data stored in two or more registers and

Shift register
instruction

Bit shift left (BSL)
instruction

Bit shift right (BSR)
instruction

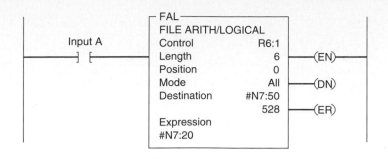

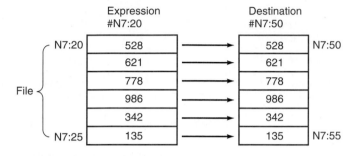

Fig. 12-75   File-to-file copy instruction.

make _____ based on the program instruction.

69. True or false? Data comparison is used in conjunction with the timer and counter instructions.

70. True or false? The limit test instruction can only be true if the test value is between or equal to the values of the high limit and low limit.

71. Set point control in its simplest form _____ an input value to a set point value.

72. Math instructions allow the PLC to perform arithmetic functions on _____ stored in memory words.

73. The four basic math functions performed by PLCs are _____, _____, _____, and _____.

74. True or false? A thumbwheel switch is a digital device that requires a word to input the data.

75. Basic word level logical instruction include _____, _____, _____, and _____ instructions.

76. A file is a group of consecutive _____.

12-6 SHIFT REGISTER AND SEQUENCER INSTRUCTIONS

The *shift register instruction* is often used with conveyors to monitor and control the flow of the individual parts. In general, it can be used in the control of machines and processes in which parts are continually shifted from one position to the next.

You can program a shift register to shift status data either right or left. This is accomplished by shifting either status or values through data files. When tracking parts on a status basis, *bit* shift registers are used. Bit shift instructions will shift bit status from a source bit address, through a data file, and out to an unload bit, one bit at a time. There are two bit shift instructions: *bit shift left (BSL),* which shifts bit status from a lower address number to a higher address number through a data file; and the *bit shift right (BSR),* which shifts data from a higher address number to a lower address number through a data file. Figure 12-76(*a*) on page 368 shows the basic concept of a shift register; a common shift pulse or clock causes each bit in the shift register to move one position to the right. At some point, the number of data bits fed into the shift register will exceed the register's storage capacity. When this happens, the first data bits fed into the shift register by the shift pulse are lost out the end of the shift register. Figure 12-76(*b*) shows a shift-right register. The status data (1 or 0) enters the register and is automatically shifted through the register from one bit address to the next on a time or event-driven basis. In the shift-left register shown in Fig. 12-76(*c*), normally, the shift register instruction is retentive. Figure 12-76(*d*) shows an 8-bit circulating shift register. Figure 12-76(*e*) is an

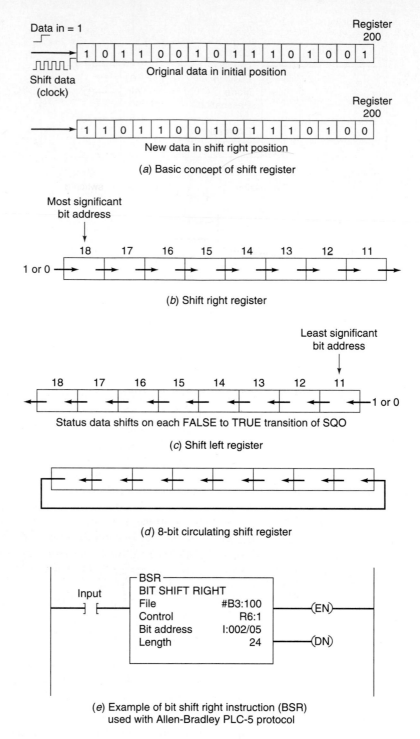

Data in = 1

Register 200

| 1 | 0 | 1 | 1 | 0 | 0 | 1 | 0 | 1 | 1 | 1 | 0 | 1 | 0 | 0 | 1 |

Shift data (clock)

Original data in initial position

Register 200

| 1 | 1 | 0 | 1 | 1 | 0 | 0 | 1 | 0 | 1 | 1 | 1 | 0 | 1 | 0 | 0 |

New data in shift right position

(*a*) Basic concept of shift register

Most significant bit address

| 18 | 17 | 16 | 15 | 14 | 13 | 12 | 11 |

1 or 0

(*b*) Shift right register

Least significant bit address

| 18 | 17 | 16 | 15 | 14 | 13 | 12 | 11 |

1 or 0

Status data shifts on each FALSE to TRUE transition of SQO

(*c*) Shift left register

(*d*) 8-bit circulating shift register

```
        ┌─ BSR ──────────────────────┐
Input   │ BIT SHIFT RIGHT            │
─┤ ├──── │ File           #B3:100     ├──(EN)
        │ Control            R6:1     │
        │ Bit address    I:002/05     │
        │ Length               24     ├──(DN)
        └────────────────────────────┘
```

(*e*) Example of bit shift right instruction (BSR) used with Allen-Bradley PLC-5 protocol

Fig. 12-76   Shift registers.

example of the bit shift right instruction (BSR) used with the Allen-Bradley PLC-5 protocol. When the instruction goes TRUE, the status of the bit address, I:002/05, is shifted into B3:101/07, which is the twenty-fourth bit in the file. The status of all of the bits in the file are shifted one position to the right, through the length of 24 bits. The status of B3:100/00 is

shifted to the unload bit, R6:1/UL. The status previously in the unload bit is lost.

Figure 12-77 illustrates a spray painting operation controlled by a shift register. The shift register's function is used to keep track of the items to be sprayed. As the parts pass along the production line, the shift register bit patterns represent the items on the conveyor

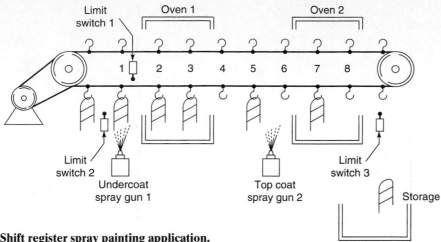

Fig. 12-77   **Shift register spray painting application.**

hangers to be painted. The logic of this operation is such that when a part to be painted and a part hanger occur together, indicated by the simultaneous operation of LS1 and LS2, a logical 1 is input into the shift register. The logical 1 will cause the undercoat spray gun to operate, and, five steps later, when a 1 occurs in the shift register, the top coat spray gun is operated. Limit switch 3 counts the parts as they exit the oven. The count obtained by limit switch 1 and limit switch 3 should be equal at the end of the spray painting run and is an indication that the parts commencing the spray painting run equal the parts that have completed it. A logical 0 in the shift register indicates that the conveyor has no parts on it to be sprayed and therefore inhibits the operation of the spray guns.

Sequencer instructions are used to control machines that operate in a sequential manner. The primary advantage of sequencer instructions is to conserve program memory. The *sequencer output instruction* functions in a manner similar to a mechanical drum switch, which is used to control output devices sequentially.

The programming of sequences will vary between programmable controller manufacturers, but the operational concepts are the same. The sequence of events controlled by the sequencer is determined by the bit pattern of each consecutive word and the number of words in the sequence.

Figure 12-78 on page 370 shows the way a typical sequencer output instruction works. In this example, 16 lights are used for outputs. Each light represents one bit address (1 through 16) of output word 050. The lights are programmed in a four-step sequence to simulate the operation of two-way traffic lights.

Data is entered into a work file for each sequencer step as illustrated. In this example, words 60, 61, 62, and 63 are used for the four-word file. Using the programmer, binary information (1s and 0s) that reflects the desired light sequence is entered into each word of the file. For ease of programming, some PLCs allow the word file data to be entered using the octal, hexadecimal, BCD, or similar number system. When this is the case, the required binary information for each sequencer step must first be converted to whatever number system is employed by the PLC. This information is then entered with the programmer into the word file.

Once the data has been entered into the word file of the sequencer, the PLC is ready to control the lights. When the sequencer is activated and advanced to *step 1*, the binary information in word 060 of the file is transferred into word 050 of the output. As a result, lights 1 and 12 will be switched ON and all the rest will remain OFF. Advancing the sequencer to *step 2* will transfer the data from word 061 into word 050. As a result, lights 1, 8, and 12 will be ON and all the rest will be OFF. Advancing the sequencer to *step 3* will transfer the data from word 062 into word 050. As a result, lights 4 and 9 will be ON and all the rest will be OFF. Advancing the sequencer to *step 4* will transfer the data from word 063 into word 050. As a result, lights 4, 5, and 9 will be ON and all the rest will be OFF. When the last step is reached, the sequencer is either automatically or manually reset to step 1.

When a sequencer operates on an entire output word, there may be outputs associated with the word that do *not* need to be controlled by the sequencer. In our example, bits

CHAPTER 12   PROGRAMMABLE LOGIC CONTROLLERS (PLCs)   **369**

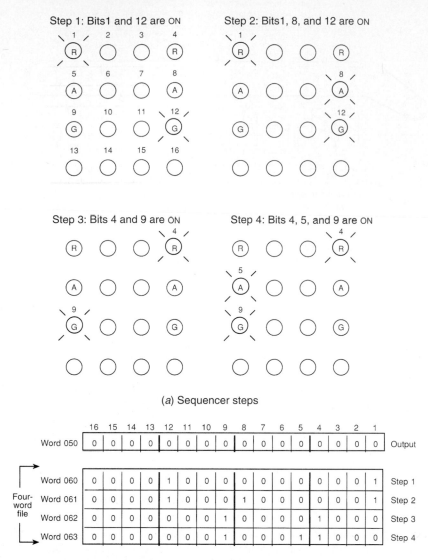

Step 1: Bits1 and 12 are ON

Step 2: Bits1, 8, and 12 are ON

Step 3: Bits 4 and 9 are ON

Step 4: Bits 4, 5, and 9 are ON

(a) Sequencer steps

	16	15	14	13	12	11	10	9	8	7	6	5	4	3	2	1	
Word 050	0	0	0	0	0	0	0	0	0	0	0	0	0	0	0	0	Output
Word 060	0	0	0	0	1	0	0	0	0	0	0	0	0	0	0	1	Step 1
Word 061	0	0	0	0	1	0	0	0	1	0	0	0	0	0	0	1	Step 2
Word 062	0	0	0	0	0	0	0	1	0	0	0	0	1	0	0	0	Step 3
Word 063	0	0	0	0	0	0	0	1	0	0	0	1	1	0	0	0	Step 4

Four-word file: Word 060, Word 061, Word 062, Word 063

(b) Binary information for each sequencer step

Fig. 12-78   Four-word sequence.

2, 3, 5, 7, 10, 11, 13, 14, 15, and 16 of output word 050 are not used by the sequencer but could be used elsewhere in the program.

To prevent the sequencer from controlling these bits of the output word, a *mask word* (040) is used. The use of a mask word is illustrated in Fig. 12-79. The mask word selectively screens out data from the sequencer word file to the output word. For each bit of output word 050 that the sequencer is to control, the corresponding bit of mask word 040 must be set to 1. All other bits of output word 050 are set to 0 and therefore can be used independently of the sequencer.

To program a sequencer instruction into a PLC, an input file is created which holds the bit pattern of the sequence read from a number of input words, and an output file is used to output the sequence to the required devices. The bit pattern of the mask word controls which outputs are made active. Sequencer instructions are usually retentive, and there is an upper limit as to the number of external outputs and steps that can be operated upon by a single instruction. Many sequencer instructions reset the sequencer automatically to step 1 upon completion of the last sequence step. Other instructions provide an individual reset control line or a combination of both. Figure 12-80 illustrates a typical block formatted sequencer instruction.

A sequencer program can be *event-driven* or *time-driven*. An event-driven sequencer operates like a mechanical stepper switch that

	16	15	14	13	12	11	10	9	8	7	6	5	4	3	2	1	
Word 050	0	0	0	0	0	0	0	0	0	0	0	0	0	0	0	0	Output
Word 040	0	0	0	0	1	0	0	1	1	0	0	1	1	0	0	1	Mask
Word 060	0	0	0	0	1	0	0	0	0	0	0	0	0	0	0	1	Step 1
Word 061	0	0	0	0	1	0	0	0	1	0	0	0	0	0	0	1	Step 2
Word 062	0	0	0	0	0	0	0	1	0	0	0	0	1	0	0	0	Step 3
Word 063	0	0	0	0	0	0	0	1	0	0	0	1	1	0	0	0	Step 4

Fig. 12-79 Using a mask word.

```
┌─ SQO ──────────────────────────┐
│  SEQUENCER OUTPUT               │
│  File                    #N7:1  │
│  Mask                    0F0F   │
│  Dest                    O:014  │
│  Control                 R6:20  │
│  Length                      4  │
│  Position                    0  │
└────────────────────────────────┘
```

Fig. 12-80 Sequencer instruction.

increments by one step for each pulse applied to it. A time-driven sequencer operates like a mechanical drum switch that increments automatically after a preset time period.

Self-Test

Answer the following questions.

77. Shift registers are often used for _____ parts on a production line.

78. Generally, there are two bit shift instructions: bit shift _____ and bit shift _____.

79. With a bit shift register, the status data (1 or 0) is automatically shifted through the register from one bit _____ to the next.

80. True or false? The advantage of sequencer programming is a large saving of memory words.

81. The sequence of events controlled by the sequencer is determined by the _____ pattern of each consecutive word and the number of _____ in the sequence.

82. The _____ word selectively screens out data from the sequencer word file to the output word.

83. A sequencer program can be either _____-driven or _____-driven.

SUMMARY

1. The programmable controller has eliminated much of the hand wiring associated with conventional relay control circuits.
2. The PLC central processing unit (CPU) is a microprocessor-based system that replaces control relays, counters, timers, and sequencers.
3. Memory is where the control plan or program is held or stored in the controller.
4. The I/O section of a PLC consists of input modules and output modules used to interface field devices with the controller.
5. Input modules are used to sense when a signal is received, convert the input signal to the correct voltage level, isolate the PLC from fluctuations in the input signal's voltage or current, and send a signal to the PLC indicating which sensor originated the signal.
6. The output interface module acts as a switch to supply power from the user power supply to operate the output.
7. The programming device allows the user to enter, edit, and monitor programs.
8. During each operating cycle or scan, the controller examines the status of input devices, executes the user program, and changes outputs accordingly.
9. PLCs work on binary numbers in one form or another.
10. The base of a number system determines the total number of different symbols or

digits used by that system—for example, decimal (base 10), octal (base 8), hexadecimal (base 16), or binary (base 2).

11. The data table stores the information needed to carry out the user program and includes such information as the status of input and output devices, timer and counter values, and data storage.

12. The input image table stores the status of digital inputs, which are connected to input interface circuits.

13. The output image table is an array of bits that controls the status of digital output devices.

14. A scan cycle is the time required for a PLC to scan its inputs and generate appropriate control responses at its outputs.

15. The three basic symbols used to translate relay control logic to contact symbolic logic are the EXAMINE ON, EXAMINE OFF, and OUTPUT ENERGIZE instructions.

16. In general, a logic ladder rung consists of a set of input conditions, represented by contact instructions, and an output instruction at the end of the rung represented by the coil symbol.

17. An internal output operates just as any output that is controlled by programmed logic; however, the output is used strictly for internal purposes.

18. Regardless of PLC model, some common operating modes are CLEAR MEMORY, PROGRAM, TEST, and RUN.

19. PLC timers provide the same types of functions offered by mechanical and electronic timing relays.

20. Associated with a timer instruction may be its address, type of timer, time base, preset value, and accumulated value.

21. A retentive timer will hold its accumulated value when the timer rung goes FALSE and will continue timing where it left off when the timer rung goes TRUE again. This timer must be accompanied by a timer RESET instruction to reset the accumulated value of the timer to zero.

22. Counter instructions allow for counting of events and controlling of other events based on the accumulated count.

23. Data transfer instructions simply involve the transfer of the contents from one word or register to another.

24. Data compare instructions compare the data stored in two or more words (or registers) and make decisions based on the program instructions.

25. Math instructions allow the PLC to perform arithmetic functions on values stored in memory words.

26. Word-level logical instructions include AND, OR, EXclusive OR (XOR), and NOT instructions.

27. A file is a group of consecutive words in the data table with a defined start and end and is used to store information.

28. The shift register instruction is often used with conveyors to monitor and control the flow of the individual parts.

29. The sequencer output instruction functions in a manner similar to that of a mechanical drum switch, which is used to control output devices sequentially.

CHAPTER REVIEW QUESTIONS

Answer the following questions.

12-1. List four advantages that PLCs offer over the conventional relay type of control system.

12-2. Explain the function of the central processing unit (CPU) of a PLC.

12-3. List four tasks performed by a PLC input module.

12-4. State the main function of the output module.

12-5. Compare discrete and analog I/O modules with respect to the type of input or output devices they can be used with.

12-6. How does the processor identify the location of a specific input or output device?

12-7. Why does a programmable controller system require two power supplies?

12-8. List three possible functions of a PLC programming device.

12-9. State the base used for each of the following number systems:
  a. Octal
  b. Decimal
  c. Binary
  d. Hexadecimal
12-10. List typical types of information that would be normally stored in a PLC data table.
12-11. Explain what a scan cycle is.
12-12. **a.** What does an EXAMINE ON or EXAMINE OFF instruction represent?
  b. What does an OUTPUT ENERGIZE instruction represent?
  c. The status bit of an EXAMINE ON instruction is examined and found to be 0. What does this mean?
  d. The status bit of an EXAMINE OFF instruction is examined and found to be 1. What does this mean?
  e. Under what condition would the status bit of an OUTPUT ENERGIZE instruction be 0?
12-13. When is the ladder rung considered TRUE, or as having logic continuity?
12-14. What is the function of an internal control relay?
12-15. Why must the output LATCH and output UNLATCH coil instructions have the same address?
12-16. Briefly describe each of the following modes of operation of PLCs:
  a. CLEAR MEMORY **c.** TEST
  b. PROGRAM **d.** RUN
12-17. In what ways may a PLC program be documented?
12-18. State five pieces of information usually associated with a PLC timer instruction.
12-19. When is the output of a programmed timer energized?
12-20. Explain the difference between the operation of a nonretentive and that of a retentive timer.
12-21. Explain how the accumulated count of programmed retentive and nonretentive timers is reset to zero.
12-22. Name the three types of PLC counters and explain the basic operation of each.
12-23. State four pieces of information usually associated with a PLC counter instruction.
12-24. When is the output of a PLC counter energized?
12-25. The counter instructions of PLCs are normally retentive. Explain what this means.
12-26. **a.** Compare the operation of a standard PLC examine-for ON contact with that of a transitional OFF-to-ON contact.
  b. What is the normal function of a transitional contact used in conjunction with a counter?
12-27. Explain the difference between a register or word and a table or file.
12-28. What is involved in a data-transfer instruction?
12-29. What is involved in a data-compare instruction?
12-30. Name five different types of data-compare instructions.
12-31. Explain the purpose of math instructions as applied to the PLC.
12-32. What are the four basic word-level logical instructions that can be programmed on some PLCs?
12-33. Give a practical example of a situation in which a shift register instruction would be used.
12-34. What type of operations are sequencers most suitable for?
12-35. Explain the purpose of a mask word when used in conjunction with the sequencer instruction.

12-1. Design a PLC program and prepare a typical I/O connection diagram and logic ladder program that will correctly execute the hard-wired control circuit as shown in Fig. 12-81.

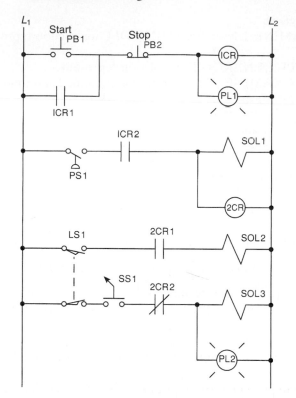

Note: PB1 and PS1 are wired NO.

PB2 is wired NC.

LS1 is wired using *one* set of NC contacts.

Fig. 12-81

12-2. Design a PLC program and prepare a typical I/O connection diagram and logic ladder program that will correctly execute the hard-wired control circuit as shown in Fig. 12-82.

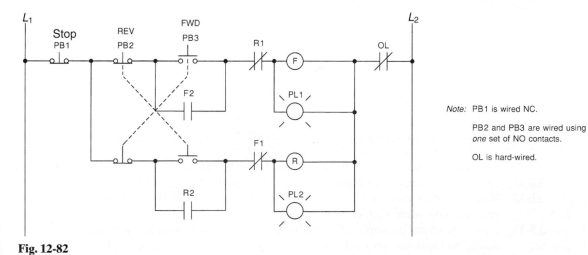

Note: PB1 is wired NC.

PB2 and PB3 are wired using *one* set of NO contacts.

OL is hard-wired.

Fig. 12-82

12-3. Design a PLC program and prepare a typical I/O connection diagram and logic ladder program that will correctly execute a hard-wired control circuit as shown in Fig. 12-83.

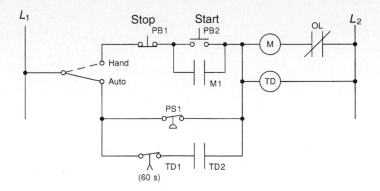

Fig. 12-83

12-4. Design a PLC program and prepare a typical I/O connection diagram and logic ladder program for the following counter specifications:
- Counts the number of times a pushbutton is closed.
- Decrements the accumulated value of the counter each time a second pushbutton is closed.
- Turns ON a light any time the accumulated value of the counter is less than 20.
- Turns ON a second light when the accumulated value of the counter is equal to or greater than 20.
- Resets the counter to zero when a selector switch is closed.

12-5 Design a PLC program and prepare a typical I/O connection diagram and logic ladder program that will correctly execute the following control circuit:
- Turns ON a nonretentive timer when a switch is closed (preset value of timer is 10 s).
- Timer is automatically reset by a programmed transitional contact when it times out.
- Counter counts the number of times the timer goes to 10 s.
- Counter is automatically reset by a second programmed transitional contact at a count of 5.
- Latches ON a light at the count of 5.
- Resets light to OFF when a selector switch is closed.

12-6 Write a program to perform the following:
- Turn ON pilot light 1 (PL1) if the thumbwheel switch value is less than 4.
- Turn ON pilot light 2 (PL2) if the thumbwheel switch value is equal to 4.
- Turn ON pilot light 3 (PL3) if the thumbwheel switch value is greater than 4.
- Turn ON pilot light 4 (PL4) if the thumbwheel switch value is less than or equal to 4.
- Turn ON pilot light 5 (PL5) if the thumbwheel switch value is greater than or equal to 4.

Answers to Self-Tests

1. relay	8. ON/OFF	15. false
2. false	9. true	16. ladder
3. processor	10. power; logic	17. instructions; addresses
4. true	11. relay; triac; transistors	18. speed
5. bits	12. false	19. true
6. true	13. true	20. binary
7. interface	14. located	21. false

22. zero; one
23. bit
24. least significant
25. false
26. true
27. 001 101 011
28. 16
29. 47
30. 11101; 00101001
31. map
32. user program
33. status data; numbers or codes
34. one
35. false
36. scan
37. false
38. symbolic
39. contact; coil
40. false
41. true
42. true

43. true
44. false
45. false
46. RUN
47. personal computers
48. true
49. coil; block
50. false
51. preset; accumulated
52. true
53. 10
54. false
55. retentive
56. true
57. true
58. transition
59. false
60. FALSE; TRUE
61. false
62. true
63. true
64. true

65. true
66. word; file
67. transfer
68. decisions
69. true
70. false
71. compares
72. values
73. ADD; SUBTRACT; MULTIPLY; DIVIDE
74. true
75. AND; OR; XOR; NOT
76. words
77. tracking
78. left; right
79. address
80. true
81. bit; words
82. mask
83. event; time

CHAPTER 13

Computer-Controlled Machines and Processes

■

CHAPTER OBJECTIVES

This chapter will help you to:

1. *Discuss* how a computer's operating system is designed to function.
2. *Explain* how a work cell functions.
3. *Compare* the methods by which computers communicate with each other.
4. *Discuss* the principle of operation of computer numerical control.
5. *Discuss* the principle of operation of robotic computer control.

Since personal computers (PCs) are now commonly used in industry, this chapter contains a brief discussion about important PC operations likely to be encountered. The use of machine interfacing circuits, along with protocol standards, is presented. This chapter also serves as an introduction to computer numerical control (CNC) and robotic computer control.

■

13-1 COMPUTER FUNDAMENTALS

Computers, particularly personal computers, are now accepted as useful devices for machine and process control. As manufacturing systems are becoming increasingly computer-based, understanding of computer fundamentals is essential.

All computers consist of two basic components: hardware and software. The computer *hardware* (Fig. 13-1 on page 378) is the physical component which makes up the computer system and includes:

Power supply. Converts the 120 V ac electricity from the line cord to dc voltages that are needed by the computer system.
Floppy drives. Allows information to be stored and read from removable floppy disks.
Hard drive. Allows information to be stored and read from nonremovable hard disks.
Motherboard. Holds and electrically inter-

connects the major sections of the entire computer system.
Microprocessor chip. Interprets the instructions for the computer and performs the required process for each of these instructions.
ROM chips. Read-only memory. Memory programmed at the factory that cannot be changed or altered by the user.
RAM chips. Read/write memory. Memory used to store computer programs and interact with them.
Peripheral cards. Allows accessory features and connects the computer to input and output devices such as drives, printers, monitors, and other external devices.
Expansion slots. Connectors used for the purpose of connecting other circuits to the motherboard.

Software is what gives the computer "life." The software resides inside the hardware. You can think of software as a computer *program.* A computer program is nothing more than a

From page 377:

Hardware

Software

On this page:

Disk drives

Floppy disk

Write-protect

Hard disk drives

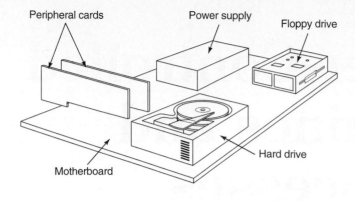

Peripheral cards | Power supply | Floppy drive | Motherboard | Hard drive

(*a*) Motherboard holds and electrically interconnects major hardware sections

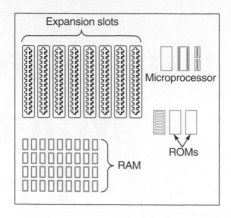

Expansion slots | Microprocessor | ROMs | RAM

(*b*) Plug-in components and cards are easily serviceable; expansion slots allow upgrading of computer capabilities

Fig. 13-1   Computer hardware.

list of instructions telling the computer what to do. The programs are usually stored on some form of mass storage system, such as a floppy disk, and loaded into the computer's random access memory (RAM) as required.

Disk drives act as both input and output devices, so they are used for getting information into the computer as well as storing information from the computer. There are actually three different kinds of disk drives popularly used in personal computers: the hard (fixed) disk drive, the $5\frac{1}{4}$-in. floppy disk drive and the $3\frac{1}{2}$-in. floppy disk drive.

The *floppy disk* (diskette) is a magnetic storage device on which information is stored for later retrieval (Fig. 13-2). The information is stored on tracks in sectors of the disk; the location of the information is configured to allow the computer access to a sector to read the information. The disk must be formatted into tracks and sectors using the disk operating system FORMAT command before any information can be stored on it.

All disks have a *write-protect*. For $5\frac{1}{4}$-in. disks, this involves merely putting an adhesive label over the notch. For a $3\frac{1}{2}$-in. disk, there is a small physical sliding tab on the lower left side which should be opened if you want to protect the disk from being written on. The smaller $3\frac{1}{2}$-in. microfloppy can generally store more information than its larger $5\frac{1}{4}$-in. counterpart. This is because of the higher quality of the magnetic surface in the $3\frac{1}{2}$-in. disk.

All disks must be formatted before they can be used. This sets the disk up according to the specifications of the computer you are using. Once the disk is inserted into the computer, the drive motor spins the disk while the step-

per motor moves the read/write head to different positions on the disk. The indicator light indicates whether the disk drive is active.

The disk is inserted in the disk drive with the slot and labels up. The disk drive door must then be shut to allow the disk reading heads to read the information. Care must be taken when handling the disk, or the information stored on the disk may be corrupted:

- Never touch the exposed disk at the head slot, or surface.
- Never bend the disk.
- Never expose the disk to heat or excessive sunlight.
- Never expose the disk to electromagnetic or electrostatic fields that are likely to occur in the industrial environment.
- Use only labels designed specifically for use with disks. These labels will not get stuck in a disk drive when you remove a disk.
- Use only soft-tip pens when labeling the disk.
- Never force the disk into the disk drive.
- Always make backup disks (copies) and store the original disks safely.

Hard disk drives (Fig. 13-3 on page 380) store and retrieve data more quickly and can hold much more information than floppy disks. The hard disk mass storage system is permanently located within the computer and can store a large number of programs simultaneously. The contents of any floppy disk can be relocated onto the hard disk, and the programs then used as required, keeping the floppy disk as backup copies. The hard disk drive is normally referred to as the C: drive (it is conventional to place a colon after the drive letter). If one floppy disk drive is used, it is called the A:

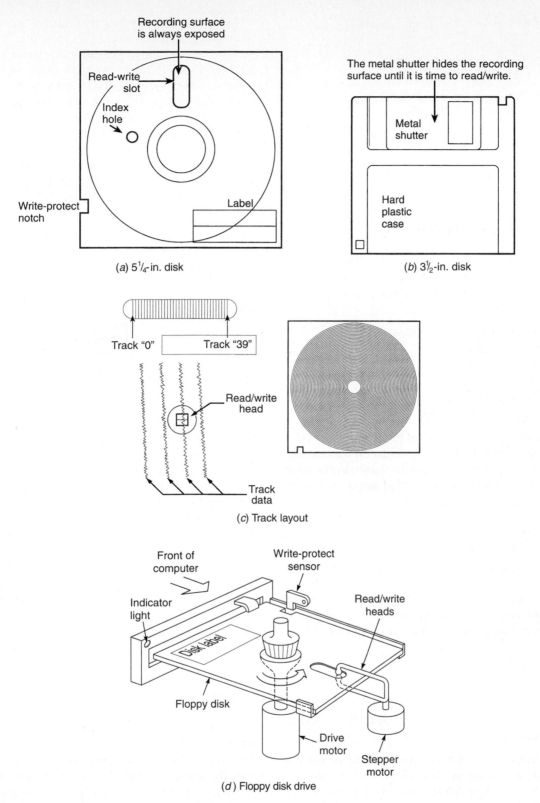

Recording surface
is always exposed

Read-write
slot

Index
hole

Write-protect
notch

Label

(a) 5¼-in. disk

The metal shutter hides the recording
surface until it is time to read/write.

Metal
shutter

Hard
plastic
case

(b) 3½-in. disk

Track "0" Track "39"

Read/write
head

Track
data

(c) Track layout

Front of
computer

Write-protect
sensor

Indicator
light

Read/write
heads

Disk label

Floppy disk

Drive
motor

Stepper
motor

(d) Floppy disk drive

Fig. 13-2 Floppy disk.

drive, and if a second floppy disk drive is used, it is called the B: drive.

There are two major categories of software: system software and application software. *System software* provides the programs that allow you to interact with your computer—to operate the disk drives, the printer, and other devices used by the computer.

The most popular operating system for personal computers is *MS-DOS:* MS refers to the

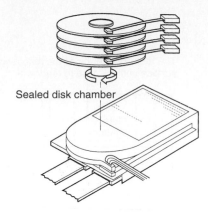

Fig. 13-3   Hard disk drive unit.

manufacturer, Microsoft, and DOS to its Disk Operating System. DOS creates a common platform for all the software you use and allows you to:

- Read, write, and edit files
- Back up files
- Organize a disk
- Manage files
- Display the contents of a disk
- View error messages

Different "versions" of DOS have evolved as people wanted the program to do more things and as technology gave us faster equipment. VER is a DOS command that will show you the version of DOS your computer is using.

A *file* is a collection of data stored under a single name. When information is stored on the disk, the file is saved or written to the disk. When the information is being used, the disk is being read or data is being loaded into the computer's memory. A computer can move, copy, rename, or delete the file at your command. Some important types of files include:

- *Executable file.* A list of instructions to the CPU
- *Data file.* A list of information
- *Text file.* A series of characters like a letter
- *Graphics file.* A picture converted to digital code

After a disk is formatted, writing or reading a file is a process that involves your software, DOS, the PC's *BIOS* (*Basic Input/Output System*), and the mechanism of the disk drive itself. The BIOS serves as a link between the computer hardware and the software programs. It is permanently stored in the computer's memory, which serves as a translator between the CPU and all the other entities it talks to.

When the computer is first turned on, the BIOS causes the computer to look into your floppy disk drives to see whether there is a disk there containing DOS. If there is, the computer will load the DOS into itself. The process of loading DOS into your computer is called *booting*. If there are no floppy disks in your drives that contain DOS, the hard drive (C: drive) will be examined for DOS. If there is no DOS (or no hard drive), then an error message will be displayed (Fig. 13-4).

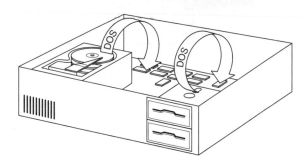

Fig. 13-4   Loading DOS into memory.

Application software involves programs written to give the computer a specific application, such as a word processor. Application software makes up the majority of the software available in the market. In order for application software to work in your computer, the system software must first be loaded into the computer's memory.

Directories are simply a way of organizing all the programs and files of the hard drive in your computer. The directory structure performs the same basic function as a filing cabinet. The first and main directory of the computer is called the *root directory*. The second level of the directory structure ideally contains the general listings of the contents of the disk. *Tree* is a DOS command that will show you the directory structure of your computer.

The *DOS prompt* (A>) is an indicator to you that DOS is loaded and that the computer is ready for you to enter a command. It will stay like this until you either enter an instruction or turn the computer off. The letter A indicates that the computer is currently ready to access Drive A. If you wish to switch to another disk drive, you must tell DOS to do so by entering the new drive letter followed by a colon and then press Enter (e.g., A>b:). The screen will then display the new prompt B>.

DOS is simply a very large list of *commands* that the computer can use and understand to

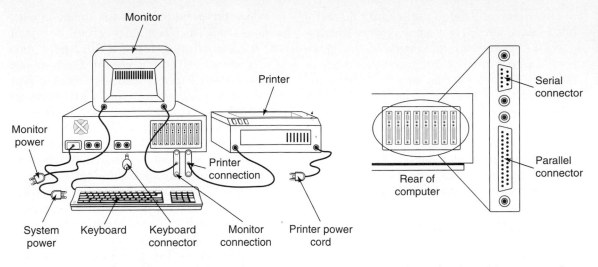

Monitor

Printer

Monitor power

Serial connector

System power

Keyboard

Keyboard connector

Printer connection

Monitor connection

Printer power cord

Rear of computer

Parallel connector

(a) Typical cable connections for computer

(b) Serial and parallel port connections

Fig. 13-5 Computer peripheral connections.

control, manage, and run other programs. Some of the commands used most often are listed below:

- *DIR command*
  - Directory command
  - Used to list the contents of any disk
- *CLS command*
  - Clear screen command
  - Used to clear the video display screen and place the prompt and cursor at the top left corner of the screen
- *COPY command*
  - Used to copy files from one location to another, for example, from Drive A to Drive B
- *REN command*
  - Rename command
  - Used to rename a file, giving it a new name without changing its contents
- *DEL command*
  - Delete command
  - Used to delete any unwanted files on a disk
- *TYPE command*
  - Used to display a file on screen, allowing you to read its contents
- *FORMAT command*
  - Used to prepare a new disk for use with DOS
- *DISKCOPY command*
  - Used to make an exact duplicate of a floppy disk
- *CD command*
  - Change directory command
  - Used to allow you to move around in the directory structure

- *CHKDSK command*
  - Check disk command
  - Used to check the status of a disk, how many files are on it, and how much of the disk is used

Microsoft's Windows is a graphics-based interface for MS-DOS. It uses graphical images to replace the verbal commands to run DOS. Instead of typing commands, you select graphic pictures (icons), using a mouse.

Windows allows for *multitasking*. No computer can actually do two things at once, but it appears to do so by very rapidly shuffling back and forth from one program to another. Programs in Windows share the CPU resources, passing control back and forth at times when the computer is not being controlled by the user. Windows' multitasking operation is one of the principal factors drawing people away from DOS application development. Instead of having to develop an application that does everything, with Windows you can use multiple specialized applications that run concurrently and work together.

Ports are the connecting devices that stick out the back end of the computer case (Fig. 13-5). They are actually the back edge of an expansion card, providing the electrical door or gateway connecting the computer and a peripheral. Ports are usually a preset number of pins to allow the proper cable connection and are referred to as male or female (with pins or holes to fit the pins). Ports are usually referred to as *serial* (for a mouse connection) or *parallel* (for a printer connection). A serial port sends data one bit at a time over a single one-way wire; a parallel port can send several bits of data across eight parallel wires simultaneously. Because of its simplicity, the serial port has been used at one time or another to make a PC communicate with just about any device imaginable. The serial port is often referred to as an *RS-232 port* (Electronics Industries Associations designation). The parallel port is often called a *Centronics port.*

Selecting a computer can be a complex task, as there are numerous vendors and scores of different configurations possible. Key specifications include:

- *Access time.* Refers to the length of time it takes for an information storage device to return a piece of information once it is requested. For a hard drive, the typical time is 9 to 28 milliseconds (thousandth of a second). For RAM, typical times are 70 to 120 nanoseconds (billionths of a second).
- *Baud.* Is a measure of information transfer speed over modems. One baud represents one signal change per second. The faster the signal, the faster the data can be transferred. Newer modems actually transfer data faster than one bit with every signal change; the transfer of information is actually measured in bps (bits per second). For example, a modem running at 1200 baud, and passing 2 bits of information with every signal change is transferring data at a rate of 2400 bps, or about 240 characters per second.
- *Bus.* Is a wire or group of wires that carries a flow of digital information. When a computer moves information from component to component, it uses a bus to do it. To increase the speed of data movement, engineers have increased the size of the bundle of wires. The earliest PCs used eight data channels, which were called 8-bit buses because they could carry eight pieces of information at one time, similar to an eight-lane highway. Newer machines use 16-bit and 32-bit buses.
- *Cache.* A memory cache is a superfast set of memory chips that tries to anticipate what data the CPU will need next and save it so it can be delivered to the CPU quickly, unlike RAM, which provides the information much more slowly. The result is that the CPU spends less time looking for instructions, and thus speeds up processing time and overall system performance.
- *Clock speed.* Refers to the rate at which the CPU clock moves. Every computer has a clock; the CPU uses pulses from the clock to pace its activities and coordinate the execution of instructions with other system components. If the clock beats faster, the system components move faster. Clock speed rates are measured in megahertz (MHz), or millions of oscillations (beats) per second. Typical clock speeds range from 4 MHz through 150 MHz.
- *Central processing unit (CPU).* PC/XT/AT/386/486/Pentium/RISC . . . With all these terms and different processors, which is the right one for your application? The different terms—PC, XT, and so on—are only marketing names. What is important is the type of processor and the speed at which it operates. In the "IBM" world, there are multiple processor types. The 8088 8-bit processor, the 80286 16-bit processor, the 80386 32-bit processor, the 80486 32-bit processor family, and several new generation 64-bit processors. Although there are many differences, the key factors are the number of instructions per clock cycle and the number of clock cycles per second of each CPU.
- *Hard disk drive.* The hard disk drive is the primary mass information storage device of the computer. Although 40 Mb of memory used to be a lot, standard drives now seem to be 80 Mb, with 600 Mb and even 1.2 GB (gigabyte or billion bytes) available. Cost and the requirements of the programs are the primary concerns in this area.
- *Monitors.* Every screen displays thousands of tiny dots (called pixels) just below the screen surface. The greater the number of dots and the closer together the dots are, the sharper the image and the more information displayed. VGA (video graphics array) and Super-VGA are the current industry standards. A VGA monitor has at least 480 rows of dots, with each row consisting of 640 dots.
- *RAM.* Random access memory is the working memory of the computer. How much RAM the computer has is closely related to performance. RAM is referred to in megabytes (Mb), and average capacities are 640 K to 16 Mb. Without enough RAM, the memory becomes cramped and programs can mysteriously fail to function because the computer has nowhere to write the new instructions.

Self-Test

Answer the following questions.

1. A computer is made up of _____ and _____ components.

2. True or false? Floppy drives allow information to be stored and read from nonremovable hard disks.

3. A floppy disk must be _____ before any information can be stored on it.

4. True or false? Exposure of a computer disk to electromagnetic fields can corrupt the information stored on the disk.

5. The hard drive of a computer is normally referred to as the _____ drive.

6. DOS is _____ software.

7. True or false? In practical applications, disk files are sorted so that every one can be accessed at the same time.

8. True or false? The process of loading DOS into your computer is called booting.

9. The first and main directory of the computer is called the _____ directory.

10. A> is known as the DOS _____.

11. True or false? DOS is simply a very large list of commands that the computer can use and understand.

12. Windows is a(n) _____-based interface.

13. True or false? The serial port is most often used for a printer connection.

14. True or false? A memory cache is used to speed up processing time.

15. Clock speed refers to the rate at which the _____ clock moves.

13-2 COMPUTERS IN PROCESS AND MACHINE CONTROL

Today automation is moving rapidly toward a true point of central control that resides in the system operator's office. It is becoming increasingly necessary for system operators to have fingertip control of the process by way of their personal computers. One application in which the computer is used to monitor and control a networked PLC system is shown in Fig. 13-6.

Computer-integrated manufacturing systems provide individual machines used in manufacturing with data communication functions and compatibility, allowing the individual devices to be integrated into a single system. This type of flexible manufacturing system area, containing machines, is termed a *work cell*. A work cell with associated machines is shown in Fig. 13-7. The computer, or cell controller, is basically a communicator between components. The main difference between the programmable controller and the cell controller is the language used to program the cell controller. The PLC programming language is simple and

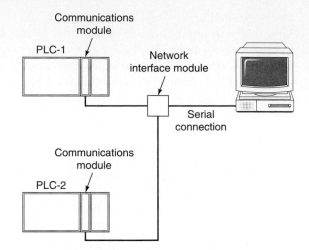

Fig. 13-6   Computer used to monitor and control a networked PLC system.

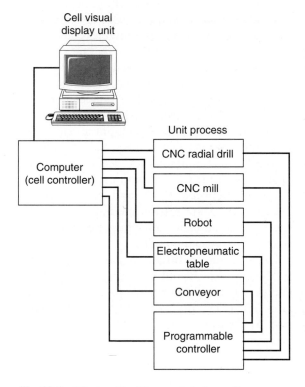

Fig. 13-7   Work cell with associated machines.

requires limited programming knowledge; the cell controller requires the programmer to have more programming knowledge.

The entire industry of cell control didn't come about until the advent of the industrial personal computer. More and more cell control programs are running on the personal computer level of hardware. Cell control involves supervising and coordinating the components and devices from simple input/output through a microprocessor-controlled robot and data collection and storage for communication within the cell and to other cells.

Cell control today involves being able to influence what happens up or down a production line. For example, picture an assembly line, and assume you produce a part and that part has to be tested. A robot can take the part, position it for the test, and, depending on the outcome, put it back on the line or reroute the part. If the specs are off, you will want to convey this information up or downstream, wherever the effects will be. You also want to reroute that part with a bill that describes the problem so that the part doesn't undergo diagnostics testing again but, rather, can be repaired or adjusted and sent back on-line. This requires communications links to avoid the problems that the defective part will cause if information is not relayed on.

Initially, *data communication systems* were provided between programmable controllers, but it was apparent that the advantages of integrated manufacture required computer-controlled and numerically controlled machines and robot controllers to be connected to the programmable controller, which in turn would interconnect with a host computer and other equipment via the factory local area network.

The fundamental job of a *local area network (LAN)* is to provide communication between PLCs or between PLCs and computers. In industry, the transmission medium most often used is coaxial cable or optical fibers because of the high noise immunity.

LANs come in three basic *topologies* (that is, the physical layout or configuration of the communications network): star, ring, or bus (Fig. 13-8). The points where the devices connect to the transmission medium are known as nodes or stations.

In a *star network*, a central control device is connected to a number of nodes. This allows for bidirectional communication between the central control device and each node. All transmission must be between the central control device and the nodes, since the central control device controls all communication.

In a *bus network*, each node is connected to a central bus. When a node sends a message on the network, the message is aimed at a particular station or node number. As the message moves along the total bus, each node is listening for its own node identification number and accepts only information sent to that number. Control can be either centralized or distributed among the nodes.

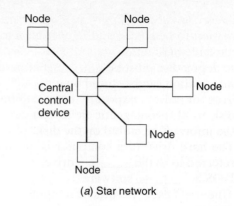

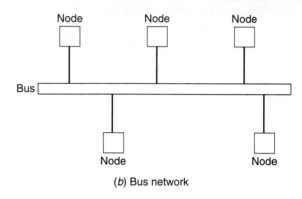

(a) Star network

(b) Bus network

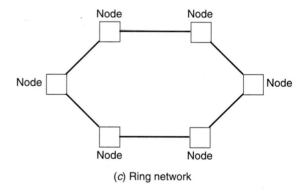

(c) Ring network

Fig. 13-8   Network topologies.

In a *ring network*, each node is connected to another node, in ring fashion. There is no end or beginning to the network. Messages are aimed at a particular node or station number, with each node listening for its own identification number. Signals are passed around the ring, and are regenerated at each node. Control can be centralized or distributed.

Protocol refers to a predetermined set of conventions which specify the format and timing of message transmission between two or more communicating devices. Several standards exist which define the signal protocol. For computer-integrated manufacturing, you need to tie all the different devices together using a common protocol. When the protocol is different, additional hardware and programming are required (Fig. 13-9). General Motors identified the need for a common protocol and developed the *MAP* (manufacturing automation protocol) specification, which is available to anyone. It has become an industry standard, and most new devices indicate that they are MAP-compatible or feature MAP interfaces.

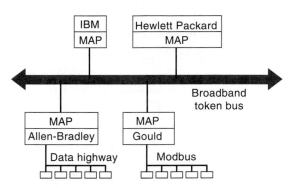

MAP gateway used to connect equipment with different protocols

Fig. 13-9   Network interconnections.

The two basic network communication formats are master-slave and peer-to-peer. A *master-slave system* (Fig. 13-10) uses a host computer to manage all network communications between network devices. The programming of the master network device incorporates routines to address each slave device individually. Direct communication between slave devices is not possible. Information to be transferred between slaves must first be sent to the network master unit, which will in turn retransmit the message to the designated slave device.

With a *peer-to-peer system* (Fig. 13-11), each network device has the ability to request use of,

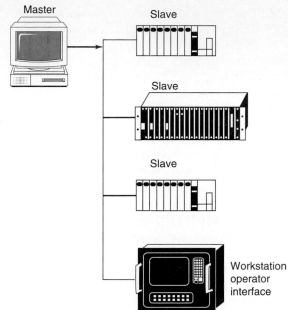

Fig. 13-10   Master-slave communication network.

and then take control of, the network for the purpose of transmitting information to or requesting information from other network devices. This type of network communication scheme is often described as a *token passing system*, since control of the network can be thought of as a token that is passed from unit to unit.

Self-Test

Answer the following questions.

16. True or false? A personal computer can be connected to monitor and control a number of different manufacturing operations.
17. True or false? PLC programming language is more complex than that required for a cell controller.

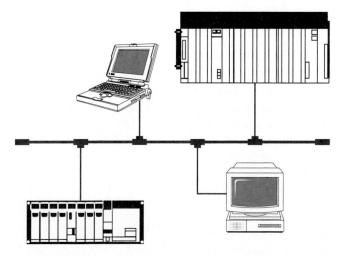

Fig. 13-11   Peer-to-peer communication network.

Master-slave system

Peer-to-peer system

Token passing system

18. True or false? A work cell could include CNC, robotic, and PLC control components.

19. Types of local area network topologies include _____, _____, and _____.

20. MAP stands for _____ _____ _____.

21. True or false? In a master-slave network communication system, direct communication between slave devices is not possible.

22. A peer-to-peer network system is often described as a(n) _____ passing system.

13-3 COMPUTER NUMERICAL CONTROL (CNC)

In general terms, *numerical control (NC)* is a flexible method of automatically controlling machines through the use of numerical values. Numerical control devices are driven over a continuous path or to various points to manufacture a component or device using a program which numerically coordinates the machine movement and operation. NC machines are program-dependent and bear a close relation to programmable controllers.

Numerical control enables an operator to communicate with machine tools through a series of numbers and symbols. A set of commands makes up the NC program and directs the machine to orientate a cutting tool with respect to a workpiece, select different tools, control cutting speed, and direct spindle rotation, as well as perform a range of auxiliary functions such as turning coolant flow on or off. Many languages exist for writing an NC program, but the one most used is called *automatically programmed tools (APTs)*. Originally NC programs were stored on punched paper tape, but in recent times they have been stored using some form of mass storage such as magnetic tape, floppy disk, or solid state memory (e.g., RAM or ROM; Fig. 13-12).

Data that is input manually [Fig. 13-12(a)] can be used to program the control system by setting the dials, switches, pushbuttons, etc. Most NC machine tools can be programmed manually, especially for setup purposes. Magnetic tape [Fig. 13-12(b)] in a tape cassette is used for some applications. The major advantage of floppy disks [Fig. 13-12(c)] is that it is much faster to retrieve information from this medium than from any other.

Figure 13-13 illustrates a typical numerical control system. The controller reads, interprets the instructions, and directs the machine to

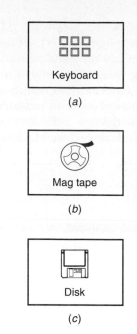

Fig. 13-12 Methods used to enter NC program.

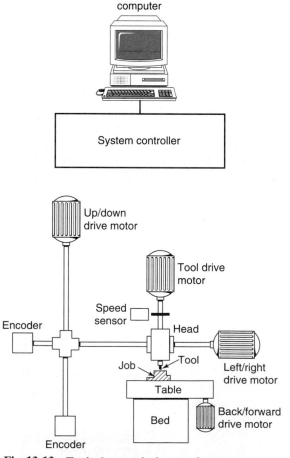

Fig. 13-13 Typical numerical control system.

perform the operations desired. The machine operator is alerted when material must be inserted, tools changed, and so on. The system controller is usually an industrial computer which electronically stores and reads the program and converts the program information into signals that drive motors to control the machine tool. The controller provides signals to either electrical or hydraulic motors which cause the machine head to be driven left or right, as well as up or down. The table can be motor driven to move the workpiece back and forth. Feedback to the controller is provided by the position encoders and speed sensor, so that the exact location and speed of the machine head is known. As the accuracy of the electronics exceeds the mechanical accuracy of the system, the repeatability, and therefore consistency, of the manufactured components are within very tight tolerances.

Numerical control is ideally suited for operations which involve the production of parts made from similar feedstock (raw material) with variations in size and shape. Even if production quantities are in small lots, NC can be economically feasible, but it is necessary that the sequence of operations be such that it can be performed on the same NC machine. However, for complete manufacturing of parts involving several sequences of operation that are dissimilar, several NC machines may be used in production.

If the controlled machine is a three-axis type, the location address of the tool is prefixed with the letter X, Y, or Z. Using X, Y, and Z coordinates, the machine can be directed to the correct location. The workpiece is located by using Cartesian coordinates, a means whereby the position of a point can be defined with reference to a set of axes at right angles to each other as illustrated in Fig. 13-14. The vertical axis is the Y axis and the horizontal axis is the X axis. The point where the two axes cross is the zero or origin point. To the left of the point of origin on the X axis and below the point of origin on the Y axis, locations are written preceded by a minus (−) symbol. Above the point of origin on the Y axis and to the right of the point of origin on the X axis, locations are written preceded by a plus (+) symbol. The Z axis of motion is parallel to the machine spindle and defines the distance between the workpiece and the machine.

In Fig. 13-14(b), X, Y, and Z words refer to coordinate movement of the machine tool for positioning or machining purposes. Using

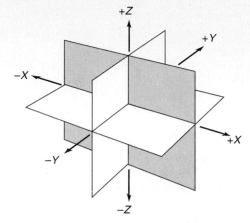

(a) Three-dimensional coordinate planes (axes) used in numerical control

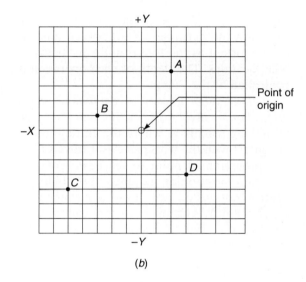

(b)

Fig. 13-14 **Locating points using Cartesian coordinates.**

Cartesian coordinates each point would be located as follows:

A is $X = +2$ and $Y = +4$
B is $X = -3$ and $Y = +1$
C is $X = -5$ and $Y = +4$
D is $X = +3$ and $Y = -3$

The programming of NC machines can be categorized into two main areas: point-to-point programming, and contour programming. *Point-to-point programming* involves straight-line movements. Figure 13-15 on page 388 shows an example of point-to-point programming in which four holes are to be drilled. The point where each hole is to be located is identified using X and Y coordinates. After each hole is drilled, the machine is instructed to move to the next point where a hole is to be drilled, and so on. The holes are drilled

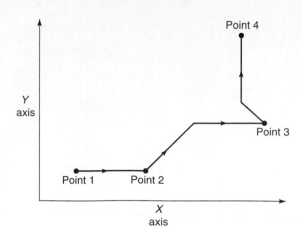

Fig. 13-15 Point-to-point programming.

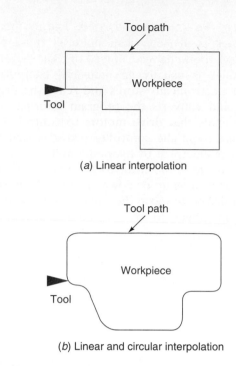

(a) Linear interpolation

(b) Linear and circular interpolation

Fig. 13-16 Contour programming.

sequentially until the program is completed. The path the machine takes between holes is not important as the tool is in the air between hole locations. The depth of each operation is controlled by the Z axis.

Contour (also known as *continuous path*) *programming* involves work like that produced on milling machines, where the cutting tool is in contact with the workpiece as it travels from one programmed point to the next. Continuous path positioning requires the ability to control motions on two or more machine axes simultaneously to keep a constant cutter-workpiece relationship. The method by which contouring machine tools move from one programmed point to the next is called *interpolation.* All contouring controls provide linear interpolation, and most are capable of linear and circular interpolation (Fig. 13-16).

Computerized numerical control (CNC) was introduced to replace the punched tape and hard-wired machine control of older NC units (Fig. 13-17). The CRT screen shows the exact position of the machine table and/or the cutting tool at every position while a part is being machined. CNC introduced a new flexibility into the manufacturing industry as the microprocessor-based control equipment brought features to NC machines, such as:

- Improved program mass storage, disk instead of punched tape
- Ease of editing programs
- Possibility of more complex contouring because of the computer's capability for mathematical manipulation
- Reusable machine pattern, which could be stored and retrieved as required
- Possibility of plantwide communication with many peripheral devices

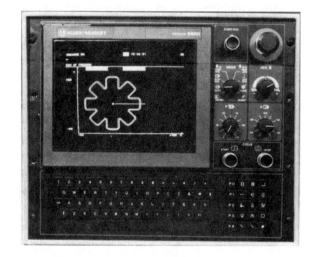

Fig. 13-17 Computer numerical control (CNC).
(Courtesy Allen-Bradley Company, Inc.)

Simple CNC programming consists of taking information from a part drawing and converting this information into a computer program. The program design can be carried out on a personal computer. A numerical control programming language must be used to design the program. This simplifies the program writing as the computer will calculate the machine coordinates. Using a computer to assist in program writing allows the program to be:

- Stored on a convenient mass storage system
- Retrieved and edited as required

- Tested off-line prior to using the program to control a machine
- Plotted using a plotter connected to the computer to assist in program de-bugging

Self-Test

Answer the following questions.

23. True or false? Not all numerical control (NC) machines rely on a program for their operation.
24. A set of _____ make up the NC program.
25. True or false? Today, most NC programs are stored on punched paper tape.
26. If an NC machine is the three-axis type, the location address of the tool is prefixed with the letter _____, _____, or _____.
27. The programming of NC machines can be categorized as being _____ or _____.
28. CNC machines use _____-based control equipment.
29. CNC programming consists of taking information from a(n) _____ _____ and converting this information into a computer program.

13-4 ROBOTICS

Robots are computer-controlled devices which perform tasks usually done by humans. The basic industrial robot in wide use today is an *arm* or *manipulator* which moves to perform industrial operations. Tasks are specialized and vary tremendously. They include:

- *Handling.* Loading and unloading components onto machines
- *Processing.* Machining, drilling, painting, and coating
- *Assembling.* Placing and locating a part in another compartment
- *Dismantling.* Breaking down an object into its component parts
- *Welding.* Assembling objects permanently by arc welding or spot welding
- *Transporting.* Moving materials and parts
- *Painting.* Spray painting parts
- *Hazardous tasks.* Operating under high levels of heat, dust, radioactivity, noise, and noxious odors

A robot is simply a series of mechanical links driven by servomotors. The area at each junction between the links is called a *joint* or *axis*. The axis may be a straight line (linear),

circular (rotational), or spherical. Figure 13-18 illustrates a 6-axis robot arm. The wrist is the name usually given to the last three joints on the robot's arm. Going out along the arm, these wrist joints are known as the *pitch joint, yaw joint,* and *roll joint.* High-technology robots have from 6 to 9 axes. As the technology increases, the number of axes may increase to 16 or more. These robots' movements are meant to resemble human movements as closely as possible.

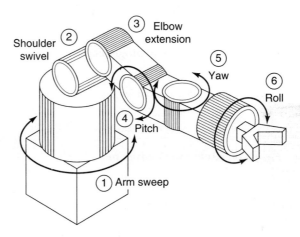

Fig. 13-18 Six-axis robot arm.

The reach of the robot is defined as the *work envelope.* All programmed points within the reach of the robot are part of the work envelope. The shape of a work envelope is determined by the major (nonwrist) types of axes a robot has (Fig. 13-19 on page 390). A robot that has two linear major axes and one rotational major axis has a cylindrically shaped work envelope. A robot that has three rotational major axes has a work envelope very much like the motion range of a human body from waist to shoulder to elbow. Being familiar with the work envelope of the robot with which you work will help you avoid personal injury or potential damage to equipment.

Most applications require that additional tooling be attached to the robot. End-of-arm tooling, commonly called the *end effector,* varies depending on the type of work the robot does. Grippers or hands are used in material handling and assembly. Spot welding and arc welding require their own tooling, as do painting and dispensing (Fig. 13-20 on page 390).

Robots usually have one of three possible sources of manipulator or muscle power: electric motors, hydraulic actuators (pistons driven by oil under pressure), or pneumatic actuators (pistons driven by compressed air).

Robots

Joint

Pitch joint

Yaw joint

Roll joint

Work envelope

End effector

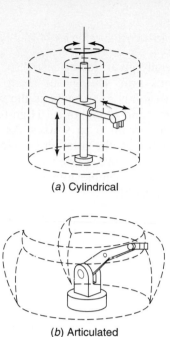

(a) Cylindrical

(b) Articulated

Fig. 13-19   Robot work envelope.

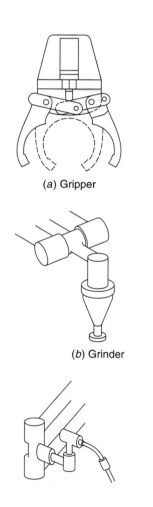

(a) Gripper

(b) Grinder

(c) Gas welding torch

Fig. 13-20   End-of-arm tooling devices.

Robots powered by compressed air are lightweight, inexpensive, and fast-moving but generally not strong. Robots powered by hydraulic fluid are stronger and more expensive but may lose accuracy if their hydraulic fluid changes temperature.

Originally all robots used hydraulic servodrives. Driven mostly by the level of service required to maintain hydraulic servosystems in these early industrial robots, engineers developed the articulated robot with dc electric servo drive motors. The industrial robot has since evolved from dc electric to ac electric. The benefits of ac servomotors over dc motors were significant. The ac servomotor incorporated brushless, maintenance-free designs, and incremental encoders for servoposition feedback.

There are two types of robot control systems: closed loop and open loop (Fig. 13-21).

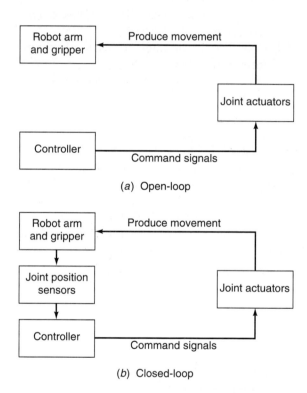

(a) Open-loop

(b) Closed-loop

Fig. 13-21   Open-loop and closed-loop robotic control systems.

⫸ **YOU MAY RECALL** that closed-loop systems have feedback and open-loop systems do not.

In the *open-loop system,* there are no sensors or feedback signals that measure how the manipulator actually moved in response to the command signals. The *closed-loop system* uses feedback signals from joint position sensors. The controller compares the actual positions of the arm joints with the programmed positions. It then issues command signals which are designed to minimize or eliminate any discrepencies or errors.

The term *servo robot* is often used to refer to a closed-loop (feedback) system, and the term *nonservo robot* to refer to an open-loop (no feedback) system.

Nonservo robots are generally small and designed for light payloads. They have only two positions for each joint (open or closed) and operate at high speed. They are often called "bang-bang" robots because of the way they bang from position to position. They are programmed for a task by setting adjustable mechanical limit stops. Nonservo robots are excellent for pick-and-place operations such as loading and unloading parts from machines.

Unlike nonservo robots, which have no control of velocity and operate with jerky motions, servo robots have smooth motions. Servo robots can control velocity, acceleration, and deceleration of each link as the manipulator goes from point to point. They can use programs which may branch to different sequences of motions depending on some condition that is measured at the time the robot is working.

Fundamentally each axis of a servo robot is a closed-loop servo control system. An example of a simple servo operation is illustrated in Fig. 13-22. In this example, the servo amplifier is responsible for amplifying the difference between the command voltage and the feedback voltage. The error signal produced is used to operate the servo motor, which is mechanically connected to the end effector and feedback potentiometer.

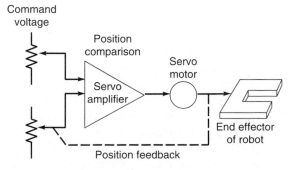

Fig. 13-22 Simple servo operation.

The *controller* contains the power supply, operator controls, control circuitry, and memory that direct the operation and motion of the robot and communication with external devices. Functionally, the controller has three major tasks to perform:

- Provide motion control signals for the manipulator unit (also known as signal processing)
- Provide storage for programmed events
- Interpret input/output signals, including operator instructions

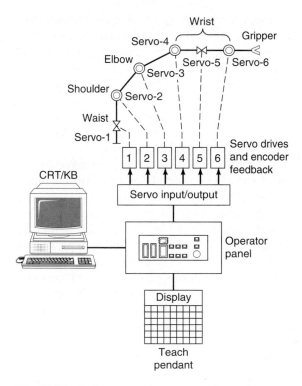

Fig. 13-23 Robot control system configuration.

Different controller configurations are used (Fig. 13-23). In general, the controller includes the following devices that are used to operate the system:

- *Operator Panel.* Comes equipped with lights, buttons, and keyswitches. Performs tasks such as powering up and powering down the system; calibrating the robot; resetting the controller after an error occurs; holding robot motion; starting or resuming automatic operation; and stopping the robot in an emergency.
- *Teach Pendant.* Comes equipped with a keypad and liquid crystal display (LCD) screen and is connected to the controller by a cable that plugs into the computer RAM board inside the controller. Performs tasks such as jogging the robot; teaching positional data; testing program execution; recovering from errors; displaying user messages, error mes-

sages, prompts, and menus; displaying and modifying positional data, program variables, and inputs and outputs; and performing some operations on computer files.

- *CRT Screen and Keyboard.* Resembles a standard computer terminal. Performs tasks such as performing computer file operations; displaying status and diagnostic information; and entering, translating and debugging programs.

The robotic controller is a microprocessor-based system that operates in conjunction with input and output cards or modules. With growing use of computers and PLCs in industry, the robot controller has become more important than the manipulator it controls. The robot controller is now required to communicate with devices outside itself such as PLCs and plant computer systems.

All major robot manufacturers have developed their own specialized program language for use with their robots. High-level language-based robot controllers are usually ASCII-based systems that are compatible with the DOS environment. Compatibility here means that robot application engineers can develop their robot programs off-line in an office environment and later down-load their program to the robot using an RS-232-C serial interface from their personal computer.

Alternate programming methods called *walk-through* and *teach-through* can also be used. In walk-through teaching, the programmer actually takes hold of the manipulator and physically moves it through the maneuvers it is intended to learn. The controller records the moves for playback later, perhaps at a much higher speed. Teach-through programming involves using a joystick or teach pendant to guide the robot along the planned path. The controller calculates a smooth path for the robot using computer-based teach software.

Self-Test

Answer the following questions.

30. True or false? The basic industrial robot is an arm which moves to perform operations.
31. The robot is a series of mechanical links driven by _____.
32. The reach of the robot can be defined as the _____ _____.
33. True or false? The end effector of a robot is always some type of gripper.
34. The power required to operate the manipulator of a robot can be _____, _____, or _____.
35. In general, a robotic controller includes the following three devices to operate the system: _____, _____, and _____.
36. True or false? Robot manufacturers normally have their own specialized program language for use with their robots.

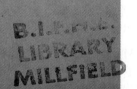
SUMMARY

1. All computers consist of two basic components: hardware and software.
2. Computer hardware includes the power supply, floppy drives, hard drive, motherboard, microprocessor chip, ROM chips, RAM chips, peripheral cards, and expansion slots.
3. A computer program is a list of instructions that give the computer instructions on what to do.
4. Disk drives act as both input and output devices.
5. The floppy disk (diskette) is a magnetic storage device which stores information on tracks in sectors of the disk.
6. A disk must be formatted before any information can be stored on it.
7. Care must be taken when handling the disk, or the information stored on the disk may be corrupted.
8. Hard disk drives store and retrieve data quicker and can hold much more information than floppy disks.
9. System software provides the programs that allow you to interact with your computer—to operate the disk drives, the printer, and other devices used by the computer.
10. The most popular operating system for personal computers is MS-DOS.
11. A file is a collection of data stored under a single name.
12. Some important types of files include executable, data, text, and graphics files.

13. The BIOS serves as a link between the computer hardware and the software programs.
14. The process of loading DOS into your computer is called booting.
15. Application software involves programs written to give the computer a specific application such as a word processor.
16. Directories are a way of organizing all the programs and files of the hard drive in your computer.
17. The first and main directory of the computer is called the root directory.
18. The DOS prompt (A>) is an indicator to you that DOS is loaded and that the computer is ready for you to enter a command.
19. DOS is simply a very large list of commands that the computer can use and understand to control, manage, and run other programs.
20. Windows uses graphical images to replace the verbal commands used to run DOS.
21. Using Windows, you can use multiple specialized applications that run concurrently and work together.
22. Ports are the back edge of expansion cards to which peripheral devices are connected.
23. A serial port sends data one bit at a time over a single one-way wire.
24. A parallel port can send several bits of data across eight parallel wires simultaneously.
25. Key computer specifications include access time, baud rate, bus channels, cache size, clock speed, CPU type, hard disk drive memory capacity, monitor type, and RAM memory capacity.
26. A work cell is a computer-integrated manufacturing system with data communication, which allows the individual devices to be integrated into a single system.
27. The cell controller requires the programmer to have more computer programming knowledge than that required for the programming of PLCs.
28. Cell control involves supervising and coordinating the components and devices.
29. The fundamental job of a local area network (LAN) is to provide communication between PLCs or between PLCs and computers.
30. LANs come in three basic configurations: start, ring, or bus.
31. Protocol refers to a predetermined set of conventions which specify the format and timing of message transmission between two or more communicating devices.
32. A master-slave system uses a host computer to manage all network communications between network devices.
33. With a peer-to-peer system, each network device has the ability to request use of and then take control of the network.
34. Numerical control machines are program-dependent and bear a close relation to programmable controllers.
35. A set of commands makes up the NC program and directs the machine to orientate a cutting tool with respect to a workpiece.
36. The NC controller is usually an industrial computer which electronically stores and reads the program and converts the program information into signals that drive motors to control the machine tool.
37. Numerical control is ideally suited for operations that involve the production of parts made from similar feedstock with variations in size and shape.
38. If the controlled machine is a three-axis type, the location address of the tool is prefixed with the letter X, Y, or Z.
39. The programming of NC machines can involve point-to-point programming and contour programming.
40. Computerized numerical control (CNC) was introduced to replace the punched tape and hard-wired machine control of older model NC units.
41. Simple CNC programming consists of taking information from a part drawing and converting it into a computer program.
42. The basic industrial robot is an arm or manipulator which moves to perform industrial operations.
43. Tasks performed by industrial robots include handling, processing, assembling, dismantling, welding, transporting, painting, and hazardous tasks.
44. A robot is simply a series of mechanical links driven by servomotors.
45. The area at each junction between the links of a robot is called a joint or axis.
46. The reach of the robot is defined as the work envelope.
47. End-of-arm tooling, commonly called the end effector, varies depending on the type of work the robot does.
48. The power that is used to operate the robotic manipulator can be hydraulic, pneumatic, or electric.
49. The term servo robot is often used to refer to a closed-loop system, and the term nonservo robot to an open-loop system.

50. A robotic controller functions to provide motion control signals for the manipulator unit. The robotic controller also provides storage for programmed events and interprets input/output signals, including operator instructions.

51. All major robot manufacturers have developed their own specialized program language for use with their robots.

52. Alternate robotic programming includes both walk-through and teach-through programming.

CHAPTER REVIEW QUESTIONS

Answer the following questions.

13-1. Explain the function of each of the following computer hardware components:

a. power supply	**f.** ROM chips
b. floppy drives	**g.** RAM chips
c. hard drive	**h.** peripheral cards
d. motherboard	**i.** expansion slots
e. microprocessor chip	

13-2. What are the kinds of disk drives popularly used in personal computers?

13-3. List five precautions to be observed when handling floppy disks.

13-4. Compare the storage capacity and speed of hard disk drives with those of floppy disks.

13-5. How are the different disk drives of a computer identified?

13-6. Compare system software and application software.

13-7. List five things DOS allows you to do.

13-8. Describe the contents of each of the following types of files:

a. executable file	**c.** text file
b. data file	**d.** graphics file

13-9. Explain the function of the computer BIOS (basic input/output system).

13-10. What does the term *booting* refer to?

13-11. Explain the function of directories.

13-12. What does the DOS prompt indicate?

13-13. Explain the function of each of the following DOS commands:

a. DIR	**f.** TYPE
b. CLS	**g.** FORMAT
c. COPY	**h.** DISKCOPY
d. REN	**i.** CD
e. DEL	**j.** CHKDSK

13-14. Explain how Windows allows for multitasking.

13-15. Name the two types of ports found on the back end of the computer case and describe how each sends and retrieves data.

13-16. Explain the importance of each of the following in the selection of a computer:

a. access time	**f.** CPU
b. baud	**g.** hard disk drive
c. bus	**h.** monitor
d. cache	**i.** RAM
e. clock speed	

13-17. Describe the make-up of an industrial work cell.

13-18. List three types of local area network (LAN) configurations or topologies.

13-19. Explain the importance of protocol in establishing communications between devices.

13-20. Compare master-slave and peer-to-peer communication formats.

13-21. Give a brief description of how numerical control devices operate.

13-22. What is the most popular NC programming language?

13-23. For what types of industrial operations is numerical control best suited?

13-24. Explain the function of Cartesian coordinates in NC programming.

13-25. Compare the motions that are involved in NC point-to-point and contour programming.

13-26. List five improvements brought about by the introduction of computerized numerical control.

13-27. What does simple CNC programming consist of?

13-28. List five specialized tasks commonly performed by an industrial robot.

13-29. Give a brief description of the make-up of a robot.

13-30. Define the robot work envelope.

13-31. List three types of power that can be used to operate the robot manipulator.

13-32. What are the major tasks performed by a robotic system controller?

13-33. Describe three ways in which robots are programmed.

CRITICAL THINKING QUESTIONS

13-1. Determine as many as possible of the following specifications about the computer available to you for lab experiments.
  a. The number of floppy disk drives
  b. The type of disk drives
  c. Label designation for each drive
  d. Storage capability of each drive
  e. Access time of each drive
  f. Baud rate
  g. Bus structure
  h. Memory cache storage
  i. Clock speed
  j. Type of processor
  k. Type of monitor
  l. RAM memory capacity
  m. Version of DOS used

Answers to Self-Tests

1. hardware; software
2. false
3. formatted
4. true
5. C:
6. system
7. false
8. true
9. root
10. prompt
11. true
12. graphical
13. false
14. true
15. CPU
16. true
17. false
18. true
19. star; ring; bus
20. manufacturing automation protocol
21. true
22. token
23. false
24. commands
25. false
26. X; Y; Z
27. point-to-point; contour
28. microprocessor
29. part drawing
30. true
31. servo motors
32. work envelope
33. false
34. hydraulic; pneumatic; electric
35. operator panel; teach pendant; CRT screen and keyboard
36. true

Index

■